PRINCIPLES AND PRACTICE OF CLINICAL ANAEROBIC BACTERIOLOGY

A SELF-INSTRUCTIONAL TEXT AND BENCH MANUAL

by

Paul G. Engelkirk, Ph.D., MT (ASCP)
Associate Professor
Program in Clinical Laboratory Science
School of Allied Health Sciences
University of Texas Health Science Center
Houston, Texas

Janet Duben-Engelkirk, Ed.D., MT (ASCP)
Associate Professor and Senior Education Coordinator
Program in Clinical Laboratory Science
School of Allied Health Sciences
University of Texas Health Science Center
Houston, Texas

V.R. Dowell, Jr., Ph.D., MT (ASCP)
Chief, Anaerobic Bacteria Branch
Hospital Infections Program
Center for Infectious Diseases
Centers for Disease Control
Atlanta, Georgia

Star

PUBLISHING COMPANY

PUBLISHING COMPANY

P.O. Box 68
Belmont, California 94002

Managing editor: Stuart Hoffman
Cover design: Douglas Hurd
Consulting Editor: Susan G. Kelley, Ph.D.
Typography: BookPrep

The authors and publisher have exerted every effort to ensure that procedures, data, analytes, reference ranges, and drug selection and dosage set forth in this text are in accord with current recommendations and practice at the time of publication. However, in view of ongoing research, changes in government regulations, and the constant flow of information relating to analytes, drug therapy, drug reactions, microbiology and medical research in general, the reader is urged to check the latest available information for each procedure, product, analyte, and drug for any change in indications, directions, and dosage and for added warnings and precautions. This is particularly important when the recommended agent is a new or infrequently employed drug.

This book was written by the authors in their private capacities. No official support or endorsement by the University of Texas or CDC is intended or should be inferred.

IN MEMORIAM

V.R. (BUD) DOWELL, JR., Ph.D. (1927-1990)

    Dr. V.R. Dowell, Jr., scientist-teacher, believed that microbiology was truly an art; the subtle nuances of microorganisms delighted and fascinated him. He was devoted to excellence in scientific investigation and teaching, especially in his chosen field of anaerobic microbiology. He relished his role as teacher and always made time for students and guest researchers. His laboratory was often filled with students actively working under his enthusiastic eyes; countless numbers of these students, encouraged by his ideals and standards, are now practicing scientists throughout the world. Dr. Dowell was especially a champion of developing countries with limited economic resources in their pursuit of public health. He often traveled to these countries, training and encouraging their microbiologists.

    During his illness, the completion of this textbook and bench-training manual was a continual source of strength. This book represents the final vehicle of his scientific principles and is, I believe, a thorough presentation of the state of the art of our science. Our scientific community has lost a great pioneer and leader. I feel most honored to have known the scientist and the man, and it is to his memory that this book is dedicated.

Suzette L. Bartley, M.S.
Microbiologist-in-Charge
Anaerobe Reference Laboratory
Anaerobic Bacteria Branch
Centers for Disease Control
Atlanta, Georgia

THIS BOOK IS ALSO DEDICATED TO MILDRED DOWELL,

MARJORIE DUBEN, AND PHYLLIS ENGELKIRK,

WITHOUT WHOSE LOVE IT COULD NEVER HAVE BEEN WRITTEN

CONTENTS

PREFACE

Anaerobic bacteria are involved in a wide variety of human diseases, playing an important, often major, role in infectious processes of the respiratory, gastrointestinal, and female genital tracts. Anaerobes are key pathogens in brain abscesses, oral/dental diseases, actinomycosis, aspiration pneumonia, lung abscesses, chronic osteo-myelitis, tetanus, botulism, gas gangrene, and numerous other infectious processes of soft tissue. Certain infectious processes are so apt to involve anaerobes that their presence should always be assumed.

Yet the involvement of anaerobes in infectious processes is frequently over-looked. This may be due in part to an insufficient understanding of, and appreciation for, the role of anaerobes in human diseases. Perhaps it is the result of inadequate emphasis on, and/or insufficient time devoted to, the subject of anaerobes in clinical microbiology courses in medical schools, graduate microbiology programs, pathology and infectious disease residencies, and clinical laboratory science (CLS) and medical laboratory technician (MLT) programs. Authorities in clinical anaerobic bacteriology have noted that clinicians and microbiologists need to be educated about the importance of these organisms and the best procedures for recovering and identifying them.

In an effort to determine the extent of anaerobic bacteriology training in CLS curricula, we conducted a survey of United States CLS education programs in 1988. Results revealed that 25% of the responding programs included fewer than three hours of didactic training in anaerobes and that 16% provided no anaerobic bac-teriology training as part of their student microbiology laboratory courses. Perhaps most shocking of all, 27% of the respondents provided fewer than three days of actual bench experience in anaerobic bacteriology during the students' clinical experience in microbiology. Sixty-six percent of the responding programs indicated an interest in a textbook that specifically addressed anaerobes, and 57% expressed interest in a self-study course on the subject. This book represents a response to these needs. In writing it, we assumed that readers would have some knowledge of general micro-biology.

Principles and Practice of Clinical Anaerobic Bacteriology is both a self-instructional text and a bench manual. The text portion can be used in either an academic or clinical laboratory setting. It may be used to supplement and reinforce lecture/laboratory instruction in clinical anaerobic bacteriology or as a self-study course on this topic. In the latter case, students would be able to learn this important subject on their own, even when time cannot be devoted to anaerobes in the curriculum. Each chapter contains learning objectives, definitions of important terms, a chapter review, and self-assessment exercises. Where appropriate, case presentations are also included.

This book may also be used in the student laboratory or clinical microbiology laboratory setting as a bench manual. It contains flow charts for specimen processing, photographs depicting cell and colony morphology, step-by-step instructions for performing key tests, and a compendium of clinically encountered anaerobes. The detailed, logical, step-by-step flow charts and other information in this book enable even small clinical laboratories to make presumptive identifications of the most commonly isolated anaerobes.

Principles and Practice of Clinical Anaerobic Bacteriology is primarily directed to students in the clinical laboratory sciences, but it is also applicable to graduate students and postdoctoral fellows in clinical microbiology, as well as pathology and infectious disease residents. Because the book is a combined textbook and bench manual, a copy should be available in those clinical microbiology laboratories where anaerobic bacteriology procedures are performed. Employees who have never received formal training in anaerobic bacteriology could learn the subject on their own by using this book.

A chapter is devoted to veterinary aspects of anaerobic bacteriology, but virtually all the information in the book is directly applicable to anaerobes of veterinary importance. Thus, the book will be of value in the education of veterinarians, veterinary microbiologists, and veterinary technologists/technicians and of use in veterinary microbiology laboratories.

We use the term "microbiologist" throughout the book in reference to the individual responsible for performing microbiological procedures in the clinical laboratory, whether in a human or veterinary setting. Therefore, the term includes CLSs, MLTs, veterinary laboratory technicians/technologists, as well as individuals with bachelors, masters, and doctoral degrees in microbiology. We use the more encompassing term "clinical laboratory professionals" to denote any and all professionals employed in the clinical laboratory. Thus, in addition to those specifically working in the microbiology section, this term includes pathologists, veterinarians, laboratory managers, and employees working in areas other than microbiology.

This book could not have been written without the assistance of very special friends and colleagues. We are deeply grateful to the following people for their numerous and varied contributions: Ms. Suzette L. Bartley, Mr. Max Crumley, Dr. Michel Delmée, Mr. Dan Edson, Dr. Sydney M. Finegold, Ms. Bridget Hahn, Dr. John G. Holt, Dr. Stanley C. Holt, Mr. Spencer Jang, Dr. Robert L. Jones, Dr. Susan Kelley, Ms. Mary Ann Lambert-Fair, Mr. Mike Leary, Dr. Lois H. Lindberg, Mr. James I. Mangels, Drs. L. V. H and W. E. C. Moore, Ms. Susan Strawn, Ms. Diane Wall, Ms. Sylvia Stein Wright, Mr. John Zlockie, corporate officials and technical representatives of the many companies that provided product information and photographs, and the publishers who granted permission for us to reproduce diagrams and tables from their publications. We extend special thanks to Mr. Mike Cox for encouraging and supporting our return to the clinically important world of anaerobes following a brief respite and to Ms. Charlene Backe for granting us permission to use materials from the third edition (1990) of our CACMLE self-study course in clinical anaerobic bacteriology. We are especially grateful and indebted to Mr. Stuart Hoffman for sharing our dream and turning it into reality.

We salute the pioneers and champions of anaerobic bacteriology education, including Drs. L. V. H. and W. E. C. Moore and their co-workers at the Virginia Polytechnic Institute; Dr. Sydney M. Finegold and his colleagues at the Wadsworth V. A. Medical Center in Los Angeles; Drs. V. R. Dowell, Jr. and George L. Lombard and others at the Centers for Disease Control in Atlanta; and Mr. Mike Cox at Anaerobe Systems in San Jose. Collectively, these dedicated individuals have presented thousands of lectures and workshops in clinical anaerobic bacteriology and have contributed hundreds of publications on this important topic. This book incorporates many of the innovations, teachings, and philosophical approaches of these distinguished and highly respected educators, especially those applicable in the cost-conscious environment of today's diagnostic microbiology laboratories.

Together with the rest of the scientific community, we were deeply saddened by the premature death of our coauthor, Dr. V. R. (Bud) Dowell. Bud and the late Dr. George L. Lombard were Paul's original instructors in anaerobic bacteriology at an outstanding CDC course in 1973. We are grateful to Bud for his valuable contributions to this book and find comfort in knowing that he is undoubtedly teaching anaerobic bacteriology in heaven at this very moment and that we shall one day be students in his classroom again.

A portion of the proceeds of this book will be donated to the American Cancer Society to assist that organization in its battle against colon cancer, the evil malady that claimed the lives of Bud Dowell, Paul's dad, Jan's grandfather, and countless others. It is to some extent ironic that clostridial bacteremias are known to occur in some cases of colon cancer and that intestinal anaerobes play roles in the generation of tumor promoters, mutagens, and carcinogens.

PGE
JDE

CHAPTER 1

INTRODUCTION

Anaerobes Defined
    Aerobes
    Microaerophiles
    Anaerobes
    Capnophiles
Why Oxygen Is Lethal to Anaerobes
    Oxidation-Reduction (Redox) Potential
The Source and Location of Anaerobes
The Importance of Anaerobes

At the conclusion of this chapter, you will be able to:

1. Describe how anaerobes differ from other categories of bacteria encountered in the clinical microbiology laboratory.
2. Define the following terms: obligate aerobe, microaerophile (microaerophilic aerobe), facultative anaerobe, aerotolerant anaerobe, microaerotolerant anaerobe, obligate anaerobe, moderate obligate anaerobe, strict obligate anaerobe, anaerobiosis, capnophile.
3. Describe the difference between "respiration" and "fermentation."
4. List three of the toxic products formed during the reduction of molecular oxygen.
5. Describe the function of enzymes in the superoxide dismutase family.
6. Describe the two-phase theory for the destructive effect of oxygen on anaerobic bacteria.
7. List four environmental niches where anaerobes are found.
8. Define the terms "endogenous anaerobe," "exogenous anaerobe," and "indigenous microflora."
9. State four reasons why anaerobes are important.

Definitions

Aerotolerant anaerobe: An organism capable of growing in atmospheres containing molecular oxygen (such as air) but growing best in an anaerobic environment

Anaerobe: An organism that does not require molecular oxygen for life and reproduction

Anaerobiosis: Life in the absence of molecular oxygen

Bactericidal: Destructive to bacteria

Bacteriostatic: Inhibiting the growth or multiplication of bacteria

Capnophile (capnophilic organism): An organism that may or may not be an anaerobe but grows best in the presence of increased concentrations of carbon dioxide

Catalase: An enzyme that catalyzes the conversion of two molecules of hydrogen peroxide (H_2O_2) to two molecules of water (H_2O) and a molecule of oxygen (O_2)

Endogenous (or indigenous) anaerobes: Anaerobes that comprise part of the indigenous microflora of animals (including humans)

Exogenous anaerobes: Anaerobes that exist in or arise from locations **outside** of or other than the bodies of animals (e.g., anaerobes that live in soil or mud)

Facultative anaerobe: An organism that multiplies equally well in the presence or absence of oxygen; sometimes called a facultative aerobe

Fermentation: An oxidation-reduction process in which electrons pass from one organic compound (an electron donor) to another organic intermediate (an electron acceptor) and molecular oxygen does not participate

Indigenous microflora (in the past called normal flora): The mixture of micro-organisms that typically reside on or in (colonize) various anatomic sites of a healthy animal body; such organisms may either be permanent or transient residents

Microaerophile (microaerophilic aerobe): An organism that requires molecular oxygen for multiplication but in concentrations lower than that found in air (about 21%)

Microaerotolerant anaerobe: An organism that grows in an anaerobic system and a microaerophilic environment (5% oxygen) but does not grow in a CO_2 incubator (15% oxygen) or in air (about 21% oxygen)

Obligate aerobe: An organism that requires an atmosphere containing 15–21% molecular oxygen (as found in a CO_2 incubator or air) for maximum growth

Obligate anaerobe: An organism growing only in an anaerobic environment, not in a microaerophilic environment, a CO_2 incubator, or air

Respiration: An oxidation-reduction process in which molecular oxygen serves as an oxidizing agent and the final electron acceptor

Superoxide anion: A highly reactive product formed during the reduction of molecular oxygen, that is toxic to cells (including anaerobic bacteria)

Superoxide dismutase: An enzyme (actually a family of enzymes) that catalyzes the conversion of two superoxide anions into a molecule of oxygen (O_2) and a molecule of hydrogen peroxide (H_2O_2)

Anaerobes Defined

In the clinical microbiology laboratory, it is useful to classify bacteria on the basis of their relationship to oxygen and carbon dioxide. This is easily accomplished by comparing an organism's ability to multiply on a solid medium (such as blood agar) in four different environments: air (about 21% oxygen), a CO_2 incubator (15% oxygen), a microaerophilic system (5% oxygen), and an anaerobic system (0% oxygen). Using this technique, a bacterial isolate can be classified into one of six groups: obligate aerobe, microaerophilic aerobe, facultative anaerobe, aerotolerant anaerobe, micro-aerotolerant anaerobe, and obligate anaerobe (see Table 1-1). Such information is useful in the preliminary identification of aerobes as well as anaerobes.

Aerobes

In order to grow and multiply, **obligate aerobes** require an atmosphere containing molecular oxygen in concentrations comparable to that found in a CO_2 incubator or room air (i.e., 15–21% O_2). Mycobacteria and certain fungi are examples of micro-organisms that are obligate aerobes. Aerobic organisms obtain most of their energy from a process known as **respiration,** in which molecular oxygen serves as an oxidizing agent in oxidation-reduction reactions and as the final electron acceptor.

Table 1-1. Grouping of Bacteria Based upon Growth in Various Concentrations of Molecular Oxygen

Group	Growth in			
	Air (~21% O_2)	CO_2 Incubator (15% O_2)	Microaerophilic Environment (5% O_2)	Anaerobic Environment (0% O_2)
Obligate aerobe	9	9	8	1
Microaerophilic aerobe	0	7	9	1
Facultative anaerobe	9	9	9	9
Aerotolerant anaerobe	4	6	8	9
Microaerotolerant anaerobe	0	0	8	8
Obligate anaerobe	0	0	0	9

Key:
0 = no growth
1 = sparse, tiny colonies
2 = sparse, small to medium colonies
3 = sparse, large colonies
4 = moderate, tiny colonies
5 = moderate, small to medium colonies
6 = moderate, large colonies
7 = abundant, tiny colonies
8 = abundant, small to medium colonies
9 = abundant, large colonies

Source: Based on reference 3.

Microaerophiles

Microaerophiles (microaerophilic aerobes) also require oxygen for multiplication, but in concentrations lower than that found in room air. *Neisseria gonorrhoeae* and *Campylobacter* species are examples of microaerophilic bacteria that prefer about 5% oxygen.

Anaerobes

Anaerobes can be defined as organisms that do not require oxygen for life and reproduction. However, they vary in their sensitivity to oxygen. The terms "obligate anaerobe," "microaerotolerant anaerobe," "aerotolerant anaerobe," and "facultative anaerobe" are used to describe the organism's relationship with molecular oxygen.

An **obligate anaerobe** is an anaerobe that grows only in an anaerobic environment. It will not grow in a microaerophilic environment (5% oxygen), a CO_2 incubator (15% oxygen), or air (about 21% oxygen). Loesche has further divided this group into **moderate anaerobes** and **strict anaerobes**.[6] A moderate anaerobe cannot multiply in an atmosphere containing more than 2–8% oxygen, and strict anaerobes cannot multiply in the presence of more than 0.5% oxygen. Moderate anaerobes (e.g., *Bacteroides fragilis*) can tolerate exposure to air for several hours on the surface of blood agar but require an anaerobic environment for multiplication. Strict anaerobes (e.g., *Clostridium novyi* type B) are killed by exposure to air for only a few minutes. Fortunately, strict anaerobes are seldom associated with human infections.

A **microaerotolerant anaerobe** is an organism capable of growth in an anaerobic system and a microaerophilic environment, but it cannot grow in a CO_2 incubator or air. An **aerotolerant anaerobe** is capable of multiplication in atmospheres containing molecular oxygen (such as air and a CO_2 incubator), but it grows best in an anaerobic environment.

Facultative anaerobes (sometimes called **facultative aerobes**) are capable of surviving in either the presence or absence of oxygen. Many of the bacteria routinely isolated from clinical specimens are facultative anaerobes (e.g., members of the family *Enterobacteriaceae*, streptococci, and staphylococci). Facultative organisms preferentially use molecular oxygen as an oxidizing agent and final electron acceptor when it is available, but they are capable of using a variety of compounds (e.g., NO_3) as sources of energy in the absence of oxygen.

Specific examples of clinically encountered obligate, microaerotolerant, and aerotolerant anaerobes are shown in Table 1-2.

Table 1-2. Examples of Clinically Encountered Obligate, Microaerotolerant, and Aerotolerant Anaerobes

---

Obligate Anaerobes (AN)

Anaerobiospirillum succiniciproducens (some strains MA)

Bacteroides spp.
- *B. caccae*
- *B. distasonis*
- *B. fragilis* (some MA strains from Mainland China)
- *B. gracilis* (most strains MA)
- *B. merdae*
- *B. ovatus*
- *B. putredinis*
- *B. stercoris*
- *B. thetaiotaomicron*
- *B. uniformis*
- *B. ureolyticus* (some strains MA)
- *B. vulgatus*

Bifidobacterium dentium (some strains AT)

Campylobacter sputorum subsp. *mucosalis*

Centipeda periodontii

Clostridium spp.
- *C. difficile*
- *C. novyi* type A
- *C. ramosum*
- *C. sphenoides*
- *C. sporogenes*
- *C. tetani*

Desulfomonas pigra

Table 1-2. Examples of Clinically Encountered Obligate, Microaerotolerant, and Aerotolerant Anaerobes (*continued*)

Desulfovibrio desulfuricans

Eubacterium spp.
  E. alactolyticum
  E. lentum
  E. limosum

Fusobacterium spp.
  F. mortiferum
  F. necrophorum
  F. nucleatum
  F. varium

Peptostreptococcus spp.
  P. anaerobius (some strains MA)
  P. asaccharolyticus
  P. indolicus
  P. magnus (some strains AT)
  P. micros
  P. prevotii
  P. tetradius

Porphyromonas spp.
  P. asaccharolytica
  P. endodontalis
  P. gingivalis

Propionibacterium spp.
  P. acnes (some strains FA, AT)
  P. propionicus (some strains FA, AT)

Sarcina ventriculi (some strains MA)

Selenomonas sputigena

Staphylococcus saccharolyticus (some strains MA)

Succinivibrio dextrinosolvens

Veillonella parvula (some strains MA)

Wolinella recta (Campylobacter rectus) (some strains MA)

Microaerotolerant Anaerobes (MA)

Campylobacter spp.
  C. concisus
  C. sputorum subsp. *sputorum*

Clostridium spp.
  C. perfringens (some strains AN)
  C. septicum (some strains AN)
  C. sordellii (some strains AN)

Wolinella curva (Campylobacter curvus)

Table 1-2. Examples of Clinically Encountered Obligate, Microaerotolerant, and Aerotolerant Anaerobes (*continued*)

Aerotolerant Anaerobes (AT)

Actinomyces spp.

        A. israelii (most strains FA, AT, or MA; some strains AN)

        A. meyeri (most strains FA)

        A. naeslundii (most strains FA)

        A. odontolyticus (most strains FA)

Clostridium tertium (most strains FA)

Key:

AN = obligate anaerobe

AT = aerotolerant anaerobe

FA = facultative anaerobe

MA = microaerotolerant anaerobe

(See text for definitions.)

Source: Based on references 1 and 2.

"**Anaerobiosis**" is a term used to describe life that occurs in the absence of molecular oxygen. Like aerobes, anaerobes obtain most of their energy from oxidation-reduction reactions. Unlike aerobes, however, they employ **fermentation** reactions rather than respiration. In fermentative processes, electrons pass from one organic intermediate (an electron donor) to another (an electron acceptor). Molecular oxygen does not participate in the reaction. Oxidation is a loss of electrons; when a substance is oxidized, it gives up electrons. Reduction is a gain of electrons. So when a substance is reduced, it accepts (or gains) electrons.

Capnophiles

Some bacteria grow better in the presence of increased concentrations of carbon dioxide (CO_2). Such organisms are referred to as **capnophiles (capnophilic organisms)**. Some anaerobes (e.g., certain *Bacteroides* and *Fusobacterium* species) are capnophiles, as are some aerobes (e.g., certain *Neisseria* and *Haemophilus* species). Capnophilic aerobes will grow in a candle extinction jar, but not in room air. A candle extinction jar (or candle jar, as it is sometimes called) generates a final atmosphere containing 12–17% oxygen and 3–5% CO_2.

Why Oxygen Is Lethal to Anaerobes

Although molecular oxygen itself can be toxic to some anaerobes, even more toxic substances are produced when oxygen becomes reduced. During oxidation-reduction reactions, molecular oxygen (O_2) is reduced in a stepwise manner by the addition of one, two, three, or four electrons (Equations 1-1 through 1-4).[7]

Equation 1-1. $O_2 + e^- = {}^-O_2$ (superoxide anion)

Equation 1-2. $O_2 + 2e^- = H_2O_2$ (hydrogen peroxide)

Equation 1-3. $O_2 + 3e^- = H_2O + OH^-$ (hydroxyl radical)

Equation 1-4. $O_2 + 4e^- = H_2O$

In the first step of the reduction process, molecular oxygen is reduced to **super-oxide anion,** a highly reactive free radical capable of inflicting severe damage on enzyme systems and cell structure. Further reduction of oxygen leads to the production of other destructive derivatives of oxygen (e.g., hydrogen peroxide and hydroxyl radicals). To further compound the potentially toxic effects, superoxide anions can react with hydrogen to form hydrogen peroxide and with hydrogen peroxide to yield hydroxyl radicals. Superoxide anion is clearly the most destructive product of aerobic metabolism.

Every living creature that uses oxygen has one or more enzymes that protect it from superoxide anions and their toxic derivatives. The most important of these is a family of enzymes known as **superoxide dismutases (SODs).** SODs are found in every type of cell that uses oxygen as a final electron acceptor as well as many that do not. As shown in Equation 1-5, SODs catalyze the conversion of two superoxide anions into a molecule of oxygen (O_2) and a molecule of hydrogen peroxide (H_2O_2).

Equation 1-5.
$$^-O_2 + {}^-O_2 \xrightarrow{\text{SOD}} O_2 + H_2O_2$$
(two super-oxide anions)   (oxygen)   (hydrogen peroxide)

In the presence of catalase, the resulting H_2O_2 can be converted to water and oxygen, as shown in Equation 1-6.

Equation 1-6.
$$2H_2O_2 \xrightarrow{\text{CATALASE}} 2H_2O + O_2$$
(two hydrogen peroxide molecules)   (2 water molecules)   (oxygen)

Thus, the presence of SODs not only reduces the concentration of superoxide anions, but also decreases the production of other toxic derivatives of oxygen. If anaerobes lacked SODs, this would explain their unique susceptibility to oxygen and its reduction products. Until recently, this has been a popular theory. Unfortunately, it is not the complete explanation. Although some anaerobes lack SODs, many obligate anaerobes produce them in varying quantities.

Likewise, the absence of catalase is not the answer. Although most obligate anaerobes do not produce catalase, some (e.g., *Bacteroides fragilis* and *Propionibacterium acnes*) produce this enzyme. In fact, the catalase test is a useful aid in identifying certain anaerobes. Because they don't produce catalase, certain anaerobes (e.g., fusobacteria) are highly susceptible to the toxic effects of hydrogen peroxide.

A popular current theory to explain the destructive effects of oxygen on anaerobes suggests that the damage occurs in a two-phase process.[7] The first phase is **bacterio-static** and the second, **bactericidal.**

In phase 1, when anaerobes are exposed to oxygen, electrons that would usually be available for metabolic functions are diverted to the reduction of molecular oxygen. There is a consequent decrease in the amount of energy available for growth and synthesis of new cell material, resulting in a slowing down or complete cessation of growth. This then is a bacteriostatic effect. If the period of exposure to oxygen is sufficiently brief, it could be reversible. Should the anaerobes be placed back into an anaerobic environment at this point, electrons would again be available for normal metabolic processes, energy production, and cell growth.

However, should the anaerobes remain in the presence of oxygen, phase 2 would occur. Phase 2 is the lethal, irreversible effect of oxygen toxicity due to the previously mentioned superoxide anions, hydroxyl radicals, and hydrogen peroxide. Phase 2, therefore, is a bactericidal effect.

Oxidation-Reduction (Redox) Potential

The mere exclusion of oxygen from a closed environment may, in itself, be insufficient for initiation of growth by some anaerobes. Strict anaerobes may also require an environment having a low oxidation-reduction potential (redox potential). This may be in part because certain enzymes, essential for bacterial growth, require fully reduced sulfhydryl (-SH) groups to be active.[7]

Redox potential (designated E_h and expressed in terms of volts or millivolts) can be defined as the tendency of a system to accept or give up electrons.[7] It can be measured using two electrodes: a measuring electrode and a hydrogen reference electrode. The pH at which the measurement is made must also be stated.

A dye called resazurin is frequently used in microbiological media to measure its E_h. Resazurin becomes reduced to a substance called resorufin, which is pink near neutrality (pH 7.0). Resorufin undergoes another reduction to a colorless compound, dihydroresorufin, near E_h -51 millivolts (mv). Thus, a shift from colorless to pink indicates that the medium has become oxidized (and thus no longer has a low E_h). Reducing agents (such as thioglycollate, cysteine, and dithiothreitol) are often added to microbiological media to obtain a low redox potential.

In vivo (in the body), bacteria have a tendency to lower the redox potential at their site of growth. Consequently, anatomic sites colonized with mixtures of organisms frequently provide conditions conducive to the growth of obligate anaerobes.

The Source and Location of Anaerobes

Based upon scientific evidence, many scientists believe that anaerobes originated about three to four billion years ago in warm, shallow waters, where they were protected from the sun's deadly ultraviolet rays.[7] It is thought that life on this planet then remained anaerobic for hundreds of millions of years.

Today anaerobes are found in specific ecological niches (Figure 1-1). They can be found in soil, in fresh- and saltwater sediments, and in the bodies of animals—including humans. Anaerobes that comprise part of the indigenous microflora of animals are referred to as endogenous (or indigenous) anaerobes. The endogenous anaerobes of humans are discussed in Chapter 3. Anaerobes that exist in or arise from locations **outside** of or other than the bodies of animals are referred to as exogenous anaerobes.

Figure 1-1.   Ecological niches of anaerobic bacteria include soil, mud, and the bodies of humans, dogs, cattle, birds, fish, and other animals.

The Importance of Anaerobes

Anaerobic bacteria are significant for a variety of reasons. They are important in human and veterinary medicine because they play a role in serious, often fatal, infectious processes and intoxications. They can be involved in infectious processes in virtually any organ or tissue of the body and consequently can be recovered from virtually any type of clinical specimen. This book describes the role of anaerobes in disease; proper techniques for selecting, collecting, transporting, and processing clinical specimens for anaerobic bacteriology; and procedures for identifying and testing the antimicrobial susceptibility of anaerobic isolates. Although other functions and applications of anaerobes are important, a thorough discussion of these aspects of anaerobic bacteriology is beyond the scope of this book.

Anaerobic bacteria, together with other gastrointestinal flora, contribute to the digestion of foods in humans and other animals. The first stomach of ruminants (e.g., antelopes, cattle, deer, goats, oxen, and sheep) is called the rumen. Anaerobes of the rumen play essential roles in digesting macromolecules such as pectin, starch, and cellulose. Cellulose-digesting anaerobes include *Bacteroides succinogenes, Butyrivibrio fibrisolvens, Eubacterium cellulosolvens,* and *Clostridium loeschii.* Anaerobes found in soil and sediments are involved in such processes as nitrate reduction, nitrogen fixation, sulfate reduction, and methanogenesis (production of methane gas).

Anaerobes are also important in industry. They have been used, to varying degrees, in the commercial production of acetic acid, acetone, butanol, butyric acid, ethanol, isopropanol, lactic acid, methane, and propionic acid. Various species of *Clostridium* are used to produce vaccines of human and veterinary importance. Examples of human vaccines include tetanus toxoid and *C. botulinum* toxoids, which protect against tetanus and botulism, respectively. Veterinary vaccines include toxoids derived from *C. chauvoei, C. haemolyticum, C. novyi,* and *C. septicum.* Toxoids, also called bacterins, are produced by treating toxins with formaldehyde, yielding proteins that have lost their toxicity but retain their antigenicity.

Perhaps the most widespread use of anaerobes is in the anaerobic digestion of sewage sludge. This process converts the noxious sludge into a less objectionable form and simultaneously generates methane, which can be used as a fuel.

Another significant use of anaerobes is in the textile industry—in a process known as "retting." Retting is used to separate cellulosic fibers of plants such as flax and hemp. Certain anaerobes (e.g., *Clostridium felsineum*) produce enzymes that degrade pectin, a plant substance that holds the cellulosic fibers together. Once separated, the fibers can be used in textile manufacture.

Some anaerobes of the genus *Propionibacterium* are used to produce Gruyere and Emmenthal cheeses. These organisms are responsible not only for the characteristic flavors of these Swiss cheeses (due to fermentation of lactate to propionate and acetate), but also for the large holes or "eyes" in the cheeses (the result of carbon dioxide formation). Propionibacteria are also employed in the production of vitamin B_{12}, used in the treatment of pernicious anemia, and as an animal feed additive.[4, 5]

Chapter in Review

- Many scientists believe that anaerobic bacteria (anaerobes) have existed for billions of years.

- Anaerobes do not require molecular oxygen for life and reproduction, but different species vary in their ability to withstand oxygen and its toxic reduction products.

- Anaerobes are found in specific environmental niches: soil, fresh- and saltwater mud, and many anatomical sites in the bodies of animals.

- Although anaerobes exist as part of the indigenous microflora of animals, they can also be involved in very serious diseases of humans and other animals.

- Important agricultural and industrial applications of anaerobes include methanogenesis, sewage treatment, retting, and the production of vaccines, cheeses, and vitamin B_{12}.

Self-Assessment Exercises

1. Define the following terms:

    a. Anaerobiosis _____

    _____

    b. Capnophile _____

    _____

    c. Obligate anaerobe _____

    _____

    d. Microaerotolerant anaerobe _____

    _____

    e. Aerotolerant anaerobe _____

    _____

2. TRUE or FALSE: Respiration involves molecular oxygen as an oxidizing agent and electron acceptor, whereas fermentation does not.

3. The toxicity of oxygen to anaerobes is thought to occur in a two-stage process: a bacteriostatic phase (phase 1) and a bactericidal phase (phase 2). Briefly describe what happens in both phases.

    a. Phase 1: _____

    _____

    _____

    b. Phase 2: _____

    _____

    _____

4. List three products formed during the reduction of molecular oxygen that are capable of causing extensive damage to cellular enzyme systems and cell structure.

   a. _____

   b. _____

   c. _____

5. List two enzymes possessed by bacteria (including some obligate anaerobes) that enable the organisms to counteract or neutralize the toxic effects of products formed during the reduction of molecular oxygen.

   a. _____

   b. _____

6. List three ecological niches where anaerobes are found.

   a. _____

   b. _____

   c. _____

7. Define the following terms:

   a. Endogenous anaerobe _____

   _____

   b. Exogenous anaerobe _____

   _____

   c. Indigenous microflora _____

   _____

8. List three industrial uses of anaerobes.

a. _____

_____

b. _____

_____

c. _____

_____

References

1. Dowell, V.R., Jr. 1988. Characteristics of some clinically encountered anaerobic bacteria. Unpublished document that accompanied a lecture presented at the IV National Symposium on Clinical Microbiology of the Sociedade Brasileira de Microbiologia, Rio de Janeiro, Brazil. Centers for Disease Control, Atlanta.
2. Dowell, V.R., Jr. 1990. Personal communication.
3. Dowell, V.R., Jr., and G.L. Lombard. 1984. Procedures for preliminary identification of bacteria. Centers for Disease Control, Atlanta.
4. Holland, K.T., J.S. Knapp, and J.G. Shoesmith. 1987. Anaerobic bacteria. Chapman and Hall, New York.
5. Ketchum, P.A. 1988. Microbiology: concepts and applications. John Wiley and Sons, New York.
6. Loesche, W.J. 1969. Oxygen sensitivity of various anaerobic bacteria. Appl. Microbiol. 18:723-727.
7. Smith, L.D.S., and B.L. Williams. 1984. The pathogenic anaerobic bacteria, 3rd ed. Charles C. Thomas, Springfield, IL.

CHAPTER **2**

CLASSIFICATION

Classification of Anaerobes
    Clostridium Species
    Nonsporeforming, Anaerobic, Gram-Positive
      Bacilli
    Nonsporeforming, Anaerobic, Gram-Negative
      Bacilli
        Bacteroides, Porphyromonas, and *Prevotella*
          Species
        Fusobacterium Species
        Curved, Motile, Anaerobic, Gram-
          Negative Bacilli
    Anaerobic Cocci
      Gram-Negative Cocci
      Gram-Positive Cocci
Anaerobes Isolated Most Frequently from Clinical
    Specimens

At the conclusion of this chapter, you will be able to:

1. Define the following terms: endospore, diphtheroid, fusiform, and pleomorphic.
2. Describe the gram reaction and cellular morphology of the following anaerobes: *Actinomyces* spp., *Bacteroides* spp., *Bifidobacterium* spp., *Clostridium perfringens, Fusobacterium nucleatum, Mobiluncus* spp., *Peptostreptococcus anaerobius, Porphyromonas* spp., *Prevotella* spp., *Propionibacterium acnes, Veillonella* spp., and *Wolinella* spp.
3. Give two distinguishing characteristics of the genus *Clostridium.*
4. Name two species of *Clostridium* that routinely stain pink in Gram-stained preparations.
5. Name the two genera of anaerobic, gram-negative bacilli most commonly isolated from clinical specimens.
6. List five species in the *Bacteroides fragilis* group and three species of *Porphyromonas.*

15

7. Name the anaerobe most commonly involved in soft tissue infectious processes and bacteremia.
8. Identify the only species in the genus *Fusobacterium* with cells that routinely appear "fusiform" in shape.
9. Name the genus of anaerobic, gram-positive cocci and the genus of anaerobic, gram-negative cocci most frequently isolated from clinical specimens.
10. List five genera of nonsporeforming, anaerobic, gram-positive bacilli.
11. Name the anaerobe most often found as a contaminant in blood cultures.
12. List the five anaerobes or groups of anaerobes isolated from two-thirds of all infectious processes involving anaerobes.

Definitions

Asaccharolytic: The inability to split sugars chemically

Beta-lactamases: Bacterial enzymes capable of inactivating beta-lactam antimicrobial agents, such as penicillins and cephalosporins

Contaminant: A member of the indigenous microflora isolated from a clinical specimen but not contributing to the infectious process

Diphtheroid: A gram-positive bacillus resembling *Corynebacterium diphtheriae* in appearance; small, club-shaped bacilli that tend to form aggregates referred to as "Chinese letters" and "picket fences"

Endospore: A thick-walled body formed within a bacterial cell capable of withstanding prolonged periods of adverse environmental conditions and germinating to form a vegetative bacterium when conditions once again become favorable

Fusiform: Long, thin, and tapered (pointed) at the ends

Nonsporeformer: A bacterial species incapable of producing endospores

Pleomorphic: When the cells of a given species of bacterium exist in a variety of forms

Protoheme: A darkly colored pigment (consisting of protoporphyrin and heme) produced by *Porphyromonas* spp. that causes colonies of these anaerobic, gram-negative bacilli to become brown to black in color

Protoporphyrin: A darkly colored pigment produced by pigmented *Prevotella* spp., resulting in the characteristic brown-black colonies of these anaerobic, gram-negative bacilli

Sporeformer: A bacterial species capable of producing endospores; examples of sporeforming anaerobes are *Clostridium* spp. and *Sarcina ventriculi*

Subterminal spore: An endospore located somewhere other than the end of the cell

Terminal spore: An endospore located at the end of the cell

Taxonomy is that branch of biology concerned with orderly classification of all living organisms into appropriate categories (taxa) on the basis of relationships among them. Some taxonomic schemes place anaerobes in the kingdom Protista together with other bacteria, algae, slime molds, fungi, and protozoa. All organisms in this kingdom are single celled. The levels of classification most commonly used for the Protista are family, tribe, genus, and species. This chapter introduces the genus and species names of anaerobic bacteria encountered in human clinical specimens.

These organisms may be present in clinical materials as indigenous microflora contaminants (Chapter 3), or they may be contributing to the disease process (Chapter 4). Many of the anaerobes encountered in human clinical materials are also recovered from veterinary specimens. Anaerobes unique to veterinary specimens are described in Chapter 13.

A set of characteristics or criteria has been established for each microorganism to establish that it belongs to a given species.[20] In the clinical microbiology laboratory, identifications of organisms are based upon a variety of traditional characteristics as well as those determined by more "modern" technologies (Table 2-1). By applying these criteria, you can classify a given microorganism and report a genus and species to the attending physician.

Table 2-1. Characteristics Used to Identify Anaerobic Isolates

---

"Traditional" Characteristics[20]

> Colony characteristics
> Gram reaction
> Cell morphology
> Differential biochemical reactions
> Motility
> Presence/absence of spores
> Growth characteristics
> > Rapidity of growth
> > Morphology in solid and liquid culture media
> > Optimal atmospheric conditions
> > Optimal temperature of incubation
> Detection of antigen or antibodies in clinical specimens

"Modern" Technologies[8]

Characteristic Analyzed	Technique(s)
Cellular proteins	Polyacrylamide gel electrophoresis [PAGE][30]; two-dimensional electrophoresis[19]
Enzyme characteristics	Multilocus enzyme electrophoresis[33]
Cellular lipids	Gas-liquid chromatography[31]
Cell wall composition	Various chemical and physical techniques[19]
Isoprenoid quinones	Thin-layer, reverse phase thin-layer, and conventional column chromatography[19]
Fermentation products	Gas-liquid chromatography[14,23] and high-pressure liquid chromatography[31]
Nucleic acid composition	DNA % guanine + cytosine[17]
Nucleic acid relatedness	DNA/DNA and DNA/rRNA hybridization[17]; 16S ribosomal RNA sequencing[6]

Source: Based on references as noted.

Classification of Anaerobes

The anaerobes isolated from clinical specimens either as true pathogens or indigenous microflora contaminants are divided into those capable of forming spores (called **sporeformers**) and those that are not (**nonsporeformers**). This taxonomic division and the genus names of clinically encountered anaerobes are shown in Figure 2-1.

Sporeformers

    Bacilli   *Clostridium* (m +/−)

    Cocci   *Sarcina*

Nonsporeformers

Gram-positive

    Bacilli   *Actinomyces, Arcanobacterium, Bifidobacterium, Eubacterium, Lactobacillus, Mobiluncus** (m), *Propionibacterium*

    Cocci   *Coprococcus, Gemella, Peptococcus, Peptostreptococcus, Streptococcus*

Gram-negative

    Bacilli   *Anaerobiospirillum* (m), *Bacteroides, Bilophila, Butyrivibrio* (m), *Campylobacter* (m), *Centipeda* (m), *Desulfomonas, Desulfovibrio* (m), *Fusobacterium, Leptotrichia, Mitsuokella, Porphyromonas, Prevotella, Selenomonas* (m), *Succinivibrio* (m), *Tissierella* (m), *Wolinella* (m)

    Spirochetes   *Treponema (m)*

    Cocci   *Acidaminococcus, Megasphaera, Veillonella*

* Technically gram-positive, but stain gram-negative
(m) = motile

Figure 2-1. Classification of clinically encountered anaerobes

All sporeforming anaerobic bacilli are classified in the genus *Clostridium* and are collectively referred to as clostridia. The spores produced by bacteria are technically known as **endospores**. They are thick-walled bodies formed within the bacterial cells—one endospore per cell. *Bacillus* spp. are examples of sporeforming aerobic bacteria.

Like other bacterial endospores, clostridial spores are capable of withstanding prolonged periods of adverse environmental conditions. When conditions once again become favorable, they germinate to form vegetative bacteria (i.e., bacteria capable of growing and multiplying).

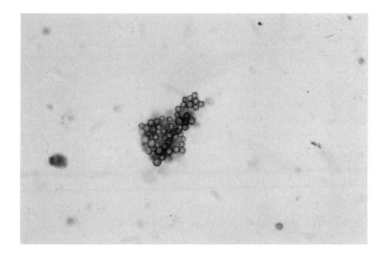

Figure 2-2. Endospore-stained appearance of *Sarcina ventriculi* (courtesy of Suzette L. Bartley, James D. Howard, and Ray Simon, Centers for Disease Control, Atlanta)

Sarcina ventriculi is an anaerobic, sporeforming coccus found in soil, mud, and animal feces. Endospores can be selectively stained using either of the spore stains described in Appendix D. Endospore-stained S. *ventriculi* are shown in Figure 2-2.

Nonsporeforming (also referred to as nonsporing) anaerobes are either gram-positive or gram-negative, and within each category are bacilli (rod-shaped bacteria) and cocci (spherical bacteria). Figure 2-1 lists some of the most important genera of nonsporeforming anaerobes.

Like other groups of bacteria, the nomenclature of anaerobes undergoes periodic change. Thus, some of the names in Tables 2-2 through 2-6 may have changed. The best sources of up-to-date information on bacterial taxonomy are recent issues of the *International Journal of Systematic Bacteriology*, available in many medical libraries, and the newsletter, *Anaerobe Abstracts*, available by subscription.

Clostridium Species

As previously mentioned, all sporeforming anaerobic bacilli are members of the genus *Clostridium*. Although all clostridia are capable of producing spores, some species do so readily and others, only under extremely harsh conditions. Thus, Gram-stained smears of clinical specimens containing clostridia may or may not contain spores. Even in the absence of spores, gram-positive bacilli observed in Gram-stained smears of clinical specimens should be suspected of being clostridia. With certain species, spores may not even be observed in Gram-stained smears of *Clostridium* colonies from an agar plate. Appendix D contains details of laboratory techniques used to induce spore formation.

Although they have a cell wall structure like other gram-positive bacteria, some clostridial species are difficult to stain, some appear gram-variable, and some routinely stain gram-negative. Thus, certain species (e.g., C. *ramosum* and C. *clostridioforme*) routinely appear as pink-staining bacilli in Gram-stained preparations. Chapter 8 contains information on special potency antimicrobial disks you should always use to determine the **true** gram reaction of a pink-staining anaerobic bacillus.

There are many ways to subdivide the clostridia. In some taxonomic schemes, they are grouped according to the location of the endospore formed within the cell (Figure 2-3). Spores are described as being either **terminal** (at the end of the cell; see Figure 2-4) or **subterminal** (at a location other than the end of the cell; see Figure 2-5). Clostridia are sometimes grouped according to various biochemical reactions or their ability to produce enzymes such as lecithinase and lipase. These enzymes are described in Chapter 8.

Clostridia with Terminal Spores

C. cadaveris
C. innocuum
C. paraputificum
C. ramosum
C. sphenoides (most strains)
C. tertium
C. tetani

Clostridia with Subterminal Spores

C. bifermentans
C. botulinum
C. butyricum
C. difficile
C. novyi type A
C. perfringens (spores rarely observed)
C. septicum
C. sordellii
C. sphenoides (some strains)
C. sporogenes

Figure 2-3. Classification of some clinically encountered clostridia by endospore location

Table 2-2 is an alphabetical listing of clinically encountered clostridia and the types of clinical specimens most apt to contain them. Virtually all the clostridia listed in Table 2-2 can be isolated from human feces, as either permanent or transient members of the gastrointestinal flora.

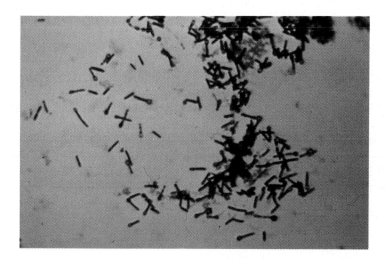

Figure 2-4. Gram-stained appearance of terminal spores of *Clostridium tetani* (courtesy of Suzette L. Bartley, James D. Howard, and Ray Simon, Centers for Disease Control, Atlanta)

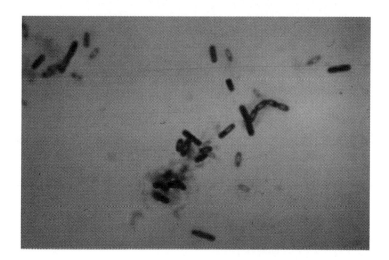

Figure 2-5. Gram-stained appearance of subterminal spores of *Clostridium sordellii* (courtesy of Suzette L. Bartley, James D. Howard, and Ray Simon, Centers for Disease Control, Atlanta)

Table 2-2. Clinically Encountered *Clostridium* Species

Clostridium Species	Clinically Relevant Information
C. aminovalericum	Bacteriuria (pregnant women)
C. argentinense	Infected wounds, bacteremia, botulism, amniotic fluid[40]
C. baratii	Infected war wounds; peritonitis; infectious processes of the eye, ear, and prostate
C. beijerinckii	Infected wounds
C. bifermentans	Infected wounds, abscesses, bacteremia
C. botulinum	Food poisoning, wound botulism, infant botulism
C. butyricum	Infectious processes of the urinary tract, lower respiratory tract, pleural cavity, abdomen; infected wounds; abscesses; bacteremia
C. cadaveris	Abscesses, infected wounds
C. carnis	Infectious processes of soft tissue, bacteremia
C. clostridioforme	Abdominal, cervical, scrotal, pleural, and other infectious processes; septicemia; peritonitis; appendicitis
C. cochlearium	No information available
C. difficile	Antibiotic-associated diarrhea, pseudomembranous colitis, bacteremia (rarely), pyogenic infectious processes (rarely) (also isolated from the hospital environment)
C. fallax	Infectious processes of soft tissues
C. ghonii	Infectious processes of soft tissues
C. glycolicum	Infected wounds, abscesses, peritonitis
C. hastiforme	Infected war wounds, bacteremia, abdominal abscesses
C. histolyticum	Infected war wounds, gas gangrene (has been isolated from the gingival plaque of institutionalized and primitive populations)
C. indolis	Infectious processes of the gastrointestinal tract
C. innocuum	Commonly isolated from infectious processes of the gastrointestinal tract, empyema
C. irregulare	Penile lesions
C. leptum	No information available
C. limosum	Bacteremia, peritonitis, pulmonary infectious processes
C. malenominatum	Various infectious processes
C. novyi	Infected wounds, gas gangrene
C. oroticum	Infectious processes of the urinary tract, rectal abscess (one report)
C. paraputrificum	Bacteremia, peritonitis, infected wounds, appendicitis

Table 2-2. Clinically Encountered *Clostridium* Species (*continued*)

Clostridium Species	Clinically Relevant Information
C. perfringens*	Infectious processes derived from colonic contents (e.g., peritonitis, intraabdominal abscesses, and infectious processes of soft tissues below the waist); involved in about 80% of cases of gas gangrene; bacteremia
C. putrefaciens	Bacteriuria (pregnant women with bacteremia)
C. putrificum	Abscesses, infected wounds, bacteremia
C. ramosum	Infectious processes of the abdominal cavity, genital tract, lung, biliary tract; bacteremia
C. sartagoforme	No information available
C. septicum	Bacteremia, suppurative infectious processes, necrotizing enterocolitis, gas gangrene
C. sordellii	Infected wounds, penile lesions, bacteremia, abscesses, infectious processes of the abdomen and vagina
C. sphenoides	Appendicitis, bacteremia, infectious processes of bone and soft tissue, intraperitoneal infectious processes, infected war wounds, visceral gas gangrene, renal abscess
C. sporogenes	Bacteremia, endocarditis, CNS and pleuropulmonary infectious processes, penile lesions, infected war wounds, other pyogenic infectious processes
C. subterminale	bacteremia (rarely); empyema; infectious processes of biliary tract, soft tissue, and bone
C. symbiosum	Liver abscesses, bacteremia, infectious processes derived from bowel flora (occasionally)
C. tertium	Appendicitis, brain abscesses, infectious processes related to the intestinal tract and soft tissue, infected war wounds, bacteremia (has been isolated from the gingival sulcus of two patients with periodontitis)
C. tetani	Infected gums and teeth, corneal ulcerations, infectious processes of mastoid and middle ear, intraperitoneal infectious processes, tetanus neonatorum, postpartum uterine infectious processes, various soft tissue infectious processes related to trauma (including abrasions and lacerations) and use of contaminated needles (has been isolated from the hospital environment)
C. thermosaccharolyticum	No information available

Note: Virtually all *Clostridium* spp. listed here have been isolated from fecal specimens of apparently healthy persons. Such isolates could represent transient rather than permanent residents of the normal colonic flora. Eighteen additional clostridial species isolated from feces but not from clinical specimens or other anatomical sites have been excluded from this list.

* *Clostridium perfringens* is the most commonly isolated clostridial species. It can be isolated from virtually any type of clinical specimen and can be involved in virtually any type of infectious process.

Source: Based on reference 38 except as otherwise noted.

Nonsporeforming, Anaerobic, Gram-Positive Bacilli

The next group of clinically significant anaerobes contains the heterogeneous, nonsporeforming, gram-positive bacilli (Table 2-3). These organisms have varying morphologies, ranging from very short rods to long, branching filaments. Although certain species of *Lactobacillus* are anaerobic, their role in infectious processes is uncertain, and they have therefore been omitted from Table 2-3.

Table 2-3. Clinically Encountered Nonsporeforming, Anaerobic, Gram-Positive Bacilli

Genus/Species	Anatomic Site(s)	Clinically Relevant Information
Actinomyces		
A. israelii	Oral cavity, tonsillar crypts, dental plaque, intestinal and female genital tracts	The principal causative agent of human cervicofacial, thoracic, and abdominal actinomycosis; infectious processes of the eye such as lacrimal canaliculitis, conjunctivitis, and dacryocystitis; cervicitis and endometritis in women using intrauterine or vaginal contraceptive devices; "sulfur granules" are produced
A. meyeri	Periodontal sulcus	Infrequently isolated from brain abscesses and pleural fluid; less often from abscesses of cervicofacial area, hip, hand, foot, spleen, and bite wounds
A. odontolyticus	Oral cavity, especially deep carious lesions, dental plaque, and calculus	Actinomycosis (rare cause); infectious processes of the eye such as lacrimal canaliculitis; periodontitis (possibly) and dental caries (possibly)
A. viscosus	Oral cavity, caries, dental plaque, and calculus	Periodontal disease; cervicovaginal secretions of women with and without IUDs; uninfected conjunctiva and cornea; cervicofacial and abdominal cases of actinomycosis (occasionally); lacrimal canaliculitis and other infectious processes of the eye; cervicitis and endometritis of women using IUDs (possibly)
Bifidobacterium		
B. bifidum	Colon, vagina	No information available
B. breve	Colon, vagina	Has been isolated from clinical specimens

Table 2-3. Clinically Encountered Nonsporeforming, Anaerobic, Gram-Positive Bacilli (*continued*)

Genus/Species	Anatomic Site(s)	Clinically Relevant Information
B. dentium	Colon, vagina, oral cavity, dental caries, dental plaque	Has been isolated from clinical specimens
B. infantis	Colon (infants), vagina	No information available
B. longum	Colon	Has been isolated from clinical specimens
B. pseudocatenulatum	Colon (infants)	No information available
Eubacterium		
E. aerofaciens	Colon	Occasionally isolated from blood cultures and various infectious processes, including subacute bacterial endocarditis, renal abscess fluid, and appendicial abscess
E. alactolyticum	Oral cavity	Frequent occurrence in infectious processes; dental calculus and gingival crevice in periodontal disease, root canals, purulent pleurisy, jugal cellulitis, post-operative wounds; abscesses of the brain, lung, intestinal tract, and mouth
E. brachy	Oral cavity	Subgingival samples and supra-gingival tooth scrapings from persons with periodontal disease; lung abscesses
E. combesii	Unknown	Has been isolated from various infectious processes
E. contortum	Colon, vagina	Blood, abdominal aortic aneurysm, wound
E. lentum	Colon	Blood, postoperative wounds, and various kinds of abscesses (brain, rectal, scrotal, pelvic)
E. limosum	Colon	Rectal and vaginal abscesses, blood, wounds
E. moniliforme	Colon	Blood, various infectious processes
E. nitritogenes	Colon	Has been isolated from various infectious processes
E. nodatum	Oral cavity	Subgingival samples and supra-gingival tooth scrapings from persons with periodontal diseases

Table 2-3. Clinically Encountered Nonsporeforming, Anaerobic, Gram-Positive Bacilli (*continued*)

Genus/Species	Anatomic Site(s)	Clinically Relevant Information
E. rectale	Colon	No information available
E. saburreum	Dental plaque and gingival crevice	No information available
E. tenue	Unknown	Abscess following abortion, knee synovial fluid, blood
E. ventriosum	Colon	Mouth abscess, infectious process of neck, purulent pleurisy, pulmonary abscesses, bronchiectasis

E. yurii
    subsp. *margaretiae*
    subsp. *schtitka*
    subsp. *yurii*
	Unknown	Periodontal pockets, subgingival dental plaque[25,26]

Mobiluncus *[39]		

M. curtisii
    subsp. *curtisii*
    subsp. *holmesii*
	Vagina	Thought to play a role in bacterial vaginosis

M. mulieris	Vagina	Thought to play a role in bacterial vaginosis
Propionibacterium		
P. acnes	Skin, colon	Acne vulgaris, wounds, blood, pus, and soft tissue abscesses
P. avidum	Moist areas of the skin (e.g., vestibule of the nose, axilla, perineum, sinuses)	Has been isolated from infected sinuses, chronically infected wounds, and submaxillary abscesses, but is probably not the primary cause of these infectious processes
P. granulosum	Oily areas of skin (e.g., forehead or between shoulder blades)	May play some part in the pathogenesis of acne but probably not otherwise pathogenic
P. propionicus	Oral cavity, cervicovaginal secretions of healthy women, secretions from uninfected conjunctiva and cornea	A cause of actinomycosis and lacrimal canaliculitis[6]

Note: Ten *Eubacterium* spp. and three *Bifidobacterium* spp. isolated from feces but not from clinical specimens or other anatomical sites have been excluded from this list.

* Although *Mobiluncus* spp. have a typical gram-positive cell wall, they stain pink with the Gram-staining procedure.

Source: Based on reference 38, except as otherwise noted.

Actinomyces spp. are straight to slightly curved bacilli ranging in size from short rods to long filaments. The shorter rods may have clubbed ends and may be seen in diphtheroid arrangements, short chains, or small clusters. Longer rods and straight or wavy filaments may be branched. Although the *Actinomyces* are gram-positive, irregular staining may cause a beaded or banded appearance. The typical branching, filamentous, Gram-stained appearance of an *Actinomyces*-like organism depicted in Figure 2-6 is referred to as "*Actinomyces*-like" in Chapter 8. Investigators at the CDC Anaerobic Bacteria Branch have recently found that members of the genus *Actinomyces* are seldom obligate anaerobes. Some are quite fastidious, however, requiring special vitamins (e.g., vitamin K_1), certain amino acids, and hemin for adequate growth.

Propionibacterium spp. are pleomorphic rods that may be coccoid, diphtheroidal, club shaped, bifurcated (forked), or branched. The term "diphtheroid" refers to an organism that morphologically resembles *Corynebacterium diphtheriae*. Because *P. acnes* is a common member of the skin microflora, it is frequently isolated from blood culture bottles as a skin **contaminant** (much like *Staphylococcus epidermidis*). However, like *S. epidermidis*, *P. acnes* can also cause subacute bacterial endocarditis and bacteremia and is thus not always a contaminant. A gram-positive, anaerobic diphtheroid that is both indole-positive and catalase-positive can be presumptively identified as *P. acnes*. The diphtheroid, Gram-stained appearance of *P. acnes* is shown in Figure 2-7; this is the morphology that is referred to as "diphtheroid appearance" in Chapter 8.

The genus *Arachnia* was originally created to accommodate organisms that had previously been designated *Actinomyces propionicus* but differed in several respects from *Actinomyces* spp., including their production of propionic acid as a major end product of glucose metabolism. *Arachnia propionica* has been reclassified in the genus *Propionibacterium* as *P. propionicus*.[6] Like *Actinomyces* spp., *P. propionicus* varies considerably in size and shape, ranging from coccoid and short, diphtheroidal rods to long, branched filaments. Individual cells may be of uneven diameter and may have distended or clubbed ends.

Eubacterium spp. may be observed as either uniform or pleomorphic, gram-positive rods. They may be coccoid, diphtheroidal, or filamentous and range in width from thin to plump.

Nonsporeforming, Anaerobic, Gram-Negative Bacilli

Tables 2-4 and 2-5 list many of the nonsporeforming, anaerobic, gram-negative bacilli involved in human infectious processes and found as members of the indigenous microflora. At first glance, the variety of such organisms appears overwhelming. Fortunately, only a few of the genera are commonly encountered in clinical specimens (*Bacteroides, Porphyromonas, Prevotella,* and *Fusobacterium* spp. and, less commonly, *Wolinella* spp.).

Bacteroides, Porphyromonas, and Prevotella Species. One of the most important genera of anaerobic, gram-negative bacilli is the genus *Bacteroides*. Organisms in this genus have received a great deal of publicity of late due to their frequent involvement in infectious processes and their resistance to antimicrobial agents. In recent years, many species of *Bacteroides* have been reclassified into other genera.

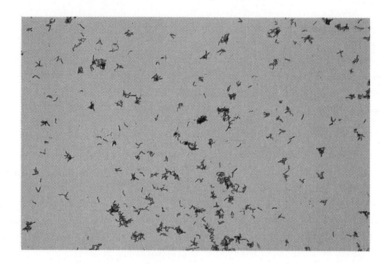

Figure 2-6. Gram-stained appearance of *Actinomyces israelii,* illustrating the term *"Actinomyces*-like" (courtesy of Suzette L. Bartley, James D. Howard, and Ray Simon, Centers for Disease Control, Atlanta)

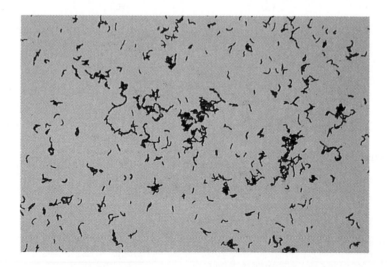

Figure 2-7. Gram-stained appearance of *Propionibacterium acnes,* illustrating the term "diphtheroid" (courtesy of Suzette L. Bartley, James D. Howard, and Ray Simon, Centers for Disease Control, Atlanta)

Table 2-4. Clinically Encountered *Bacteroides, Porphyromonas,* and *Prevotella* Species

Species	Anatomic Site(s)	Clinically Relevant Information
Bile tolerant		
***Bacteroides fragilis* group**		
B. caccae	Colon	Rarely isolated from clinical specimens[18]
B. distasonis	Colon (common)	Occasionally isolated from clinical specimens
B. fragilis	Colon	The most common species of anaerobic bacteria isolated from infectious processes of soft tissue
B. merdae	Colon[18]	No information available
B. ovatus	Colon	Occasionally isolated from clinical specimens
B. stercoris	Colon[18]	No information available
B. thetaiotaomicron	Colon (common)	Frequently found in clinical specimens
B. uniformis	Colon	Various clinical specimens
B. vulgatus	Colon (common)	Occasionally isolated from infectious processes
Others		
B. eggerthii	Colon	Occasionally isolated from clinical specimens
B. splanchnicus	Colon, vagina	Occasionally isolated from clinical specimens
Bile sensitive		
Pigmented		
***Prevotella*[35]**		
P. corporis	Unknown	Various clinical specimens
P. denticola	Gingival crevice	Various clinical specimens
P. intermedia	Gingival crevice	Specimens from head, neck, and pleural infectious processes; occasionally isolated from blood, abdominal, and pelvic sites
P. loescheii	Gingival crevice	No information available
P. melaninogenica	Gingival crevice	Various clinical specimens

Table 2-4. Clinically Encountered *Bacteroides, Porphyromonas,* and *Prevotella* Species (*continued*)

Species	Anatomic Site(s)	Clinically Relevant Information
Porphyromonas[34]		
P. asaccharolytica	Unknown	Various clinical specimens
P. endodontalis	Unknown	Dental root canals[44]
P. gingivalis	Mouth	No information available
Nonpigmented		
Pitting		
B. gracilis	Gingival crevice	No information available
B. ureolyticus	Buccal cavity, intestinal and urogenital tracts	Infectious processes of the respiratory and intestinal tracts; has been isolated from blood following tooth extractions
Nonpitting		
Bacteroides		
B. capillosus	Colon, mouth	Cysts and wounds
B. coagulans	Colon, urogenital tract	Occasionally isolated from clinical specimens
B. forsythus	Oral cavity[43]	No information available
B. galacturonicus	Colon[15]	No information available
B. pectinophilus	Colon[15]	No information available
B. pneumosintes	Gingival crevice	Nasopharyngeal washings from healthy individuals; secondary infectious processes of the upper respiratory tract; periodontal pockets, blood, various infectious processes of the head and neck; brain abscesses
B. putredinis	Colon, mouth (rarely)	Abdominal and rectal abscesses, cases of acute appendicitis
Prevotella[35]		
P. bivia	Vagina	Infectious processes of the urogenital or abdominal region; occasionally isolated from mouth, blood, chest fluid, breast abscesses

Table 2-4. Clinically Encountered *Bacteroides, Porphyromonas,* and *Prevotella* Species (*continued*)

Species	Anatomic Site(s)	Clinically Relevant Information
P. buccae	Gingival crevice	Chest drainages, blood, sinus asprirate (sinusitis), peritoneal fluid, mandibular cyst
P. buccalis	Oral cavity	No information available
P. disiens	Vagina, mouth	Abdominal and urogenital infectious processes
P. heparinolytica	Unknown	Periodontitis lesions[32]
P. oralis	Gingival crevice	Infectious processes of the oral cavity, upper respiratory, and genital tracts
P. oris	Gingival crevice	Systemic infectious processes; non-oral isolates from face, neck, and chest abscesses and drainages; abdominal wound drainages; peritoneal fluid; blood; spinal fluid
P. oulora	Gingival crevice[36]	No information available
P. veroralis	Oral cavity	No information available
P. zoogleoformans	Gingival sulcus	No information available

Source: Based on reference 21, except as otherwise noted.

The genus *Bacteroides* can first be divided into species that are bile-tolerant and those that are bile-sensitive. The various methods used in the laboratory to demonstrate bile tolerance are described in Chapter 8 and Appendix E. In the past, bile-sensitive species were further subdivided into pigmented and nonpigmented species (Table 2-4). Recently, however, the pigmented species of bile-sensitive *Bacteroides* have been reclassified into the genera *Porphyromonas* and *Prevotella*.

As shown in Table 2-4, the bile-tolerant *Bacteroides* include members of the *B. fragilis* group plus two additional species. Although *B. fragilis* is the most common species of anaerobic bacteria isolated from infectious processes of soft tissue and anaerobic bacteremia, this species accounts for less than 1% of the human intestinal microflora. *Bacteroides vulgatus, B. thetaiotaomicron,* and *B. distasonis* are among the most common species of bacteria isolated from human feces. The Gram-stained appearance of a typical *Bacteroides* species is shown in Figure 2-8.

Penicillin-resistant strains of *B. fragilis* and other members of the *B. fragilis* group have been encountered for a number of years, but there are also recent reports of resistance to tetracycline, cefotaxime, cefoperazone, moxalactam, and clindamycin

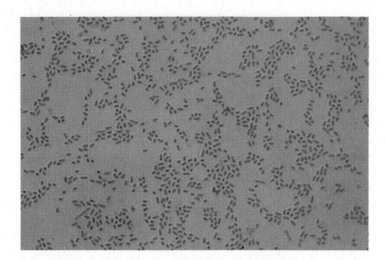

Figure 2-8. Gram-stained appearance of *Bacteroides thetaiotaomicron,* illustrating the typical appearance of *Bacteroides* spp. (courtesy of Suzette L. Bartley, James D. Howard, and Ray Simon, Centers for Disease Control, Atlanta)

among isolates of this group. It is important to identify members of the *B. fragilis* group to the species level because of species-to-species variability in both virulence and drug resistance.

As described in Chapter 10, anaerobes other than members of the *B. fragilis* group are also resistant to a number of antimicrobial agents. Members of the *B. fragilis* group were the first anaerobes reported to produce **beta-lactamases,** enzymes capable of inactivating penicillins and cephalosporins. It is now known that many other anaerobes are also capable of producing beta-lactamases, including isolates of the following species: *B. coagulans, B. splanchnicus, Prevotella oralis, P. bivia, P. disiens, P. oris,* and *P. buccae.* Beta-lactamase production by anaerobes and their resistance to antimicrobial agents are described in Chapter 10.

Some species of *Prevotella* (e.g., *P. corporis, P. intermedia, P. loescheii, P. melaninogenica,* and some species of *P. denticola*) produce a dark pigment (**protoporphyrin**) that causes their colonies to become brown to black. Pigmentation of colonies may not become evident until two to three weeks of incubation. Plated media containing laked blood are frequently used in an attempt to enhance pigmentation.

Three asaccharolytic, pigmented species of *Bacteroides* (namely, *B. asaccharolyticus, B. gingivalis,* and *B. endodontalis*) were reclassified into a genus called *Porphyromonas.*[34] The genus name, *Porphyromonas,* comes from the Greek adjective porphyreos, meaning purple. Although the name refers to the dark pigment produced by *Porphyromonas* spp., the pigment is actually **protoheme** rather than protoporphyrin.[34] *Porphyromonas* spp. (*P. asaccharolytica, P. gingivalis,* and *P. endodontalis*) produce a dark brown to black pigment when grown for six to ten days on blood agar plates. All three species are obligate anaerobes, and most strains require hemin and menadione (a vitamin K derivative) for growth. They are often associated with human opportunistic infections, and the pathogenic potential of some has been clearly demonstrated in experimental infections. *P. asaccharolytica* is widely distributed in human tissues and fluids and has been isolated from feces, the cervix, ear tissue, the umbilical cord, amniotic fluid, blood, empyema, peritoneal and pelvic abscesses, endometritis, and bite wound infections. *P. gingivalis* appears to be positively associated with several periodontal diseases. Its common ecological niche is the gingiva or periodontal pocket. *P. endodontalis* has been isolated only from mixed oral infections, predominantly pyogenic infections of odontogenic or dental root canal infections. Its clinical importance is unknown, but its isolation from oral infections suggests a possible pathogenic role. All three species are susceptible to penicillin, clindamycin, erythromycin, metronidazole, and tetracycline, less susceptible to vancomycin, spiramycin, and chloramphenicol, and resistant to gentamicin. Because most strains are susceptible to vancomycin, they will not grow on media containing 5 μg or more of vancomycin per milliliter.[27]

A listing of the nonpigmented, bile-sensitive *Bacteroides* spp., including those capable of pitting or corroding the agar surface, can be found in Table 2-4. At one time, pitting, anaerobic, gram-negative bacilli were called *Bacteroides corrodens*, but use of that name has been discontinued. Microaerophilic, agar-pitting, gram-negative bacilli are now called *Eikenella corrodens*, whereas pitting, gram-negative anaerobes have been placed in the *Bacteroides ureolyticus* group. This group consists of *Wolinella* spp., *B. ureolyticus*, and *B. gracilis* (see Table 2-4). As described in Chapter 8, these species are easily differentiated using tests to detect motility and urease (an enzyme capable of splitting urea). Not all strains of these organisms will pit agar, and even with strains that do, not all colonies will appear to be pitting. Thus, they may resemble a mixed culture.

Fusobacterium **Species.** The term **fusiform** (meaning long, thin, and tapered at the ends) is frequently associated with the genus *Fusobacterium*. It is important to note, however, that only one species has cells consistently fusiform in shape and bacteria that are fusiform in shape are not necessarily fusobacteria.

Fusobacterium nucleatum is the only *Fusobacterium* species that **routinely** exists as long, thin, gram-negative rods, tapered (pointed) at the ends. The Gram-stained appearance of *F. nucleatum* is depicted in Figure 2-9. Although some cells of certain other fusobacteria (e.g., *F. naviforme*, *F. necrophorum*) may be fusiform, these and most other *Fusobacterium* spp. listed in Table 2-5 are **pleomorphic** (occurring in a variety of shapes). In fact, certain species (such as *F. mortiferum*) can be extremely pleomorphic, exhibiting globular forms, swellings, and other bizarre shapes. The pleomorphism of *F. mortiferum*, for example, is depicted in Figure 2-10. Organisms other than fusobacteria may also have fusiform shaped cells; examples include *Bacteroides gracilis*, *B. forsythus*, and microaerophilic *Capnocytophaga* spp.

Curved, Motile, Anaerobic, Gram-Negative Bacilli. Certain "*Vibrio*-like" anaerobes or "anaerobic vibrios" (curved, motile, gram-negative staining bacilli) have been reclassified as *Anaerobiospirillum*, *Butyrivibrio*, *Campylobacter*, *Desulfovibrio*, *Selenomonas*, *Succinivibrio*, *Mobiluncus*, and *Wolinella* spp. Of these, *Wolinella* and *Mobiluncus* are encountered most frequently in clinical materials - the former from oral, gastrointestinal, and vaginal specimens and the latter from vaginal specimens.

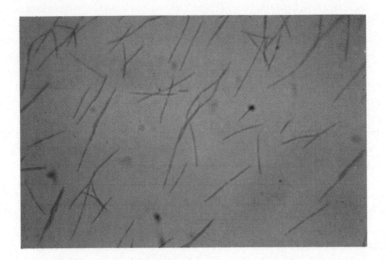

Figure 2-9. Gram-stained appearance of *Fusobacterium nucleatum*, illustrating the fusiform morphology of this organism (courtesy of Suzette L. Bartley, James D. Howard, and Ray Simon, Centers for Disease Control, Atlanta)

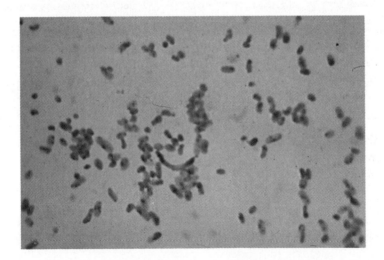

Figure 2-10. Gram-stained appearance of *Fusobacterium mortiferum*, illustrating pleomorphism (courtesy of Suzette L. Bartley, James D. Howard, and Ray Simon, Centers for Disease Control, Atlanta)

Table 2-5. Clinically Encountered Anaerobic, Gram-Negative Bacilli (Other Than *Bacteroides, Prevotella,* and *Porphyromonas* Species)

Genus/Species	Anatomic Site(s)	Clinically Relevant Information
Anaerobiospirillum		
A. succiniciproducens	Unknown	Blood, fecal specimens from patients with diarrhea[24,28]
Anaerorhabdus		
A. furcosus	Colon (infrequent)	Infected appendix, lung and abdominal abscesses
Bilophila		
B. wadsworthia	Colon	Intraabdominal specimens from patients with gangrenous and perforated appendicitis[2]
Butyrivibrio		
B. crossotus	Colon	No information available
B. fibrisolvens	Colon	Infectious processes of the eye
Campylobacter		
C. concisus	Oral cavity	Periodontal disease, wound isolate[16]
Centipeda		
C. periodontii	Oral cavity	Periodontal disease[22]
Desulfomonas		
D. pigra	Colon	Peritoneal fluid, pylonidal cyst abscess, ruptured sigmoid colon
Desulfovibrio		
D. vulgaris	Colon	Pleural fluid[16]
Fusobacterium		
F. alocis	Gingival sulcus	Submandibular abscess, mouth ulcer[5]
F. gonidiaformans	Intestinal and uro-genital tracts	Various types of infectious processes
F. mortiferum	Colon	Blood and various other clinical specimens
F. necrogenes	Colon	No information available
F. necrophorum	Body cavities	Necrotic lesions, abscesses, blood

F. nucleatum
    subsp. *nucleatum*
    subsp. *polymorphum*
    subsp. *vincentii*[9]
	Gingival margin and sulcus	Infectious processes of the upper respiratory tract and pleural cavity, occasionally from wounds and other infectious processes

Table 2-5. Clinically Encountered Anaerobic, Gram-Negative Bacilli (Other Than *Bacteroides, Prevotella,* and *Porphyromonas* Species) (*continued*)

Genus/Species	Anatomic Site(s)	Clinically Relevant Information
F. periodonticum	Oral cavity	No information available
F. prausnitzii	Colon (very common)	No information available
F. pseudonecrophorum[37]	Unknown	No information available
F. russii	Colon	No information available
F. sulci	Gingival sulcus[5]	No information available
F. ulcerans	None	Cutaneous tropical ulcers[1]
F. varium	Colon	Purulent infectious processes (upper respiratory tract, surgical wounds, peritonitis)
Leptotrichia		
L. buccalis	Oral cavity, female periurethral region	No information available
Megamonas		
M. hypermegas	Colon	No information available
Mitsuokella		
M. dentalis	Oral cavity	Dental root canals[13]
M. multiacida	Colon	Occasionally isolated from clinical specimens
Selenomonas		
S. artemidis	Gingival crevice[27]	No information available
S. dianae	Gingival crevice[29]	No information available
S. flueggei	Gingival crevice[29]	No information available
S. infelix	Gingival crevice[29]	No information available
S. noxia	Gingival crevice[29]	No information available
S. sputigena	Gingival crevice	Transtrachial aspirate and pleural fluid[16]; relationship to periodontal disease unknown
Succinimonas		
S. amylolytica	Unknown	Wound drainage[16]
Succinivibrio		
S. dextrinosolvens	Oral cavity (rare), colon (rare)	Septicemia[16]
Tissierella		
T. praeacuta	Colon	Infrequently from lung abscesses, gangrenous lesions, blood[7,16]

Table 2-5. Clinically Encountered Anaerobic, Gram-Negative Bacilli (Other Than *Bacteroides, Prevotella,* and *Porphyromonas* Species) (*continued*)

Genus/Species	Anatomic Site(s)	Clinically Relevant Information
Wolinella		
W. curva (*Campylobacter curvus*)	Oral cavity	Lesions in oral cavity, blood culture; pathogenicity unknown[16,42]
W. recta (*Campylobacter rectus*)	Gingival crevice	Periodontal pockets, necrotic dental root canals; pathogenicity unknown

Source: Based on reference 21, except as otherwise noted.

Growth of *Wolinella* spp. is stimulated by formate-fumarate, whereas growth of *Mobiluncus* spp. is not. Although cells of *W. curva* are curved, *W. recta* cells are straight rods.[42] *Wolinella* spp. are capable of pitting the agar surface like *B. ureolyticus* and *B. gracilis*, but they can be readily differentiated from these *Bacteroides* spp. by their motility. Not all strains of *Wolinella* will pit agar, and even pitting strains will produce a variety of colony types on a given plate.

Mobiluncus spp. are curved, motile bacilli that stain pink or gram-variable with the Gram staining technique. They are not gram-negative, however. Their cell walls are structurally similar to gram-positive organisms and lack lipopolysaccharide (LPS). Like gram-positive organisms, *Mobiluncus* spp. are susceptible to vancomycin and resistant to colistin.[4] For these reasons, *Mobiluncus* spp. are considered gram-positive organisms in this book and are included in Table 2-3. They are found in the healthy vagina but increase dramatically in number during and are thought to play a role in bacterial vaginosis. The Gram-stained appearance of *Mobiluncus* is depicted in Figure 2-11.

Anaerobic Cocci

Gram-Negative Cocci. Although several genera of anaerobic, gram-negative cocci (Table 2-6) are found in the indigenous microflora, only *Veillonella* spp. are implicated as pathogens. *Veillonella* are very small cocci (0.3–0.5 μm in diameter) that inhabit the oral cavity. Other genera of gram-negative cocci (*Acidaminococcus* and *Megasphaera*) live in the gastrointestinal tract but are only rarely isolated from clinical specimens. Tiny, nitrate-positive, gram-negative, anaerobic cocci may be reported as *Veillonella* spp. (Chapter 8). Some strains of *Veillonella* are nitrate-negative, however, and these require a more extensive biochemical workup for identification (Chapter 9). The Gram-stained appearance of *Veillonella* is depicted in Figure 2-12.

Gram-Positive Cocci. Most of the gram-positive, anaerobic cocci previously classified as *Peptococcus* spp. have been renamed *Peptostreptococcus* spp. The exception is *Peptococcus niger*, which produces tiny black colonies that become light gray when exposed to air and is only rarely isolated from clinical specimens. Media

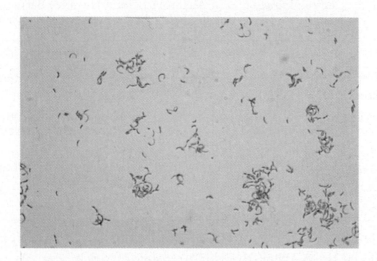

Figure 2-11. Gram-stained appearance of *Mobiluncus curtisii* subsp. *curtisii,* illustrating the curved morphology of this organism (courtesy of Suzette L. Bartley, James D. Howard, and Ray Simon, Centers for Disease Control, Atlanta)

Figure 2-12. Gram-stained appearance of *Veillonella* spp. (from reference 41; used with permission of Star Publishing Co.)

Table 2-6. Clinically Encountered Anaerobic Cocci

Genus/Species	Anatomic Site(s)	Clinically Relevant Information
Gram-negative cocci		
Acidaminococcus		
A. fermentans	Colon	Rarely isolated from clinical specimens
Megasphaera		
M. elsdenii	Colon	Rarely isolated from clinical specimens
Veillonella		
V. atypica *V. dispar* *V. parvula*	Oral cavity	Head, neck, dental, and pulmonary infectious processes; bite wounds[41]
Gram-positive cocci		
Coprococcus	Three species have been isolated from feces but not from clinical specimens	No information available
Gemella	Unknown	No information available
Peptococcus		
P. niger	Umbilicus, vaginal area	Only occasionally isolated from human clinical specimens; more commonly isolated from veterinary specimens
Peptostreptococcus		
P. anaerobius	Vagina, colon	Isolated from a wide variety of clinical specimens, including abscesses of brain, jaw, pleural cavity, ear, pelvic region, urogenital area, and abdominal region; also from blood, spinal fluid, joint cultures, and specimens from osteomyelitis; gingival crevice of persons with gingivitis or periodontal diseases
P. asaccharolyticus	Vagina	Vaginal discharge, skin abscess, peritoneal abscess
P. hydrogenalis	Unknown	Has been isolated from feces and vaginal discharge[10]

Table 2-6. Clinically Encountered Anaerobic Cocci (*continued*)

Genus/Species	Anatomic Site(s)	Clinically Relevant Information
P. magnus	Genital tract, rare in colon or gingival crevice	Wounds, abscesses of abdominal, peritoneal, appendiceal, and urogenital sites; more frequent in anatomic sites below the diaphragm
P. micros	Genital tract, infrequently in healthy gingival crevice and intestinal tract	Frequently isolated from clinical specimens, including brain, lung, jaw, head, neck and bite abscesses, spinal fluid, blood, and abscesses at other body sites; often a major component of the flora of the gingival sulcus in periodontal disease
P. prevotii	Skin, vagina, tonsils	No information available
P. productus	Colon (one of the more predominant members of the fecal flora)	No information available
P. tetradius	Vagina	Vaginal discharge and various purulent secretions
Ruminococcus	Six species have been isolated from feces but not from clinical specimens	No information available
Sarcina		
S. ventriculi	Colon	Diseased stomach
Staphylococcus		
S. saccharolyticus	Skin	Endocarditis (one case) [45]
Streptococcus		
S. hansenii *S. pleomorphus*	Colon	No information available

Source: Based on reference 38, except as otherwise noted.

preparation, age, and storage conditions are critical for the isolation of clinically significant, anaerobic, gram-positive cocci. The size and arrangement (e.g., pairs, tetrads, chains, clusters) of the gram-positive, anaerobic cocci vary with growth conditions and are, therefore, not reliable criteria for identification. The various species names and anatomical sites of this group of anaerobes are contained in Table 2-6. Most of the anaerobic, gram-positive cocci isolated from clinical specimens are in the genus *Peptostreptococcus*. The Gram-stained appearance of a *Peptostreptococcus* species is shown in Figure 2-13. Other anaerobic, gram-positive cocci encountered in clinical materials are *Coprococcus, Gemella, Sarcina,* and *Streptococcus* spp.

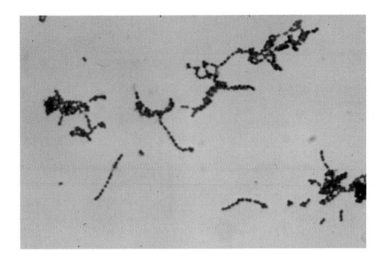

Figure 2-13. Gram-stained appearance of a *Peptostreptococcus* sp., illustrating the chain formation that occurs with some species (courtesy of Suzette L. Bartley, James D. Howard, and Ray Simon, Centers for Disease Control, Atlanta)

Anaerobes Isolated Most Frequently from Clinical Specimens

Although many different species of anaerobes can potentially be isolated from human clinical specimens, the number of species **routinely** isolated is relatively small. Finegold has reported that two-thirds of clinically significant, anaerobe-associated, infectious processes involve the following five anaerobes or groups of anaerobes:[11]

- Members of the *Bacteroides fragilis* group

- Pigmented species of *Bacteroides* (all have been reclassified as *Porphyromonas* and *Prevotella* spp.)

- *Fusobacterium nucleatum*

- *Clostridium perfringens*

- Anaerobic cocci

Consequently, these are the most critical anaerobes that microbiologists must isolate from clinical materials, identify, and test for susceptibility/resistance to appropriate antimicrobial agents.

The relative frequencies of isolation of specific anaerobes from clinical specimens were described by Brook.[3] He cited data compiled at two large military hospitals over a 12-year period. The most commonly isolated anaerobes were *Bacteroides* spp. (43% of all isolates), anaerobic, gram-positive cocci (26%), *Propionibacterium* spp. (13%), *Clostridium* spp. (7%), and *Fusobacterium* spp. (4%). Collectively, these five groups represented 93% of all anaerobic isolates (see Figure 2-14). Of the *Bacteroides* isolates, members of the *B. fragilis* group accounted for 44%, and pigmented species (all of which have since been renamed) accounted for 21%. Within the *B. fragilis* group, *B. fragilis* accounted for 63% and *B. thetaiotaomicron* for 14%. *Bacteroides melaninogenicus* (*Prevotella melaninogenica*) accounted for 42% of the pigmented species. *Peptostreptococcus magnus* (18%), *P. asaccharolyticus* (17%), *P. anaerobius* (16%), and *P. prevotii* (13%) were the most frequently isolated peptostreptococci, collectively representing 64% of the gram-positive cocci. *Clostridium perfringens* accounted for 48% of the clostridia isolated. The most commonly isolated *Fusobacterium* species was *F. nucleatum.*

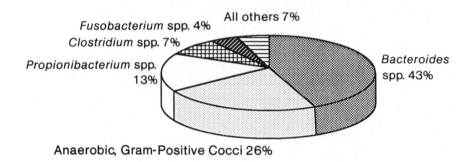

Figure 2-14. Frequency of isolation of anaerobes from clinical specimens (source: based on reference 3)

Goldstein and Citron reported data regarding the frequency of isolation of members of the *Bacteroides fragilis* group at two California community hospitals.[12] Of the anaerobic bacteria isolated from clinical sources at one of these hospitals during 1987, 34.6% were members of this group. Combining data from both hospitals gave the following relative frequencies of isolation: *B. fragilis,* 61%; *B. thetaiotaomicron,* 17%; *B. distasonis,* 7%; *B. vulgatus,* 6%; *B. ovatus,* 5%; and *B. uniformis,* 4% (see Figure 2-15). The authors concluded that 33 to 45% of the *B. fragilis* group isolates were species other than *B. fragilis,* and recommended that clinical laboratories routinely identify *B. fragilis* group members to the species level and perform *in vitro* susceptibility tests on all such isolates.

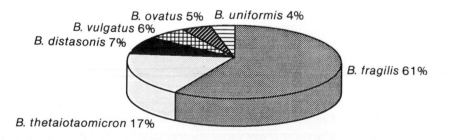

Figure 2-15. Frequency of isolation of members of the *Bacteroides fragilis* group from clinical specimens (source: based on reference 12)

Chapter in Review

- Anaerobes are taxonomically divided into those that produce endospores and those that do not. All sporeforming, anaerobic bacilli are members of the genus *Clostridium*. Although all clostridia form spores, spores are not always seen in clinical specimens containing clostridia or pure cultures of these organisms.

- Although clostridia have typical gram-positive cell walls, some species (e.g., *C. ramosum* and *C. clostridiiforme*) routinely stain pink with the Gram staining technique. Thus, techniques are need in the laboratory to determine the **true** gram reaction of pink-staining, anaerobic bacilli.

- There are many ways to subdivide the clostridia, including endospore location, biochemical reactions, and enzyme production. Spores are described as being terminal or subterminal in location.

- Nonsporeforming, anaerobic, gram-positive bacilli that are encountered most often in clinical materials are *Actinomyces*, *Bifidobacterium*, *Eubacterium*, *Lactobacillus*, and *Propionibacterium* spp.

- *Bacteroides* and *Fusobacterium* spp. are the most commonly encountered anaerobic, gram-negative bacilli. Traditionally, *Bacteroides* spp. were subdivided on the basis of tolerance of or resistance to 20% bile, their ability to produce brown-black pigmented colonies, and whether they are saccharolytic or asaccharolytic. Recently, however, most pigmented species of *Bacteroides* have been reclassified as *Porphyromonas* or *Prevotella* spp.

- Members of the *Bacteroides fragilis* group are especially pathogenic and drug resistant. It is important to identify members of the *B. fragilis* group to the species level because of species-to-species variability in both virulence and drug resistance.

- Members of the *B. fragilis* group include *B. caccae*, *B. distasonis*, *B. fragilis*, *B. merdae*, *B. ovatus*, *B. stercoris*, *B. thetaiotaomicron*, *B. uniformis*, and *B. vulgatus*. *B. fragilis* is the most common species of anaerobic bacteria isolated from soft tissue infectious processes and bacteremia.

- *Porphyromonas* spp. are asaccharolytic, bile-sensitive, gram-negative bacilli that produce brown-black pigmented colonies. Species of *Porphyromonas* are *P. asaccharolytica*, *P. endodontalis*, and *P. gingivalis*.

- *F. nucleatum* is the only *Fusobacterium* species that routinely exists as long, thin, gram-negative bacilli with tapered ends. Although some cells of certain other fusobacteria (e.g., *F. gonidiaformans*, *F. naviforme*, and *F. necrophorum*) may be fusiform, these species and most other *Fusobacterium* spp. are **pleomorphic**.

- Two of the most commonly encountered motile, curved, gram-negative bacilli, once lumped together as "anaerobic vibrios," are *Wolinella* and *Mobiluncus* spp. Not all *Wolinella* are curved, but most strains produce colonies that pit the agar surface. Although *Mobiluncus* spp. stain pink when Gram stained, they have typical gram-positive cell walls, lack LPS, are susceptible to vancomycin, and resistant to colistin. Thus, they are taxonomically classified as gram-positive organisms.

- Most anaerobic gram-positive and gram-negative cocci encountered in clinical specimens are *Peptostreptococcus* spp. and *Veillonella* spp., respectively.

Self-Assessment Exercises

1. Give the gram reaction (positive or negative) and morphologic appearance of each of the following anaerobes (e.g., *Propionibacterium acnes* is a gram-positive diphtheroid).

	Gram Reaction	Morphology
a. *Bacteroides fragilis*	_____	_____
b. *Clostridium perfringens*	_____	_____
c. *Fusobacterium nucleatum*	_____	_____
d. *Peptostreptococcus anaerobius*	_____	_____
e. *Veillonella* spp.	_____	_____
f. *Wolinella* spp.	_____	_____
g. *Actinomyces* spp.	_____	_____
h. *Mobiluncus* spp.	_____	_____
i. *Bifidobacterium* spp.	_____	_____

2. TRUE or FALSE: All sporeforming, anaerobic bacilli are classified as *Clostridium* spp.

3. Two *Clostridium* spp. that routinely stain pink in Gram-stained preparations are C. _____ and C. _____.

4. TRUE or FALSE: It is easy to recognize *Clostridium* spp. in Gram-stained smears of clinical specimens because they always contain spores.

5. The only *Fusobacterium* species that consistently has cells that are "fusiform" in shape (long, thin rods of variable length, with pointed ends) is F._____.

6. Define the following terms:

   Diphtheroid _____

   _____

   Pleomorphic _____

   _____

7. The two genera of anaerobic, gram-negative bacilli most commonly isolated from clinical specimens are _____ and _____.

8. Name the anaerobe most commonly involved in soft tissue infectious processes and bacteremia; give both genus and species. _____

9. Most anaerobic, gram-positive cocci isolated from clinical materials are species of _____, and most gram-negative cocci are species of _____.

10. List five genera of nonsporeforming, anaerobic, gram-positive bacilli.

    a. _____

    b. _____

    c. _____

    d. _____

    e. _____

11. List five species in the *Bacteroides fragilis* group.

    a. _____

    b. _____

    c. _____

    d. _____

    e. _____

12. Name three species in the genus *Porphyromonas*.

    a. _____

    b. _____

    c. _____

References

1. Adriaans, B., and H. Shah. 1988. *Fusobacterium ulcerans* sp. nov. from tropical ulcers. Int. J. Syst. Bacteriol. 38:447–448.

2. Baron, E. J., P. Summanen, J. Downes, M. C. Roberts, H. Wexler, and S. M. Finegold. 1989. *Bilophila wadsworthia*, gen. nov. and sp. nov., a unique gram-negative anaerobic rod recovered from appendicitis specimens and human faeces. J. Gen. Microbiol. 135:3405–3411.

3. Brook, I. 1988. Recovery of anaerobic bacteria from clinical specimens in 12 years at two military hospitals. J. Clin. Microbiol. 26:1181–1188.

4. Carlone, M., M. L. Thomas, R. J. Arko, G. O. Guerrant, C. W. Moss, J. M. Swenson, and S. A. Morse. 1986. Cell wall characteristics of *Mobiluncus* species. Int. J. Syst. Bacteriol. 36:288–296.

5. Cato, E. P., L. V. H. Moore, and W. E. C. Moore. 1985. *Fusobacterium alocis* sp. nov. and *Fusobacterium sulci* sp. nov. from the human gingival sulcus. Int. J. Syst. Bacteriol. 35:475–477.

6. Charfreitag, O., M. D. Collins, and E. Stackebrandt. 1988. Reclassification of *Arachnia propionica* as *Propionibacterium propionicus* comb. nov. Int. J. Syst. Bacteriol. 38:354–357.

7. Collins, M. D., and H. N. Shah. 1986. Reclassification of *Bacteroides praeacutus* Tissier (Holdeman and Moore) in a new genus, *Tissierella*, as *Tissierella praeacuta* comb. nov. Int. J. Syst. Bacteriol. 36:461–463.

8. Dowell, V. R., Jr. 1988. Characteristics of some clinically encountered anaerobic bacteria. Unpublished document that accompanied a lecture presented at the IV National Symposium on Clinical Microbiology of the Sociedade Brasileira de Microbiologia, Rio de Janeiro, Brazil. Centers for Disease Control, Atlanta.

9. Dzink, J. L., M. T. Sheenan, and S. S. Socransky. 1990. Proposal of three subspecies of *Fusobacterium nucleatum* Knorr 1922: *Fusobacterium nucleatum* subsp. *nucleatum* subsp. nov., comb. nov.; *Fusobacterium nucleatum* subsp. *polymorphum* subsp. nov., nom. rev., comb. nov.; and *Fusobacterium nucleatum* subsp. *vincentii* subsp. nov., nom. rev., comb. nov. Int. J. Syst. Bacteriol. 40:74–78.

10. Ezaki, T., S. -L. Liu, Y. Hashimoto, and E. Yabuuchi. 1990. *Peptostreptococcus hydrogenalis* sp. nov. from human fecal and vaginal flora. Int. J. Syst. Bacteriol. 40:305–306.

11. Finegold, S. M. 1987. Anaerobic bacteria: their role in infection and their management. Postgrad. Med. 81:141–147.

12. Goldstein, E. J. C., and D. M. Citron. 1988. Annual incidence, epidemiology, and comparative in vitro susceptibilities to cefoxitin, cefotetan, cefmetazole, and ceftizoxime of recent community-acquired isolates of the *Bacteroides fragilis* group. J. Clin. Microbiol. 26:2361–2366.

13. Haapasalo, M., H. Ranta, H. Shah, K. Ranta, K. Lounatmaa, and R. M. Kroppenstedt. 1986. *Mitsuokella dentalis* sp. nov. from dental root canals. Int. J. Syst. Bacteriol. 36:566–568.

14. Holdeman, L. V., E. P. Cato, and W. E. C. Moore (eds.). 1977 (with 1987 update). VPI Anaerobe Laboratory Manual, 4th ed. Virginia Polytechnic Institute and State University, Blacksburg.

15. Jensen, N. S., and E. Canale-Parola. 1986. *Bacteroides pectinophilus* sp. nov. and *Bacteroides galacturonicus* sp. nov. : two pectinolytic bacteria from the human intestinal tract. Appl. Environ. Microbiol. 52:880–887.

16. Johnson, C. C., and S. M. Finegold. 1987. Uncommonly encountered, motile, anaerobic gram-negative bacilli associated with infection. Rev. Inf. Dis. 9:1150–1162.

17. Johnson, J. L. 1984. Bacterial classification III: nucleic acids in bacterial classification, p. 8–11. *In* N. R. Krieg and J. C. Holt (eds.), Bergey's manual of systematic bacteriology, vol. 1. William and Wilkins, Baltimore.

18. Johnson, J. L., W. E. C. Moore, and L. V. H. Moore. 1986. *Bacteroides caccae* sp. nov., *Bacteroides merdae* sp. nov., and *Bacteroides stercoris* sp. nov. isolated from human feces. Int. J. Syst. Bacteriol. 36:499–501.

19. Jones, D., and N. R. Krieg. 1984. Bacterial classification V: serology and chemotaxonomy, p. 15–18. In N. R. Krieg and J. C. Holt (eds.), Bergey's manual of systematic bacteriology, vol. 1. William and Wilkins, Baltimore.

20. Koneman, E. W., S. D. Allen, V. R. Dowell, Jr., W. M, Janda, H. M. Sommers, and W. C. Winn, Jr. 1988. Color atlas and textbook of diagnostic microbiology, 3rd ed. J. B. Lippincott, Philadelphia.

21. Krieg, N. R., and J. G. Holt (eds.). 1984. Bergey's manual of systematic bacteriology, vol. 1. Williams and Wilkins, Baltimore.

22. Lai, C. -H., B. M. Males, P. A. Dougherty, P. Berthold, and M. A. Listgarten. 1983. *Centipeda periodontii* gen. nov., sp. nov. from human periodontal lesions. Int. J. Syst. Bacteriol. 33:628–635.

23. Lombard, G. L., and V. R. Dowell, Jr. 1982. Gas liquid chromatography and analysis of the acid products of bacteria. Centers for Disease Control, Atlanta.

24. Malnick, H., K. Williams, Phil-Ebosie, J., and A. S. Levy. 1990. Description of a medium for isolating *Anaerobiospirillum* spp., a possible cause of zoonotic disease, from diarrheal feces and blood of humans and use of the medium in a survey of human, canine, and feline feces. J. Clin. Microbiol. 28:1380–1384.

25. Margaret, B. S., and G. N. Krywolap. 1986. *Eubacterium yurii* subsp. *yurii* sp. nov. and *Eubacterium yurii* subsp. *margaretiae* subsp. nov.: test tube brush bacteria from subgingival dental plaque. Int. J. Syst. Bacteriol. 36:145–149.

26. Margaret, B. S., and G. N. Krywolap. 1988. *Eubacterium yurii* subsp. *schtitka subsp.* nov.: test tube brush bacteria from subgingival dental plaque. Int. J. Syst. Bacteriol. 38:207–208.

27. Mayrand, D., and S. C. Holt. 1988. Biology of asaccharolytic black-pigmented *Bacteroides* species. Microbiol. Rev. 52:134–152.

28. McNeil, M. M., W. J. Martone, and V. R. Dowell, Jr. 1987. Bacteremia with *Anaerobiospirillum succiniciproducens*. Rev. Inf. Dis. 9:737–742.

29. Moore, L. V. H., J. L. Johnson, and W. E. C. Moore. 1987. *Selenomonas noxia* sp. nov., *Selenomonas flueggei* sp. nov., *Selenomonas infelix* sp. nov., *Selenomonas dianae* sp. nov., and *Selenomonas artemidis* sp. nov., from the human gingival crevice. Int. J. Syst. Bacteriol. 37:271–280.

30. Moore, W. E. C., D. E. Hash, L. V. Holdeman, and E. P. Cato. 1980. Polyacrylamide slab gel electrophoresis of soluble proteins for studies of bacterial flora. Appl. Environ. Microbiol. 39:900–907.

31. Moss, C. W. 1985. Use of gas-liquid chromatography and high pressure liquid chromatography in clinical microbiology, p. 1029–1036. In E. H. Lennette, A. Balows, W. J. Hausler, Jr., and H. J. Shadomy (eds.), Manual of clinical microbiology, 4th ed. American Society for Microbiology, Washington, D. C.

32. Okuda, K., T. Kato, J. Shiozu, I. Takazoe, and T. Nakamura. 1985. *Bacteroides heparinolyticus* sp. nov. isolated from humans with periodontitis. Int. J. Syst. Bacteriol. 35:438–442.

33. Selander, R. K., D. A. Caugant, H. Ochman, J. M. Musser, M. N. Gilmour, and T. S. Whittam. 1986. Methods of multilocus enzyme electrophoresis for bacterial population genetics and systematics. Appl. Environ. Microbiol. 51:873–884.

34. Shah, H. N., and M. D. Collins. 1988. Proposal for reclassification of *Bacteroides asaccharolyticus, Bacteroides gingivalis,* and *Bacteroides endodontalis* in a new genus, *Porphyromonas.* Int. J. Syst. Bacteriol. 38:128–131.

35. Shah, H. N., and M. D. Collins. 1990. *Prevotella,* a new genus to include *Bacteroides melaninogenicus* and related species formerly classified in the genus *Bacteroides.* Int. J. Syst. Bacteriol. 40:205–208.

36. Shah, H. N., M. D. Collins, J. Watabe, and T. Mitsuoka. 1985. *Bacteroides oulorum* sp. nov., a nonpigmented saccharolytic species from the oral cavity. Int. J. Syst. Bacteriol. 35:193–197.

37. Shinjo, T., K. Hiraiwa, and S. Miyazato. 1990. Recognition of biovar C of *Fusobacterium necrophorum* (Flugge) Moore and Holdeman as *Fusobacterium pseudonecrophorum* sp. nov., nom. rev. (ex Prevot 1940). Int. J. Syst. Bacteriol. 40:71–73.

38. Sneath, P. H. A., N. S. Mair, M. E. Sharpe, and J. G. Holt (eds.). 1986. Bergey's manual of systematic bacteriology, vol. 2. Williams and Wilkins, Baltimore.

39. Spiegel, C. A., and M. Roberts. 1984. *Mobiluncus* gen. nov., *Mobiluncus curtisii* subsp. *curtisii* sp. nov., *Mobiluncus curtisii* subsp. *holmesii* subsp. nov., and *Mobiluncus mulieris* sp. nov., curved rods from the human vagina. Int. J. Syst. Bacteriol. 34:177–184.

40. Suen, J. C., C. L. Hatheway, A. G. Steigerwalt, and D. J. Brenner. 1988. *Clostridium argentinense* sp. nov.: a genetically homogeneous group composed of all strains of *Clostridium botulinum* toxin type G and some nontoxigenic strains previously identified as *Clostridium subterminale* or *Clostridium hastiforme.* Int. J. Syst. Bacteriol. 38:375–381.

41. Sutter, V. L., D. M. Citron, M. A. C. Edelstein, and S. M. Finegold. 1985. Wadsworth anaerobic bacteriology manual, 4th ed. Star Publishing, Belmont, CA.

42. Tanner, A. C. R., M. A. Listgarten, and J. L. Ebersole. 1984. *Wolinella curva* sp. nov.: "*Vibrio succinogenes*" of human origin. Int. J. Syst. Bacteriol. 34:275–282.

43. Tanner, A. C. R., M. A. Listgarten, J. L. Ebersole, and M. N. Strzempko. 1986. *Bacteroides forsythus* sp. nov., a slow-growing, fusiform, *Bacteroides* sp. from the human oral cavity. Int. J. Syst. Bacteriol. 36:213–221.

44. van Steenbergen, T. J. M., A. J. van Winkelhoff, D. Mayrand, D. Grenier, and J. deGraaf. 1984. *Bacteroides endodontalis* sp. nov., an asaccharolytic black-pigmented *Bacteroides* species from infected dental root canals. Int. J. Syst. Bacteriol. 34:118–120.

45. Westblom, T. U., G. J. Gorse, T. W. Milligan, and A. H. Schindzielorz. 1990. Anaerobic endocarditis caused by *Staphylococcus saccharolyticus.* J. Clin. Microbiol. 28:2818–2819.

CHAPTER 3

ENDOGENOUS ANAEROBES

Beneficial Aspects of Endogenous Anaerobes
Nonbeneficial Aspects of Endogenous Anaerobes
Anaerobes Associated with Specific Anatomic
    Sites
        Upper Respiratory Tract
        Skin
        Urethra
        Vagina
        Colon

At the conclusion of this chapter, you will be able to:

1. Define the terms "opportunistic pathogen," "sterile site," and "nonsterile site."
2. State three ways in which endogenous anaerobes are beneficial to humans.
3. List five anatomical locations of the healthy human body that would be expected to harbor anaerobic bacteria as part of the microflora at those sites.
4. Name three sterile site specimens.
5. State the anatomic site where each of the following anaerobes is especially common as a member of the microflora: *Bacteroides thetaiotaomicron* and *B. vulgatus, Porphyromonas* spp., *Prevotella bivia* and *P. disiens.*
6. Name the most common genus of anaerobe in the microflora of the skin and colon.
7. List three genera of anaerobes found in the microflora of the mouth, skin, urethra, vagina, and colon.
8. Explain the importance of careful collection of clinical specimens to be submitted to the anaerobic bacteriology laboratory.

Definitions

Nonsterile site: An anatomic site usually inhabited by (colonized with) members of the indigenous microflora (e.g., the skin, oral cavity, distal urethra, vagina, and gastrointestinal tract) (Improperly collected specimens from nonsterile sites are unsuitable for anaerobic bacteriology.)

Opportunistic pathogens: Microorganisms that, under ordinary circumstances, cause no harm but can cause disease under certain conditions (e.g., after immunosuppressive therapy or when the organisms gain access to a usually sterile body site)

Sterile site: An anatomic site usually devoid of microorganisms (e.g., blood, cerebrospinal fluid, synovial fluid, and healthy tissue)

Great fleas have little fleas
upon their backs to bite 'em;
and little fleas have lesser fleas,
and so on, ad infinitum.
— Anonymous

Many anatomic sites of the healthy human body are inhabited by a variety of microorganisms, including anaerobes. The total number of bacterial species inhabiting the human body probably will exceed 500 once they are all classified and named. Collectively, these are referred to as the **indigenous microflora** of those sites. Animal species other than humans also possess indigenous microflora, varing from one animal species to another and from one anatomic site to another. Specific microorganisms may be either permanent or transient residents.

Anaerobes outnumber aerobes at mucosal surfaces (such as the linings of the oral cavity, gastrointestinal tract, and genitourinary tract). These heavily colonized surfaces are the usual portals of entry into the tissues and bloodstream for endogenous anaerobes. For these reasons, clinicians and microbiologists should suspect anaerobe involvement in infectious processes that occur at or near mucosal surfaces. Specimens improperly collected at these sites, such as by swab, are apt to contain anaerobes as **contaminants.** Microbiologists must not only keep these important points in mind as they attempt to interpret culture results, but they must also educate others by presenting this information at in-services and rounds.

Beneficial Aspects of Endogenous Anaerobes

In their usual anatomic niches, many anaerobes of the indigenous microflora are beneficial and play an active role in maintaining the health of humans and other animals. Anaerobes, together with other microorganisms, provide a natural barrier to colonization of mucous membranes by pathogenic organisms.[1] Within the gastrointestinal tract, anaerobes provide a source of fatty acids, vitamins, and cofactors that are utilized by the host and degrade potentially toxic and/or oncogenic (cancer-causing) compounds. Anaerobes also play a role in maturation of the immune system during early development of neonates.

Nonbeneficial Aspects of Endogenous Anaerobes

Under ordinary circumstances, microorganisms that are members of our indigenous microflora do not cause us problems. Indeed, many can actually be beneficial. However, when certain of these organisms (including certain anaerobes) gain access to usually sterile areas of the body (e.g., brain, lungs, etc.), they can cause serious, even fatal, diseases. Most endogenous anaerobes do not cause disease under ordinary circumstances but can cause serious illnesses when conditions permit. Examples of such conditions are discussed in Chapter 4. Opportunistic pathogens can be thought of as usually harmless organisms capable of seizing an opportunity to become pathogenic.

Anaerobes Associated with Specific Anatomical Sites

Knowledge of the composition of the microflora at specific anatomic sites is useful for predicting the particular organisms most apt to be involved in infectious processes that arise at or adjacent to those sites. Because some anaerobes have fairly predictable susceptibility patterns, such knowledge may also be of value to physicians considering empirical antimicrobial therapy.

Furthermore, the finding of site-specific organisms at a distant and/or unusual site can serve as a clue to the underlying origin of an infectious process. For example, the isolation of oral anaerobes from a brain abscess may suggest communication between an oral lesion and the bloodstream. Figure 3-1 summarizes the variety of endogenous anaerobes that may be found at specific body sites.

Upper Respiratory Tract

In the upper respiratory tract, the number of anaerobes equals or exceeds that of aerobic organisms in nasal washings, saliva, and gingival and tooth scrapings. Ninety percent of the bacteria present in saliva are anaerobes. Expectorated (coughed up) sputum becomes contaminated with indigenous microflora of the oral cavity as it passes through the throat and mouth and mixes with saliva. Because of the large numbers of anaerobes that live in the oral cavity, virtually all oral lesions involve anaerobes, as do the majority of cases of aspiration pneumonia (i.e., pneumonia that follows aspiration of oral contents into the lungs).

A wide variety of anaerobes lives in the oral cavity, although their concentrations and relative proportions vary from one microenvironment to another. Anaerobes occurring in the highest numbers are gram-negative bacilli (*Bacteroides, Fusobacterium,* and *Porphyromonas* spp.) and anaerobic cocci (*Peptostreptococcus* and *Veillonella* spp.). Therefore, these particular anaerobes should be suspected as participants in any infectious process occurring in the oral cavity. So strong is the association between certain anaerobes (e.g., *Fusobacterium nucleatum* and pigmented *Porphyromonas* spp.) and the oral cavity that an oral nidus (origin) should be suspected whenever these organisms are recovered from the bloodstream or from abscesses located far from the oral cavity.

Motile, anaerobic, gram-negative bacilli observed in or isolated from specimens from oral lesions may represent species of *Campylobacter, Centipeda, Selenomonas, Succinivibrio, Wolinella,* or various empirical groups not yet classified and named.

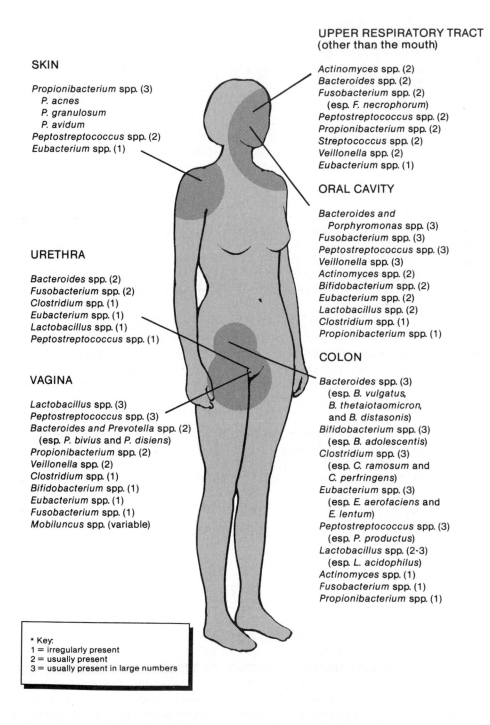

SKIN

Propionibacterium spp. (3)
   P. acnes
   P. granulosum
   P. avidum
Peptostreptococcus spp. (2)
Eubacterium spp. (1)

URETHRA

Bacteroides spp. (2)
Fusobacterium spp. (2)
Clostridium spp. (1)
Eubacterium spp. (1)
Lactobacillus spp. (1)
Peptostreptococcus spp. (1)

VAGINA

Lactobacillus spp. (3)
Peptostreptococcus spp. (3)
Bacteroides and Prevotella spp. (2)
   (esp. *P. bivius* and *P. disiens*)
Propionibacterium spp. (2)
Veillonella spp. (2)
Clostridium spp. (1)
Bifidobacterium spp. (1)
Eubacterium spp. (1)
Fusobacterium spp. (1)
Mobiluncus spp. (variable)

UPPER RESPIRATORY TRACT
(other than the mouth)

Actinomyces spp. (2)
Bacteroides spp. (2)
Fusobacterium spp. (2)
   (esp. *F. necrophorum*)
Peptostreptococcus spp. (2)
Propionibacterium spp. (2)
Streptococcus spp. (2)
Veillonella spp. (2)
Eubacterium spp. (1)

ORAL CAVITY

Bacteroides and
   Porphyromonas spp. (3)
Fusobacterium spp. (3)
Peptostreptococcus spp. (3)
Veillonella spp. (3)
Actinomyces spp. (2)
Bifidobacterium spp. (2)
Eubacterium spp. (2)
Lactobacillus spp. (2)
Clostridium spp. (1)
Propionibacterium spp. (1)

COLON

Bacteroides spp. (3)
   (esp. *B. vulgatus,*
   B. thetaiotaomicron,
   and *B. distasonis*)
Bifidobacterium spp. (3)
   (esp. *B. adolescentis*)
Clostridium spp. (3)
   (esp. *C. ramosum* and
   C. perfringens)
Eubacterium spp. (3)
   (esp. *E. aerofaciens* and
   E. lentum)
Peptostreptococcus spp. (3)
   (esp. *P. productus*)
Lactobacillus spp. (2-3)
   (esp. *L. acidophilus*)
Actinomyces spp. (1)
Fusobacterium spp. (1)
Propionibacterium spp. (1)

* Key:
1 = irregularly present
2 = usually present
3 = usually present in large numbers

Source: Based on references 1 and 2.

Figure 3-1.   The most common endogenous anaerobes of humans

Other anaerobes commonly found among the indigenous microflora of the oral cavity are species of *Actinomyces, Bifidobacterium,* and *Propionibacterium.* Anaerobic spirochetes in the genus *Treponema* also inhabit the oral cavity.

The predominant anaerobes inhabiting upper respiratory tract sites other than the mouth are *Actinomyces, Bacteroides, Eubacterium, Fusobacterium* (especially *F. necrophorum*), *Propionibacterium, Peptostreptococcus,* and *Veillonella* spp.[1, 2]

Some oral anaerobes, such as *Fusobacterium* spp. and pigmented species of *Porphyromonas,* produce volatile and foul-smelling metabolic by-products that undoubtedly contribute significantly to the odor associated with aspirates and exudates from oral lesions and aspiration pneumonia, as well as the phenomenon of "bad breath."

Expectorated sputum and oral specimens collected by swab are unacceptable for anaerobic bacteriology because they would virtually always be contaminated with anaerobes of the oral cavity and no one would be able to differentiate organisms causing the disease from the plethora of contaminants in the specimen. Proper techniques for collecting oral and pulmonary specimens for anaerobic culture are described in Chapter 5.

Skin

The indigenous microflora of the skin consists primarily of bacteria within the following genera: *Staphylococcus, Micrococcus, Corynebacterium, Propionibacterium, Brevibacterium,* and *Acinetobacter.*[1] Yeasts of the genus *Pityrosporum* are also present. The anaerobes present in highest numbers are gram-positive bacilli in the genus *Propionibacterium* (especially *P. acnes*). Gram-positive cocci (*Peptostreptococcus* spp.) and gram-positive bacilli in the genus *Eubacterium* are present in the skin flora of some individuals.

Should a venipuncture site be inadequately disinfected before collection of a specimen for blood culture, the specimen could become contaminated with skin flora, including anaerobes. For this reason, *P. acnes* is commonly isolated from positive blood cultures. As with *Staphylococcus epidermidis,* recovery of *P. acnes* from a single blood culture bottle usually represents "contamination" from the patient's skin. This is especially true when growth in the bottle does not appear until after several days of incubation, indicating that only a small number of organisms was inoculated into the bottle. However, like *S. epidermidis, P. acnes* can cause endocarditis, and its recovery could represent a true bacteremia. Superficial wound or abscess specimens aspirated by needle and syringe are much better specimens for anaerobic bacteriology than material collected by swabs because the latter are often contaminated with anaerobes of the skin microflora.

Urethra

The bacteria found colonizing the distal urethra include *Staphylococcus* spp., nonhemolytic *Streptococcus* spp., diphtheroids, and occasionally members of the family *Enterobacteriaceae.*[1] The anaerobes present in the greatest numbers are gram-negative bacilli—*Bacteroides* and *Fusobacterium* spp.[2] Other anaerobes (e.g., peptostreptococci, lactobacilli, eubacteria, and clostridia) can be found in fewer numbers. These organisms would be recovered if voided or catheterized urine was cultured for anaerobes. Although anaerobes can be a rare cause of infectious diseases of the urinary tract, their presence in such specimens generally indicates contamination of the urine with organisms flushed from the distal urethra. Thus, voided or catheterized urine specimens are unacceptable for anaerobic bacteriology. The proper method of collecting urine for anaerobic culture is described in Chapter 5.

Vagina

About 50% of the bacteria in cervical and vaginal secretions are anaerobes. Thus, a variety of anaerobes (Figure 3-1) will usually be cultured from vaginal or cervical swabs, the most common being lactobacilli (not all of which are anaerobes) and gram-positive cocci (*Peptostreptococcus* spp.). *Prevotella bivia* and *P. disiens* are frequently found as vaginal flora.

Curved, motile, anaerobic, gram-positive bacilli in the genus *Mobiluncus* are also common members of the vaginal flora. These organisms appear as curved, pink bacilli in Gram-stained preparations. Their numbers increase dramatically during bacterial vaginosis, and they are thought by some authorities to play a role in that disease. *Clostridium* spp. from the colon can also be found in the indigenous microflora of the vagina because of its proximity to the anus.

Whenever anaerobes are recovered from vaginal and cervical swabs, neither the microbiologist nor the physician can distinguish the indigenous microflora contaminants from organisms actually contributing to the patient's infectious process. For this reason, genitourinary (GU) tract swabs are unacceptable for anaerobic bacteriology. Proper methods of obtaining GU specimens are described in Chapter 5.

Colon

Of the estimated 500 species of bacteria that inhabit the human body, approximately 300 to 400 live in the colon. Microflora studies have found that anaerobes outnumber aerobes by a factor of 100 to 1000 to one.[1] Anaerobes occurring in the highest numbers in intestinal flora are *Bacteroides, Bifidobacterium, Clostridium, Eubacterium, Lactobacillus,* and *Peptostreptococcus* spp.[2] Of these, *Bacteroides* is the predominant genus.[1] It has been estimated that there are a thousand *Bacteroides* organisms for each *Escherichia coli* cell in the large intestine. The most common *Bacteroides* spp. are *B. vulgatus* and *B. thetaiotaomicron,* although *B. distasonis, B. fragilis,* and *B. ovatus* are also quite common.[1] The most common anaerobic, gram-positive bacilli are *Bifidobacterium adolescentis, Eubacterium aerofaciens, E. lentum,* and *Lactobacillus acidophilus.*[1] The most common species of *Clostridium* are *C. ramosum,* and *C. perfringens,* and the most common gram-positive coccus is *Peptostreptococcus productus.*

Stool specimens are not routinely cultured for anaerobes. However, if a patient is suspected of having pseudomembranous colitis and/or antibiotic-associated diarrhea caused by *Clostridium difficile,* diarrheal specimens from that patient should be examined. Diseases caused by *C. difficile* are described in Chapter 4. Quantitative and/or qualitative cultures of feces should also be performed to confirm foodborne disease (food poisoning) due to *C. perfringens.*

Chapter in Review

- Many anatomic sites of the healthy human body are inhabited by microorganisms. Collectively, these microorganisms are referred to as the indigenous microflora of those sites. The total number of bacterial species in the indigenous microflora of humans probably exceeds 500.

- Although many species of endogenous anaerobes are beneficial in their usual anatomic niches, they can cause serious illnesses when conditions permit. Thus, they are considered to be opportunistic pathogens.

- Anaerobes outnumber aerobes at mucosal surfaces, which are the usual portals of entry into the tissue and bloodstream for endogenous anaerobes. Clinicians and microbiologists should suspect anaerobe involvement in infectious processes at or near mucosal surfaces. Specimens improperly collected at these sites, such as by swab, are apt to contain contaminating anaerobes.

- Knowledge of the composition of the microflora at specific anatomic sites is useful for predicting the particular organisms most apt to be involved in infectious processes that arise at or adjacent to those sites, as well as the antimicrobial agents most likely to be of value in empirical antimicrobial therapy. The finding of site-specific organisms at a distant and/or unusual site can serve as a clue to the underlying origin (nidus) of an infectious process.

- Anaerobes occurring in the highest numbers in the microflora of the oral cavity are gram-negative bacilli (*Bacteroides, Fusobacterium,* and *Porphyromonas* spp.) and anaerobic cocci (*Peptostreptococcus* and *Veillonella* spp.). An oral nidus should be suspected when oral anaerobes such as *Fusobacterium nucleatum* or pigmented species of *Porphyromonas* are recovered from the bloodstream or from abscesses far distant from the oral cavity.

- The predominant anaerobes inhabiting various sites in the upper respiratory tract other than the mouth are *Actinomyces, Bacteroides, Eubacterium, Fusobacterium, Propionibacterium, Peptostreptococcus,* and *Veillonella* spp.

- The anaerobes present in highest numbers as skin flora are *Propionibacterium* spp. (especially *P. acnes*) and, less often, gram-positive cocci.

- The anaerobes colonizing the distal urethra in the greatest numbers are gram-negative bacilli—*Bacteroides* and *Fusobacterium* spp.

- About 50% of the bacteria in cervical and vaginal secretions are anaerobes, the most common being lactobacilli (not all of which are anaerobes) and gram-positive cocci. *Prevotella bivia, P. disiens,* and *Mobiluncus* spp. are common members of the vaginal flora.

- Of the 500 or so estimated species of bacteria that inhabit the human body, approximately 300 to 400 live in the colon. The anaerobes present in the highest numbers in the intestinal flora are species of *Bacteroides, Bifidobacterium, Clostridium, Eubacterium, Lactobacillus,* and *Peptostreptococcus*. *Bacteroides* spp. are the most common anaerobes in the colon, outnumbering *Escherichia coli* by about a thousand to one. In the colon, the most common *Bacteroides* spp. are *B. vulgatus* and *B. thetaiotaomicron*.

Self-Assessment Exercises

1. Define the following terms:

   a. Sterile site _____

   _____

   b. Nonsterile site _____

   _____

   c. Opportunistic pathogen _____

   _____

2. List three ways in which endogenous anaerobes are beneficial to the human body.

   a. _____

   _____

   b. _____

   _____

   c. _____

   _____

3. Which of the following clinical materials would you expect to contain endogenous anaerobes?

   a. Scrapings from gingival crevices
   b. Voided urine
   c. Expectorated sputum
   d. Cervical swab
   e. Throat swab

4. What genus of anaerobic, gram-positive bacillus is most prevalent in the indigenous microflora of the skin?

   _____

5. Name two species of *Prevotella* that are especially prevalent in the microflora of the human vagina.

    a. _____

    b. _____

6. What genus of pigmented, anaerobic, gram-negative bacilli is especially prevalent in the indigenous microflora of the oral cavity?

    _____

7. Which genus of anaerobe is most prevalent in the colon?

    _____

8. TRUE or FALSE:　The most common species of *Bacteroides* in the colon is *B. fragilis*.

9. Which of the following clinical materials are considered sterile site specimens?

    a. Blood
    b. Catheterized urine
    c. Cerebrospinal fluid
    d. Synovial fluid
    e. Fluid aspirated from an oral lesion

10. TRUE or FALSE:　Unless they are carefully collected, specimens from many anatomical locations can become contaminated with endogenous anaerobes.

References

1. Hentges, D. J. 1989. Anaerobes as normal flora, p. 37–53. In S. M. Finegold and W. L. George (eds.), Anaerobic infections in humans. Academic Press, San Diego.
2. Sutter, V. L., D. M. Citron, M. A. C. Edelstein, and S. M. Finegold. 1985. Wadsworth anaerobic bacteriology manual, 4th ed. Star Publishing, Belmont, CA.

CHAPTER 4

ANAEROBE-ASSOCIATED DISEASES

Diseases Caused by Exogenous Anaerobes
Diseases Caused by Endogenous Anaerobes
Virulence Factors of Anaerobic Bacteria
Indications of Anaerobe Involvement in Human
    Disease

At the conclusion of this chapter, you will be able to:

1. Differentiate between "exogenous" and "endogenous" anaerobes.
2. Categorize each of the following anaerobes as "exogenous" or "endogenous": *Clostridium botulinum, C. tetani, Bacteroides fragilis, Fusobacterium nucleatum.*
3. Discriminate between the terms "infection," "infectious disease," and "intoxication" as they relate to anaerobes.
4. Distinguish between the following clostridial diseases with respect to how they are acquired: foodborne botulism, infant botulism, wound botulism.
5. Differentiate between the following clostridial diseases with respect to etiologic agents and how the diseases are acquired: tetanus, tetanus neonatorum, gas gangrene, pseudomembranous colitis, clostridial food poisoning.
6. Describe the role of *Clostridium difficile* in nosocomial infections.
7. Give five examples of conditions that predispose an individual to infection by endogenous anaerobes.
8. Name five infectious processes or diseases that virtually always involve anaerobes of endogenous origin.
9. State three examples of virulence factors possessed by anaerobes.
10. List three signs, symptoms, or other indications of the involvement of anaerobes in a particular infectious process or disease.
11. Match the names of the most commonly isolated and/or clinically significant anaerobes with specific clinical specimens, symptoms, or diseases.

Definitions

Actinomycosis: A chronic, granulomatous, infectious disease caused by various *Actinomyces* spp. or closely related organisms. The disease is characterized by the development of sulfur granules (small colonies of bacteria) and fistulae, which erupt and drain pus to the surface. Actinomycosis occurs in many wild and domestic animals. Cervicofacial actinomycosis (lumpy jaw) is common in animals, especially cattle. Most cases of human actinomycosis are due to *Actinomyces israelii.* *A. bovis* causes actinomycosis in cattle but not in humans. Examinations of a wet mount and a Gram-stained preparation of pus from draining sinuses are useful diagnostic procedures for demonstrating the nonsporeforming, gram-positive bacilli that frequently exhibit branching in clinical materials.

Antibiotic-associated diarrhea: Diarrhea that develops in a patient who has received antimicrobial therapy and is a result of that therapy

Aspiration pneumonia: Pneumonia that follows the inhalation/aspiration of saliva and/or food particles into the respiratory passages

Botulism: A neuroparalytic illness affecting cholinergic nerves that is caused by botulins (neurotoxins) produced by *Clostridium botulinum* or certain other clostridia (*C. butyricum, C. barati*). Three clinical forms (foodborne, wound, infant) of botulism are now recognized. Foodborne botulism develops after ingestion of preformed botulin in food contaminated with *C. botulinum,* an obligately anaerobic, sporeforming bacillus found in soil and marine environments throughout the world. Wound botulism develops after multiplication of *C. botulinum* and elaboration of botulin in the contaminated tissues of a traumatic wound. Infant botulism is a toxicoinfectious form of botulism resulting from botulin(s) produced *in vivo* in the intestine after multiplication of bacteria (*C. botulinum, C. barati, C. butyricum*) capable of producing botulin. Seven botulins (A, B, C, D, E, F, G) are produced by *C. botulinum,* but only four (A, B, E, F) cause botulism in humans.

Endotoxin (also called **bacterial pyrogen**): A lipopolysaccharide complex (LPS) found in the cell envelope of some gram-negative bacteria. LPS is pyrogenic (fever producing) and causes increased capillary permeability.

Enterotoxin: A microbial toxin specific for cells of the intestinal mucosa; produces gastroenteritis

Exotoxin: A bacterial toxin secreted by living bacterial cells (e.g., *Clostridium perfringens* alpha toxin)

Gas gangrene (myonecrosis): An acute, severe, and painful condition characterized by necrosis of muscle and subcutaneous tissue in which the necrotic tissues become filled with a serosanguinous fluid and exudate. *Clostridium perfringens* is the most common etiologic agent of gas gangrene. Other histotoxic clostridia capable of producing gas gangrene in humans or animals include *C. septicum, C. novyi* type A, *C. sordellii,* and *C. chauvoei* (animals only).

Infant botulism: A form of botulism resulting from ingestion of spores of *Clostridium botulinum,* with subsequent germination, vegetative growth, and elaboration of botulinal toxin *in vivo.* Although more common in infants, a similar condition can also occur in adults whose intestinal flora are unable to inhibit colonization by *C. botulinum.*

Infection: The term "infection" is used th1oughout this book to indicate tissue invasion by and subsequent multiplication of microorganisms. Some infections are clinically inapparent; others, referred to as **infectious diseases** or **infectious processes,** result in clinical signs and symptoms. Some authors use the terms "infection" and "infectious disease" synonymously.

Intoxication: An illness caused by a toxic compound. Foodborne botulism is an example of an intoxication. The illness is due to ingestion of botulin elaborated by *C. botulinum* in food contaminated with the microorganism and is not an infectious disease.

Nosocomial infection: A hospital-acquired infection, as distinquished from a community-acquired infection; an infection the patient did not have upon entering the hospital

Pili (also called **fimbriae**): Minute, filamentous appendages possessed by certain bacterial cells. They are considerably smaller and less rigid than flagella and associated with antigenic properties of the cell surface. Certain types of pili (**sex pili**) enable bacterial cells to conjugate ("mate"). Other types enable bacteria to adhere to surfaces.

Polymicrobial: Characterized by the presence of more than one species of bacteria

Protoplasmic toxin: A toxin found in the protoplasm, within the bacterial cell membrane, and released after autolysis of cells (e.g., botulins of *C. botulinum, C. difficile* toxins A and B, and *C. tetani* toxin [tetanospasmin])

Pseudomembranous enterocolitis (also called **pseudomembranous colitis** or **PMC**): An acute inflammation of the colonic mucosa with formation of pseudomembranous plaques overlying an area of superficial ulceration and passage of the pseudomembranous material in the feces. Colitis following antimicrobial therapy is referred to as **antibiotic-associated colitis.**

Purulent: Pus producing

Spore-associated toxin: A toxin released after sporulation of cells (e.g., *C. perfringens* enterotoxin)

Tetanus: An intoxication caused by a neurotoxin (tetanospasmin) produced by *Clostridium tetani.* Tetanus develops after germination of spores, multiplication of bacterial cells, and release of tetanospasmin in traumatized tissue—usually a traumatic wound contaminated with *C. tetani* spores from soil. Clinical findings in tetanus may include spasms of the head and neck muscles ("lockjaw"), arching of the back, respiratory failure, and cardiac arrest. Tetanus is primarily a disease of the unimmunized. Therefore, the highest incidence is in elderly people and neonates.

Tetanus neonatorum: Tetanus of the newborn, usually resulting from infection of the umbilicus (navel) with *Clostridium tetani;* occurs most commonly in parts of the world where mud, animal feces, or other contaminated materials are applied to the infant's umbilicus shortly after birth

Virulence factor: A phenotypic characteristic thought to contribute to the virulence (degree of pathogenicity) of a microorganism (e.g., polysaccharide capsules, pili, and certain exoenzymes)

Wound botulism: Botulism that develops in a patient with a wound contaminated with *C. botulinum* spores. Botulin is produced *in vivo*. After germination of the spores, multiplication of the cells, autolysis of the cells, and release of botulin in the traumatized tissue, the botulin is absorbed by the regional lymphatic nodes. The toxin makes its way via blood to the target nerves, where it is absorbed, and development of disease ensues.

Prospective payment policies make it more important than ever to rapidly determine the etiology of a patient's infectious disease, to administer appropriate therapy, and to release the patient from the hospital as expeditiously as possible. Clinical laboratory professionals must ensure that administrators do not use DRGs and related cost-containment policies as convenient excuses for reducing or eliminating measures necessary for successful anaerobic bacteriology. The initial patient workup must include a thorough anaerobic bacteriology workup if this is found to be appropriate. Neither the patient nor the clinician can afford to wait until everything else has been tried first. If anaerobes are involved in the patient's infectious disease, that information must be provided to the clinician immediately.

Diseases Caused by Exogenous Anaerobes

Anaerobes involved in diseases of humans may originate either **outside** of (**exogenous** origin) or **within** the body (**endogenous** origin). Those of exogenous origin are usually members of the genus *Clostridium,* many species of which are found in soil. Exogenous anaerobes or their spores usually enter the body via the mouth (ingestion) or an open wound. Certain diseases (called **intoxications**) follow **ingestion of toxins** produced by exogenous anaerobes. Foodborne botulism is an example of an intoxication resulting from ingestion of botulin produced in food contaminated with *C. botulinum.*

For many years, the only anaerobic bacteria recognized as human pathogens were primarily *Clostridium* spp. originating outside of the body (exogenous anaerobes). These were the clostridia that cause diseases such as tetanus, gas gangrene, botulism, and *C. perfringens* food poisoning. It is now known that some *Clostridium* spp. also inhabit certain body sites (e.g., the colon and vagina) as part of the indigenous microflora, and the origin of clostridial disease is not always outside the body. Some of the classical diseases caused by clostridia, such as botulism, tetanus, gas gangrene (myonecrosis), and *C. perfringens* food poisoning (foodborne intoxication), have been known for many years.

In tetanus and gas gangrene, clostridial spores enter through open wounds and germinate *in vivo*. The vegetative bacteria then multiply and produce toxins. With *C. tetani,* a neurotoxin (tetanospasmin) produces a rigid type of paralysis; in gas gangrene, exotoxins (such as alpha toxin) produced by the causative organisms cause necrosis of the tissue and allow deeper penetration by the organisms.

As a result of the widespread use of DPT (diphtheria-pertussis-tetanus) vaccine, tetanus is no longer a common disease in the United States (Figure 4-1). During 1989, for example, only 53 cases were reported to the Centers for Disease Control (CDC).[14] The state reporting the highest number of cases of tetanus during 1989 was California, with 10 cases or 19% of the total.

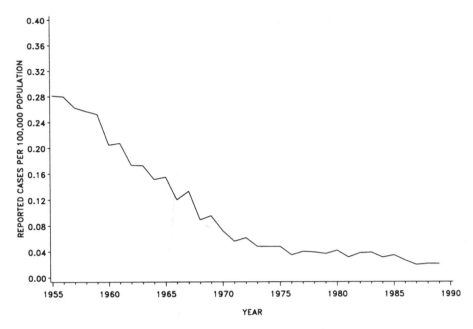

Figure 4-1. Tetanus in the United States (1955–1989) [14]

In classical foodborne botulism, characteristic signs and symptoms develop following ingestion of food contaminated with botulin, produced in the food by *C. botulinum*. Clostridial food poisoning usually follows ingestion of *C. perfringens–*contaminated food, and subsequent spore and enterotoxin production *in vivo*. Although *C. perfringens* food poisoning is generally a relatively mild and self-limited gastrointestinal illness, botulism is a very serious disease that can result in a flaccid type of paralysis and death.

Wound botulism is the result of contamination of wounds with spores of *C. botulinum*, with subsequent germination, multiplication, and production of toxins *in vivo*. Infant botulism, which is not necessarily confined to infants, follows ingestion of *C. botulinum* spores, germination, colonization of the colon, and *in vivo* elaboration of toxins. Thus, both wound botulism and infant botulism differ from classical food-borne botulism in that toxins are produced and released *in vivo* rather than *in vitro*.

During 1989, 89 cases of botulism were reported to CDC.[14] Of these, 23 cases were foodborne botulism (Figure 4-2), 60 were infant botulism (Figure 4-3), and six were either wound or unspecified botulism. The state reporting the greatest number of cases of foodborne botulism in 1989 was Washington, with 10 cases or 43% of the total. California reported the highest number of cases of infant botulism in 1989, with 35 cases or 58% of the total.

Table 4-1 summarizes data from a CDC report describing 24 confirmed outbreaks of *C. perfringens* food poisoning in the United States during the five-year period, 1983–1987.[13] An outbreak was defined as an incident in which two or more persons experienced a similar illness after ingesting a common food and epidemiologic analysis implicated the food as the source of the illness. The incubation period for *C. perfringens* food poisoning is 8 to 14 hours.

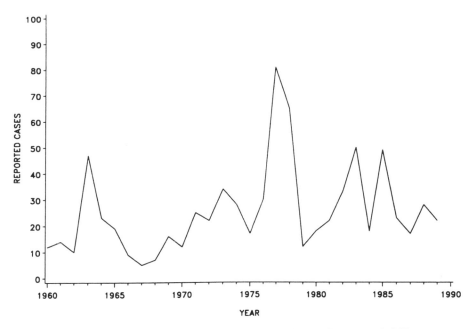

Figure 4-2. Foodborne botulism in the United States (1960–1989) [14]

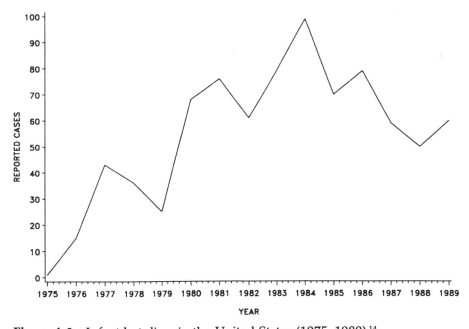

Figure 4-3. Infant botulism in the United States (1975–1989) [14]

Table 4-1. Confirmed Foodborne Disease Outbreaks Due to *Clostridium perfringens* (1983-1987)

Year	Outbreaks	Cases	Deaths	Implicated Foods
1983	5	353	0	Beef (1), Mexican food (3), unknown (1)
1984	8	882	2	Chicken (1), turkey (1), salad (1), Mexican food (2), multiple foods (2), unknown (1)
1985	6	1016	0	Beef (1), turkey (1), Mexican food (1), multiple foods (3)
1986	3	202	0	Turkey (1), multiple foods (1), unknown (1)
1987	2	290	0	Beef (1), poultry/fish/or egg salad (1)
Totals	24	2743	2	

Contributing Factors*

	Improper Holding Temperatures	Inadequate Cooking	Contaminated Equipment	Food from Unsafe Source	Poor Personal Hygiene	Other
1983	5	4	1	1	1	0
1984	8	3	0	0	0	3
1985	3	3	2	0	2	0
1986	3	2	0	0	1	0
1987	2	1	1	0	0	0

* Multiple contributing factors were identified in some outbreaks.

Source: Based on reference 13.

Clostridium difficile is the most common (but not the sole) cause of antibiotic-associated diarrhea and pseudomembranous colitis. This organism is found as part of the gastrointestinal (GI) flora of many individuals. Following antimicrobial therapy, many organisms of the GI flora other than *C. difficile* are killed, thus allowing *C. difficile* to multiply and produce abundant quantities of toxin (both enterotoxin and cytotoxin). In addition, *C. difficile* is a common cause of nosocomial (hospital-acquired) infection. The organism is frequently transmitted among hospitalized patients and is often present on the hands of hospital personnel who are caring for such patients.[12]

Using rabbit antisera and slide agglutination, *C. difficile* can be differentiated into ten serogroups, creating a typing scheme that has proven useful for clinical and epidemiological investigations.[3] Certain serogroups (A, D, G, H, and K) possess peritrichous flagella, but others (B, C, F, I, and X) do not. The flagella of one *C. difficile* serogroup are shown in Figure 4-4. Investigations have shown that cross-agglutinations due to flagellar antigens (flagellin) can be suppressed by simple shearing of the flagella.[3] An entire book has been devoted to *C. difficile*.[15]

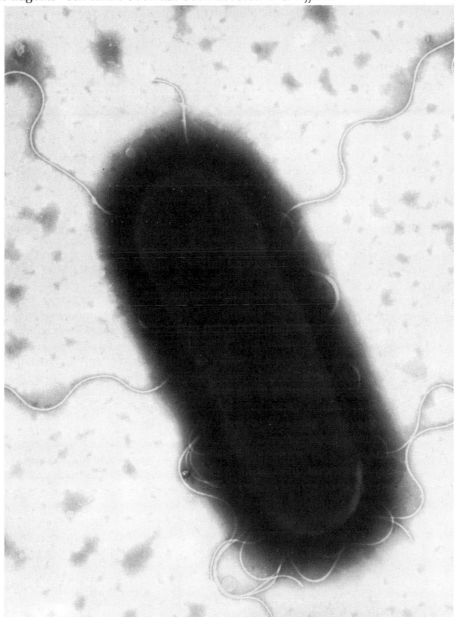

Figure 4-4. Transmission electron micrograph of *Clostridium difficile* (W1194, ATCC 49594, serogroup A), negatively stained with 3% phosphotungstic acid, pH 6 (courtesy of Michel Delmée, M.D., Guy Burtonboy, M.D., and the American Society for Microbiology)

Case Presentation 4-1. Antibiotic-Associated Diarrhea

History and Physical Examination

An 87-year-old male nursing home resident was hospitalized with a history of fever of 37.8° to 38.4°C for the preceding five days. He had been able to dress himself and move about the home until a week before his hospital admission, at which time he had felt very weak and could not get out of bed. On physical examination, he was unable to give a history and could not answer questions about his current condition. His temperature was 38.4°C, his other vital signs were normal, and, except for mild dehydration, there were no localizing physical findings.

Laboratory Results

Initial laboratory evaluation revealed a hemoglobin of 15, a hematocrit of 49, and a white blood cell count of 11,000, with 80% polymorphonuclear leukocytes (PMNs). Urinalysis was unremarkable, and liver function tests, BUN, and electrolytes were within normal limits. A chest x-ray and an electrocardiogram were unchanged from previous studies. The diagnosis of fever of unknown origin (FUO) was made, and cultures of blood and urine were ordered.

Subsequent Findings

On bedside rounds the next morning, the attending physician was informed by the staff nurse that the patient had had two loose bowel movements during the night. There had been no mention of the patient's bowel habits in the referring history sheet. A telephone call to the nursing home soon revealed that (1) one or two loose stools per day was not unusual for the patient, (2) approximately four weeks earlier the patient had been treated for seven days with ampicillin for an apparent urinary tract infection, and (3) many patients at the nursing home were experiencing diarrhea. The attending physician suspected antibiotic-associated diarrhea and requested that a stool specimen be sent to the microbiology laboratory.

Question 4-1-1: What clues led the physician to suspect antibiotic-associated diarrhea?

Answer 4-1-1:   A history of loose stools following seven days treatment with ampicillin and a "diarrhea problem" at the nursing home. Virtually all antimicrobial agents have been implicated in antibiotic-associated diarrhea, but the most common are cephalosporins, ampicillin, and clindamycin (the order reflecting the frequency of use of these drugs in clinical practice).

Question 4-1-2: What etiologic agent do you suspect in this case? Why?

Answer 4-1-2:   *Clostridium difficile*—a sporeforming, anaerobic, gram-positive bacillus. This organism is the most common cause of antibiotic-associated diarrhea. Recent evidence suggests that *C. difficile* is the most common cause of diarrhea in nursing home settings. The organism can be transmitted in hospitals and chronic care facilities by person-to-person contact or through environmental contamination.

Question 4-1-3: How does Clostridium difficile cause antibiotic-associated diarrhea?

Answer 4-1-3: Many people harbor the organism as part of their indigenous intestinal flora, where it tends to remain in a dormant state in low numbers. The sequence of events in antibiotic-related *C. difficile* disease is thought to begin with the suppression of indigenous flora by the antimicrobial agent, with persistence of the spore form of *C. difficile*. Sometime during or after antibiotic administration, the spores germinate and grow in large numbers. The organism produces its major toxins (A and B) in the lumen of the large intestine. Toxin A is primarily an enterotoxin and toxin B, a cytotoxin. The current view is that both toxins act synergistically to injure the intestinal mucosa.

Question 4-1-4: How might the laboratory establish a diagnosis of Clostridium difficile–associated disease?

Answer 4-1-4: The simplest and most reliable means of diagnosing *C. difficile* infection is to measure the toxin in stool. The preferred laboratory method is a tissue culture assay for toxin B. Toxin B produces a cytopathic change in the cultured cells, but the change can be neutralized by a specific antitoxin. The organism can also be cultured from stool specimens, but culturing cannot distinguish between carriers and patients with active disease. Additional methods for detecting *C. difficile* toxins and antigens are available.

Treatment and Outcome

Within 24 hours, the laboratory provided evidence of *Clostridium difficile* toxins in the stool specimen. Specific treatment with metronidazole was initiated. The patient became afebrile within 36 hours and was returned to the nursing home, without further laboratory investigations, within 72 hours.

Question 4-1-5: Was the therapy appropriate for C. difficile disease?

Answer 4-1-5: Most patients with antibiotic-associated diarrhea have a self-limited course without specific treatment. Simply stopping the antibiotic usually causes symptoms to subside, generally within one to three days. Specific antimicrobial treatment is indicated for patients with identified *C. difficile* disease and severe symptoms and for those in whom the condition started after the incriminated drug was stopped (as in this case). The preferred antibiotics are metronidazole or vancomycin, used as oral medications. The former drug is preferred because of its lower cost.

Additional Information

C. difficile can be isolated from environmental sources, particularly in hospitals. Under adverse environmental conditions, it reverts to its highly resistant spore form, which can be cultured from floors, bedpans, and toilets in rooms occupied by patients colonized with *C. difficile*. Spores can also be cultured from the hands and clothing of attending personnel. The spores are virtually indestructible, giving the organism a kind of microbial immortality and enabling it to threaten other patients in the area. In addition to the potential for nosocomial infections due to *C. difficile* within primary care facilities, there exists a need to maintain vigilance for this organism as a cause of fever and diarrhea in elderly patients living in nursing homes.

Source: Based on Reference 7.

Diseases Caused by Endogenous Anaerobes

In recent years, the anaerobes most frequently isolated from infectious processes in humans have been those of endogenous origin (members of the patient's own indigenous microflora that somehow gained entrance to an area of the body they don't usually inhabit). There are several reasons for the apparent shift from predominantly exogenous to predominantly endogenous anaerobes:

- More information is available today regarding the role of endogenous anaerobes in infectious processes. Therefore, physicians and microbiologists are more likely to suspect them.

- More efficient means are now available for isolating anaerobes from clinical specimens, including better collection techniques, improved specimen transport devices, special media for isolating anaerobes, holding systems for use in processing of specimens, and improved techniques for anaerobic incubation of cultures.

- Increasing numbers of patients are immunosuppressed for one reason or another. They demonstrate decreased host resistance to a variety of microorganisms, including those comprising their own indigenous microflora.

Although clostridia are no longer the predominant anaerobes isolated in today's microbiology laboratory, they are still very important and may represent a serious condition when isolated. However, because certain *Clostridium* spp. may constitute part of a patient's genitourinary (GU), GI, or skin flora, mere recovery of clostridia from clinical specimens is not in itself proof of an infectious process. This is true, of course, for any microorganism found as part of the indigenous microflora.

Anatomic sites inhabited by indigenous microflora are referred to as **nonsterile sites,** and specimens collected from such sites are sometimes called **nonsterile specimens.** A member of the indigenous microflora recovered from an inappropriate or improperly collected specimen from a nonsterile site could represent either of two things: a pathogen contributing to the infectious process or a contaminant introduced during specimen collection. Neither the microbiologist nor the physician would know which of these is the case.

Anaerobes of endogenous origin can contribute to an infectious disease in virtually any tissue, anatomic region, or organ of the body, provided suitable conditions exist for colonization and penetration of the bacteria. For example, *Actinomyces* spp. and related anaerobic bacteria (e.g., bifidobacteria, eubacteria, propionibacteria) of the indigenous flora can cause disease (actinomycosis) in the brain, orofacial region, pleuropulmonary region, or genital organs. Bacterial vaginosis is another example. Some investigators think this disease involves endogenous anaerobes of the vagina—curved, motile, gram-positive bacteria of the genus *Mobiluncus* (especially *M. curtisii* subsp. *curtisii*).

Many of the infectious processes involving anaerobes are **polymicrobial** (a variety of organisms is involved in the process). Generally, these are mixtures of obligate anaerobes or mixtures of obligate or microaerotolerant anaerobes and facultative organisms. Symbiotic relationships frequently exist between some of the bacteria involved in polymicrobial infections, which can act synergistically in the production of disease.[1,9,17]

Infectious processes involving anaerobes are usually **purulent,** meaning they contain **pus.** The absence of leukocytes does not, however, rule out the possibility that anaerobes are contributing to the process. Some of the more serious infectious processes in humans are caused by anaerobes that produce cytotoxins and other histotoxic virulence factors that contribute to the necrotizing process by destroying neutrophils, macrophages, and other cells.

The relationships between anaerobic bacteria and cancer in humans is uncertain. A correlation has been reported between *Clostridium septicum* bacteremia and malignancy or other diseases of the colon.[16] Also, the number of *Eubacterium lentum* in the bowel is reduced over a hundredfold in recently diagnosed cases of colon cancer.[10] Certain members of the *Bacteroides fragilis* group, commonly found in human feces, produce substances with mutagenic activity.[6] The significance of these findings is not known.

Factors that commonly predispose to infection with endogenous and exogenous anaerobes include trauma of mucous membranes or skin, vascular stasis, tissue necrosis, and a decrease in the oxidation-reduction (redox) potential of tissue. Vascular stasis (blockage of blood flow) prevents oxygen from entering the site, resulting in an environment conducive to multiplication of any anaerobes that might be present at that site. Examples of conditions that predispose a patient to anaerobic infections are listed in Table 4-2.

Table 4-2. Examples of Conditions Predisposing a Patient to Anaerobe-Associated Infectious Processes/Diseases

Predisposing Condition	Explanation
Human or animal bite wounds	Endogenous anaerobes of the oral cavity enter the wound.
Aspiration of oral contents into the lungs after vomiting	Endogenous anaerobes of the oral cavity and other materials (mucous, stomach contents, etc.) enter the lungs.
Tooth extraction, oral surgery, or traumatic puncture of the oral cavity	Endogenous anaerobes of the oral cavity gain entrance to traumatized tissue and the bloodstream.
Gastrointestinal tract surgery or traumatic puncture of the bowel	Endogenous anaerobes of the gastro-intestinal tract gain access to traumatized tissue and the bloodstream.
Genital tract surgery or traumatic puncture of the genital tract	Endogenous anaerobes of the genital tract gain access to traumatized tissue and the bloodstream.
Introduction of soil into a wound	Clostridial spores in the soil enter the tissues via a traumatic wound, resulting in colonization and multiplication of the bacteria.

A typical scenario for infections involving anaerobes might be as follows:

1. First, there is some condition that predisposes the patient to infection, such as an automobile accident, a tooth extraction, a ruptured appendix, or oral, gynecologic, or abdominal surgery. Any of these events could cause the initial tissue damage.
2. A primary infection then develops, not necessarily involving obligate or microaerotolerant anaerobes. Facultative anaerobes contribute to a decrease in the redox potential at the site and to the production of anaerobic conditions. Obligate anaerobes and anaerobes of other groups (in relationship to oxygen) could then flourish at the site.
3. Bacteria could next gain access to the bloodstream, where they would be disseminated hematogenously (through the circulatory system) throughout the body.
4. At locations possibly distant from the original infection site, small clots might develop in tiny capillaries, stopping the flow of blood and creating anaerobic conditions.
5. A septic thrombus (clot) could develop at one or more of these sites.
6. An abscess would form, and organisms could then be disseminated hematogenously to additional sites.
7. Further complications could develop, such as intravascular hemolysis, tissue necrosis, toxemia, shock, intravascular coagulation, vascular collapse, and even death.

Virulence Factors of Anaerobic Bacteria

The precise mechanisms by which anaerobic bacteria cause disease are not known. However, anaerobes produce or possess a variety of enzymes, capsules, and adherence factors that are thought to play a role in pathogenicity. These are collectively referred to as **virulence factors**. **Polysaccharide capsules**, such as those possessed by *Bacteroides fragilis, Porphyromonas gingivalis,* and some other anaerobic, gram-negative bacilli, are thought to promote abscess formation and serve an antiphagocytic function.[9] The capsule of *P. gingivalis* is shown in Figure 4-5.[11]

The **ability to adhere** to cell surfaces is considered necessary for the initiation of infection by certain microorganisms. Adherence often results from the presence of **pili** (fimbriae, fibrils) that protrude from the bacterial surface. Pili may serve as virulence factors in several *Bacteroides* and *Porphyromonas* spp., including *B. fragilis* and *P. gingivalis* in humans and *B. nodosus* in sheep.[18] Pili and surface vesicles ("blebs") of *P. gingivalis* are illustrated in Figure 4-6. Some investigators believe that the vesicles also play a role in pathogenesis or virulence, perhaps by competing for antibodies or serving as a vehicle for toxins and/or proteolytic enzymes.[11]

Clostridium spp. produce a variety of toxins that are responsible for many of the characteristic signs and symptoms of the diseases caused by these organisms (Table 4-3). Examples include **neurotoxins, cytotoxins, necrotizing toxins,** and **enterotoxins.**[8] Enzymes capable of breaking down hyaluronic acid (**hyaluronidase**), collagen (**collagenase**), mucopolysaccharides (e.g., **chondroitin sulfatase**), immunoglobulins (**proteases**), deoxyribonucleic acid (**DNase**), glycoproteins (e.g., **neuraminidase**), fibrin (**fibrinolysin**), heparin (**heparinase**), and beta-lactam antibiotics

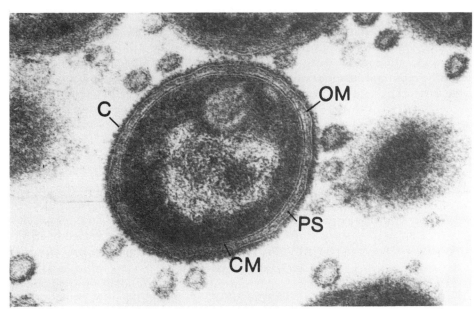

Figure 4-5. Transmission electron micrograph of *Porphyromonas gingivalis* (ATCC 33277), depicting the ruthenium red-stained capsule (C) that covers the outer cell membrane (OM). The unit cytoplasmic membrane (CM) and particulate periplasmic space (PS) can also be seen. (courtesy of S. C. Holt, Ph.D., D. Guerrero, and the American Society for Microbiology)

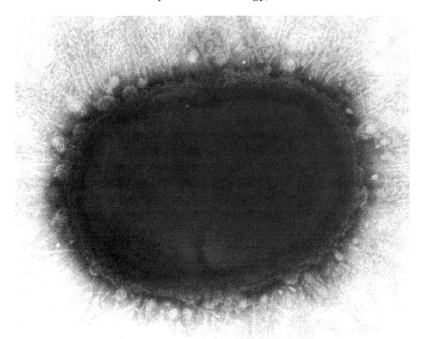

Figure 4-6. Transmission electron micrograph of *Porphyromonas gingivalis* CO, negative-stained using 2% ammonium molybdate. Many vesicles ("blebs") are present, and numerous pili can be seen emanating from the cell surface. (courtesy of S. C. Holt, Ph.D., D. Guerrero, and the American Society for Microbiology)

(**beta-lactamases**) have all been suggested as possible virulence factors of anaerobes.[1,9] **Hemolysins** and **leukocidins** are also thought to play a role in pathogenesis, as is the **endotoxin** of gram-negative anaerobes. Possible virulence factors of
clinically important anaerobes are shown diagrammatically in Figure 4-7 and listed in
Table 4-4.

Table 4-3. Examples of Toxins Produced by *Clostridium* spp.

Toxins	Effects
Collagenases	Enzymes that catalyze the degradation of collagen
Cytotoxins	Substances that have toxic action against specific types of cells
DNases	Enzymes capable of destroying DNA
Enterotoxins	Substances specifically toxic to cells of the intestinal mucosa
Hemolysins	Substances that liberate hemoglobin from red blood cells by interrupting their structural integrity (i.e., by lysing the cells)
Hyaluronidases	Enzymes that catalyze the hydrolysis of hyaluronic acid, the cement substance of tissues
Lipases	Enzymes that catalyze the hydrolysis of ester linkages between the fatty acids and glycerol of triglycerides and phospholipids
Necrotizing toxins	Substances that cause necrosis (death) of cells
Neuraminidases	Enzymes that destroy neuraminic acid, a sialic acid found on cell surfaces
Neurotoxins	Substances that are poisonous or destructive to nerve tissue
Permeases (e.g., ADP-ribosyltransferases)	Enzymes (often enterotoxins) that alter the physiology of cells in such a way as to cause severe fluid loss and ionic imbalance
Phospholipases (lecithinases)	Enzymes that catalyze the splitting of phospholipids
Proteases	Enzymes that are proteolytic (i.e., capable of splitting proteins by hydrolysis of peptide bonds) (Certain proteases destroy immunoglobulin molecules.)
Proteinases	Enzymes that split the interior peptide bonds of proteins (endopeptidases)

Source: Based on reference 8.

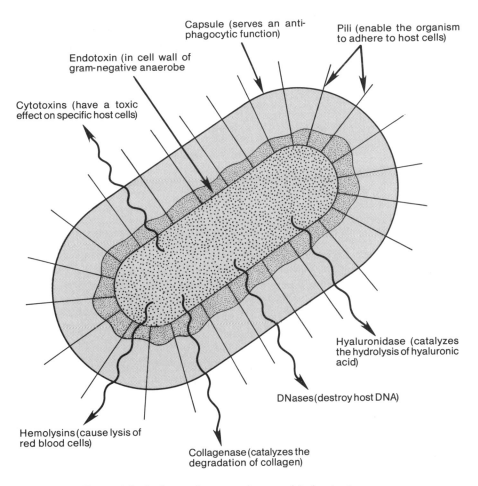

Figure 4-7. Potential virulence factors of anaerobic bacteria

Indications of Anaerobe Involvement in Human Disease

It has been stated that certain infectious diseases are so likely to involve anaerobes as significant pathogens that they should be regarded as anaerobic or mixed anaerobic-facultative infectious diseases until proved otherwise.[5] Examples of such infectious processes and diseases are shown in Figure 4-8 and include brain abscess, oral and dental infectious processes, aspiration pneumonia, lung abscess, peritonitis, intraabdominal abscess, infectious processes/diseases that follow bowel surgery or trauma to the bowel, endometritis, tuboovarian abscess, perirectal abscess, gas-forming/necrotizing infectious processes of soft tissue or muscle, and infectious processes of the lower extremities in diabetics. Table 4-5 describes specific anaerobes most commonly involved in these infectious processes.

Table 4-4. Potential Virulence Factors of Specific Anaerobes Causing Human Disease

Anaerobe	Source	Disease	Potential Virulence Factors
Clostridia			
C. botulinum	Exogenous	Botulism, wound botulism	Neurotoxin (botulin)
C. tetani	Exogenous	Tetanus	Spasmogenic neurotoxin, hemolytic tetanolysin
C. perfringens	Exogenous	Food poisoning	Enterotoxin
	Exogenous or endogenous (GI/GU flora)	Gas gangrene, infectious processes involving wounds	Phospholipase, hemolysins, necrotizing toxins, collagenase
C. septicum	Exogenous or endogenous (GI flora)	Gas gangrene	Deoxyribonuclease, hyaluronidase, hemolysins
C. difficile	Endogenous (GI flora)	Antibiotic-associated diarrhea and pseudomembranous colitis	Toxin, cytotoxin
Gram-negative bacilli			
B. fragilis	Endogenous	See Table 4-5	Capsule, collagenase, chondroitin sulfatase, fibrinolysin, heparinase, hyaluronidase, neuraminidase
P. asaccharolytica	Endogenous	See Table 4-5	Ability to adhere, collagenase, deoxyribonuclease, IgA protease, IgG protease, IgM protease
P. gingivalis	Endogenous	Periodontitis	Ability to adhere, capsule, collagenase, IgA protease, IgG protease
F. necrophorum	Endogenous	See Table 4-5	Ability to adhere, endotoxin

Source: Based on references 2, 9, and 16.

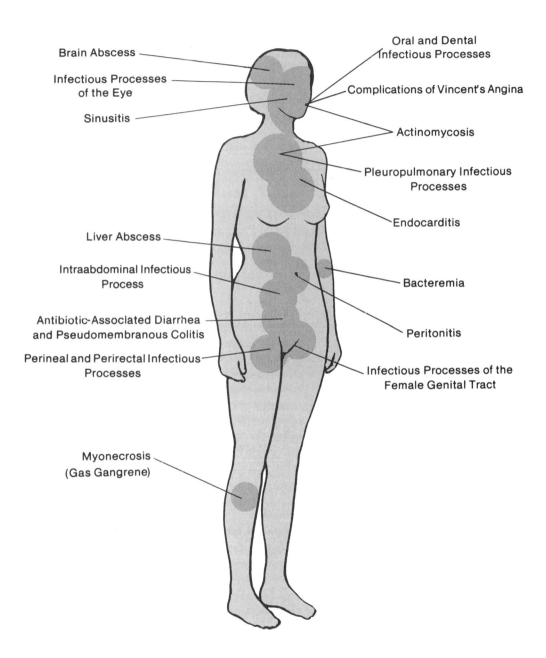

Figure 4-8. Human diseases that commonly involve anaerobes

Table 4-5. Endogenous Anaerobes Commonly Involved in Human Infectious Diseases

Infectious Processes	Anaerobes Commonly Involved
Actinomycosis	*A. israelii*, other *Actinomyces* spp., *Propionibacterium propionicus*
Antibiotic-associated diarrhea and pseudo-membranous colitis	*C. difficile, C. perfringens*
Bacteremia	Anaerobes are recovered from 10–25% of all positive blood cultures; *Propionibacterium* spp. most often represent contamination of the blood culture bottle by skin flora; 65–75% of the clinically significant anaerobic bacteremias involve *Bacteroides* and *Fusobacterium* spp.; 80–90% of these are due to the *B. fragilis* group; peptostreptococci are also frequently isolated
Brain abscess	Often polymicrobial; anaerobes frequently involved; *Bacteroides* spp., *Fusobacterium* spp., *Peptostreptococcus* spp., clostridia (less often)
Complications of Vincent's angina (necrotizing ulcerative gingivitis)	*F. necrophorum*
Endocarditis	Anaerobes are an uncommon cause. *Bacteroides* (especially *B. fragilis* group), gram-positive cocci, and some nonsporeforming, gram-positive bacilli (*P. acnes, Actinomyces,* and related bacteria)
Infectious processes of the eye	*Peptostreptococcus* spp., clostridia, *Bacteroides* spp., *Actinomyces* spp.
Infectious processes of the female genital tract	Anaerobes may cause disease in virtually every type of gynecologic or obstetric infectious process. Gram-positive cocci, *Bacteroides* spp., clostridia, and many other anaerobes
Intraabdominal infectious processes	Anaerobes are frequently encountered. Polymicrobic; *B. fragilis* group, other *Bacteroides* spp., *Fusobacterium* spp., *C. perfringens*, other clostridia, *Peptostreptococcus* spp.
Liver abscess	Gram-positive cocci, *B. fragilis* group, other *Bacteroides*, *F. necrophorum*, other fusobacteria, clostridia, *Actinomyces* spp.
Myonecrosis (gas gangrene)	*C. perfringens* (80–95% of the cases), *C. novyi, C. septicum*
Oral and dental infectious processes (e.g., periodontitis)	Oral anaerobes are almost always involved. *Peptostreptococcus* spp., *Porphyromonas* spp., *Wolinella* spp., *Fusobacterium* spp.

Table 4-5. Endogenous Anaerobes Commonly Involved in Human Infectious
Diseases (*continued*)

Infectious Processes	Anaerobes Commonly Involved
Perineal and perirectal infectious processes	Anaerobes are commonly involved. *B. fragilis* group, other *Bacteroides* spp., *Fusobacterium* spp., clostridia, gram-positive cocci, *Eubacterium* spp., *Actinomyces* spp.
Peritonitis	*Bacteroides* spp., *Peptostreptococcus* spp., *F. necrophorum*, clostridia
Pleuropulmonary infectious processes	Anaerobes are involved in 90% of cases. *Porphyromonas* spp., *F. nucleatum*, *Peptostreptococcus* spp., *B. fragilis* group, *Actinomyces* spp., *Eubacterium* spp.
Sinusitis	Anaerobes are commonly involved in chronic sinusitis, often polymicrobial. *Bacteroides* spp., *Peptostreptococcus* spp., *Fusobacterium* spp.

Source: Based on reference 5.

Generally, infectious diseases involving anaerobic bacteria follow some type of
trauma to protective barriers such as the skin and mucous membranes, thus allowing
anaerobes of the indigenous microflora (or soil anaerobes, in some cases) to gain
access to deeper tissues. A physician would usually assume that an infectious process
that arises following oral, GU, or GI surgery involves anaerobes and, pending microbiology laboratory results, would probably empirically treat the patient based upon
that assumption.

Table 4-6 contains a list of additional indications of involvement of anaerobes in
infectious processes. Although any of these should alert the physician to **possible**
involvement of anaerobes, most are **not specific** for anaerobes. For example, a foul
odor to the specimen could be absent in some infectious processes involving
anaerobes. But large quantities of gas in the specimen might be due to organisms
other than anaerobes (e.g., *Escherichia coli* or a mixture of enteric bacteria with or
without other organisms).

Table 4-6. Indications of Involvement of Anaerobes in Infectious Processes

Indication	Brief Explanation
Infectious process is in close proximity to a mucosal surface.	Anaerobes are the predominant microflora at mucosal surfaces.
Infectious process persists despite aminoglycoside therapy.	Aminoglycosides are ineffective against most anaerobes.
Specimen has a foul odor.	Certain anaerobes (e.g., *Porphyromonas* and *Fusobacterium* spp.) produce foul-smelling metabolic end products.

Table 4-6. Indications of Involvement of Anaerobes in Infectious Processes (*continued*)

Indication	Brief Explanation
Specimen contains a large quantity of gas.	Certain anaerobes (e.g., *Clostridium* spp.) produce large quantities of gas during metabolism.
Specimen has a black discoloration or fluoresces brick-red under long-wave ultraviolet light (e.g., under a Wood's lamp).	Pigmented species of *Porphyromonas* and *Prevotella* produce a pigment that fluoresces brick-red under ultraviolet light before becoming dark brown to black.
Specimen contains "sulfur granules."	"Sulfur granules" are often present in specimens from patients with actinomycosis.
Organisms exhibit distinct morphologic features in Gram-stained preparations.	The morphology of certain anaerobes is somewhat (but not completely) distinctive (e.g., certain *Bacteroides* and *Fusobacterium* spp. are quite pleomorphic; *F. nucleatum* is "fusiform" in shape; *Clostridium* spp. are often large, gram-positive rods that may or may not contain spores).

Source: Based on reference 5.

Case Presentation 4-2. Aspiration Pneumonia

History and Physical Examination

A 63-year-old man presented to the hospital with complaints of fever, cough, mild shortness of breath, and weight loss. The onset of his illness had been gradual, and he had not felt very ill until just a few days before entering the hospital. At first, the patient thought he had a chest cold, but the symptoms persisted and worsened. He stated that he had coughed up thick, foul-smelling, yellow material. There was no history of chest pain, and a chest x-ray, taken six months earlier, was normal.

The patient stated that he had consumed anywhere from a pint to a quart of liquor daily for many years and had smoked a pack of cigarettes per day for the past 30 years. There was no unusual history of travel, exposure to potential causes of pneumonia, tuberculosis, or known exposure to tuberculosis.

Physical examination revealed a temperature of 38.3°C, extensive gingivitis, foul-smelling breath, and some abnormal chest and breath sounds. Chest x-rays showed an extensive right upper lobe infiltrate in the posterior segment with numerous small excavations.

Question 4-2-1: What three clinical features in this case suggest aspiration pneumonia?

Answer 4-2-1: (1) The location of the patient's pulmonary process (the segments most commonly involved in pulmonary infection secondary to aspiration are the posterior segments of the upper lobes and the superior segments of the lower lobes); (2) gingivitis or periodontal disease is a common factor in aspiration pneumonia; (3) the transtrachial aspirate was foul smelling, which is a specific clue to the involvement of anaerobes.

Laboratory Results

The patient's white blood cell count was 16,200, with a mild shift to the left (increase in immature forms). Other hematologic and blood chemistry tests were within normal limits.

His sputum was grossly purulent and foul smelling. A Gram stain of the sputum revealed numerous PMNs, 30 squamous epithelial cells (EPIs) per low-power field (LPF), and mixed bacterial flora with no dominance by any particular morphotype. The sputum was not cultured. Transtracheal aspiration was performed, and the material obtained (the aspirate) was immediately placed in an anaerobic transport container and rapidly transported to the microbiology laboratory. The aspirate was foul smelling, and a Gram stain of the material revealed numerous PMNs and 0–1 EPIs per LPF. Also present were numerous pale-staining, gram-negative coccobacilli, numerous pale-staining, thin, fusiform, gram-negative bacilli in pairs (end-to-end), and a few tiny, gram-positive cocci in short chains. Three blood cultures failed to yield any growth.

Question 4-2-2: Why do you think this patient's sputum wasn't cultured?

Answer 4-2-2: His sputum specimen would not have been reliable for culture because it contained significant salivary contamination, as evidenced by the large number of squamous cells present. Applying Mayo Clinic standards, a satisfactory sputum specimen contains > 25 PMNs or < 10 EPIs per LPF (preferably both). Even if this patient's sputum specimen met these criteria, sputum specimens are not cultured for anaerobes due to contamination of the specimen with anaerobes of the oral flora. The indigenous microflora of saliva typically consists of 10^8 anaerobes/ml and 10^7 aerobes/ml.

Question 4-2-3: Of what value was the transtracheal aspirate?

Answer 4-2-3: Transtracheal aspiration bypasses the flora of the upper respiratory tract, permitting reliable culture—both aerobic and anaerobic. The aspirate provided immediate valuable information regarding the infecting organisms. The morphology of the organisms was distinctive enough to permit prediction of what organisms would eventually be cultured from the specimen.

Question 4-2-4: Based upon the Gram stain results, what anaerobes might you expect to be cultured from the transtracheal aspirate?

Answer 4-2-4: The presence of pale-staining, gram-negative bacilli or coccobacilli suggests *Bacteroides* or *Porphyromonas* spp. Pale-staining, thin, fusiform, gram-negative bacilli in pairs (end-to-end) suggest the presence of *Fusobacterium nucleatum*. The presence of tiny, gram-positive cocci in short chains suggests the presence of streptococci or peptostreptococci.

Treatment and Outcome

The patient was empirically treated with IV penicillin G and oral metronidazole. He received fluids intravenously and oxygen by nasal catheter. Within 48 hours, his temperature was lower, he looked and felt better, his oxygenation was improved, and his sputum had decreased in volume, was less purulent, and was no longer foul smelling. He became afebrile on the eighth day of therapy. At that point, he was placed on oral amoxicillin in place of penicillin, and his metronidazole therapy was maintained.

Further Laboratory Results

Aerobic culture of the transtrachial aspirate yielded a light growth of *Streptococcus intermedius*. Anaerobic culture yielded heavy growth of *Porphyromonas intermedia*, heavy growth of *Fusobacterium nucleatum*, and light growth of *Streptococcus intermedius*. The final culture results were not available until two weeks after the culture was set up.

Question 4-2-5: Did the culture results confirm the Gram stain observations and the appropriateness of the empirical therapy?

Answer 4-2-5: Yes on both counts.

Question 4-2-6: Do you think that two weeks was a reasonable period of time to wait for final culture results in this case?

Answer 4-2-6: Absolutely not! Whenever a properly selected, collected, and transported specimen is promptly processed, using appropriate media and anaerobic incubation conditions, final culture results should be available within three to four days.

Additional Information

The principal causes of anaerobic or mixed anaerobic-aerobic pulmonary infectious processes are aspiration of oral or gastric contents and the bacteria present in these materials. Indigenous oral flora and organisms involved in periodontal disease are most apt to be involved in community-acquired disease. In the hospital setting, *Staphylococcus aureus*, various members of the *Enterobacteriaceae* family, and *Pseudomonas* spp. may also be involved. Pneumonia following aspiration is the most common type of hospital-acquired pneumonia and, as such, is a major cause of death and disability in hospitalized patients. Circumstances predisposing to aspiration include reduced level of consciousness, dysphagia (difficulty in swallowing), nasogastric feeding, tracheostomy, and endotracheal intubation. The major conditions that heighten risk of aspiration pneumonia include alcoholism, drug addiction, seizure disorders, cerebrovascular accidents, esophageal disease, and use of general anesthesia. Periodontal disease, gingivitis, and use of antacids are important background factors.

Source: Based on Reference 4.

Chapter in Review

- Anaerobes that are involved in human diseases may originate either **outside** of the body (**exogenous** origin) or **within** the body (**endogenous** origin). Those of exogenous origin are usually members of the genus *Clostridium*, many species of which are found in soil. Anaerobes of endogenous origin are members of the indigenous microflora—organisms living on or in the human body.

- Diseases caused by exogenous *Clostridium* spp. include foodborne botulism, infant botulism, wound botulism, tetanus, gas gangrene (myonecrosis), and food poisoning.

- Endogenous clostridia can also cause disease. *Clostridium difficile*, for example, is a common etiologic agent of antibiotic-associated diarrhea and pseudomembranous colitis. *Clostridium perfringens*, commonly part of the GI or GU flora, can cause disease when it gains access to the bloodstream or other tissues of the body.

- Anaerobes of endogenous origin can contribute to infectious processes and diseases in virtually any tissue or organ of the body. These processes are usually **polymicrobial** and **purulent.** However, some necrotic processes involve histotoxic anaerobes that produce cytotoxins capable of causing necrosis of neutrophils, macrophages, and cells of various other tissues.

- Anaerobic bacteria are usually involved in infectious diseases that follow some type of mucous membrane or skin trauma, vascular stasis, tissue necrosis, and a decrease in the oxidation-reduction (redox) potential of tissue.

- Some anaerobes produce or possess a variety of potential **virulence factors,** including endotoxins, enzymes, polysaccharide capsules, and adherence factors, thought to play a role in pathogenicity. *Clostridium* spp. produce a variety of **exotoxins,** such as **neurotoxins, cytotoxins, necrotizing toxins,** and **enterotoxins,** which are responsible for the characteristic signs and symptoms of the diseases caused by these organisms.

- Certain infectious processes/diseases virtually always involve anaerobes. These include brain and lung abscesses, oral and dental infectious processes, aspiration pneumonia, peritonitis, intraabdominal abscess, infectious processes that follow bowel surgery or trauma to the bowel, endometritis, tuboovarian abscess, perirectal abscess, gas-forming/necrotizing infectious processes of soft tissue or muscle, and infectious processes of the lower extremities of diabetic patients.

- A physician would usually assume that an infectious process that arises following oral, GU, or GI surgery involves anaerobes and, pending microbiology laboratory results, would probably treat the patient empirically with antimicrobial agents known to be effective against the most likely anaerobic pathogens.

Self-Assessment Exercises

1. TRUE or FALSE: The term "exogenous anaerobes" refers to anaerobes found on or in the body.

2. TRUE or FALSE: Most infectious diseases involving anaerobes involve endogenous anaerobes.

3. TRUE or FALSE: Endogenous anaerobes can cause infectious diseases in virtually any tissue or organ of the body.

4. TRUE or FALSE: Infectious diseases involving anaerobes are rarely polymicrobial or purulent.

5. TRUE or FALSE: Foodborne botulism is acquired by ingesting food containing **spores** of *Clostridium botulinum,* whereas infant botulism results from the ingestion of food containing **toxins** of *C. botulinum.*

For questions 6–9, indicate whether the organism is considered an endogenous or exogenous anaerobe.

6. *Bacteroides fragilis* _____

7. *Clostridium tetani* _____

8. *Fusobacterium nucleatum* _____

9. *Clostridium botulinum* _____

10. Define the term "intoxication" as it refers to diseases caused by anaerobic bacteria.

_____

_____

11. Give five examples of conditions that predispose an individual to infections by endogenous anaerobes:

a. _____

b. _____

c. _____

d. _____

e. _____

12. Name the anaerobe most often involved in each of the following infectious processes/diseases:

a. Actinomycosis (give genus and species) _____

_____

b. Bacterial vaginosis (give genus) _____

_____

c. Gas gangrene (give genus and species) _____

_____

d. Infectious processes in soft tissues (give genus and species) _____

_____

e. Antibiotic-associated diarrhea and pseudomembranous colitis (give genus and species) _____

_____

13. Give three indications of possible involvement of anaerobes in an infectious process, other than predisposing factors:

a. _____

b. _____

c. _____

14. List four examples of virulence factors possessed by anaerobes:

a. _____

b. _____

c. _____

d. _____

15. Differentiate between the terms "infection" and "infectious disease" as they are used in this book. _____

_____

_____

16. Why is the presence of a *Clostridium difficile* carrier of special importance in a hospital setting? _____

_____

_____

References

1. Bergan, T. 1984. Pathogenicity of anaerobic bacteria. Scand. J. Gastroent. Suppl. 91:1–11.
2. Bjornson, H. S. 1984. Enzymes associated with the survival and virulence of gram-negative anaerobes. Rev. Infect. Dis. 6 (Suppl 1):S21–S24.
3. Delmée, M., V. Avesani, N. delferriere, and G. Burtonboy. 1990. Characterization of flagella of *Clostridium difficile* and their role in serogrouping reactions. J. Clin. Microbiol. 28:2210–2214.
4. Finegold, S. M. 1989. Clinical experience: anaerobic pulmonary infection. Hosp. Practice 24:103–133.
5. Finegold, S. M. , W. L. George, and M. E. Mulligan. 1986. Anaerobic infections. Year Book Medical Publishers, Chicago.
6. Goldin, B. R. , and S. L. Gorbach. 1989. Impact of anaerobic bowel flora on metabolism of endogenous and exogenous compounds, p. 691–714. In S. M. Finegold and W. L. George (eds.), Anaerobic infections in humans. Academic Press, San Diego.
7. Gorbach, S. L. 1989. Clinical experience: *Clostridium difficile* settles in a nursing home. Hosp. Practice 24:145–160.
8. Hatheway, C. L. 1990. Toxigenic clostridia. Clin. Microbiol. Rev. 3:66–98.
9. Hofstad, T. 1984. Pathogenicity of anaerobic gram-negative rods: possible mechanisms. Rev. Infect. Dis. 6:189–199.
10. Holland, K. T. , J. S. Knapp, and J. G. Shoesmith. 1987. Anaerobic bacteria. Chapman and Hall, New York.
11. Mayrand, D. , and S. C. Holt. 1988. Biology of asaccharolytic black-pigmented *Bacteroides* species. Microbiol. Rev. 52:134–152.
12. McFarland, L. V. , M. E. Mulligan, R. Y. Kwok, and W. E. Stamm. 1989. Nosocomial acquisition of *Clostridium difficile* infection. N. Engl. J. Med. 320:204–210.
13. MMWR. 1990. CDC surveillance summaries. Morbidity and mortality weekly report, vol. 39, no. SS–1. Centers for Disease Control, Atlanta.
14. MMWR. 1990. Summary of notifiable diseases, United States, 1989. Morbidity and mortality weekly report, vol. 38, no. 54. Centers for Disease Control, Atlanta.
15. Rolfe, R. D. , and S. M. Finegold. 1988. *Clostridium difficile:* its role in intestinal disease. Academic Press, San Diego.
16. Smith, L. D. S. , and B. L. Williams. 1984. The pathogenic anaerobic bacteria, 3rd ed. Charles C. Thomas, Springfield, IL.
17. Turgeon, D. K. , and V. R. Dowell, Jr. 1989. In vitro evidence for synergy between *Fusobacterium mortiferum* and *Escherichia coli*. Mil. Med. Lab. Sci. 18:89–93.
18. Zaleznik, D. F. , and D. L. Kasper. 1989. Role of bacterial virulence factors in pathogenesis of anaerobic infections, p. 81–95. In S. M. Finegold and W. L. George (eds.), Anaerobic infections in humans. Academic Press, San Diego.

CHAPTER 5

SPECIMEN QUALITY

Selection and Collection of Specimens for
Anaerobic Bacteriology
Acceptable Specimens
Unacceptable Specimens
Proper Transport of Specimens
Aspirates
Swabs
Tissue
Blood

At the conclusion of this chapter, you will be able to:

1. Describe three ways in which clinical laboratory professionals can influence the quality of clinical specimens submitted for anaerobic culture.
2. List six acceptable and six unacceptable clinical specimens for anaerobic bacteriology.
3. Explain why expectorated sputum, voided urine, and feces are unacceptable specimens for routine anaerobic culture.
4. Name three important considerations in the transport of clinical specimens for anaerobic bacteriology.
5. Define the term "PRAS media."

Definitions

Culdocentesis: Aspiration of fluid from the rectouterine excavation following needle puncture of the vaginal wall. The rectouterine excavation is a sac or recess formed by a fold of the peritoneum where it dips down between the rectum and uterus.

Percutaneous transtracheal aspiration: A technique used to aspirate fluid from the lung. A catheter is fed into the affected lung via a needle inserted through the wall of the trachea. When performed correctly, this technique provides a lower respiratory specimen free of contamination by extraneous oropharyngeal flora.

Suprapubic bladder aspiration: A technique used for aspiration of urine from the urinary bladder. A needle is inserted through the abdominal wall and into the bladder. This technique provides a urine specimen free of contamination by the indigenous microflora of the distal urethra.

Thoracentesis: Surgical puncture of the chest wall for the purpose of aspirating fluid from the pleural cavity

---

Chapter 4 emphasized that infectious processes/diseases involving anaerobes are both serious and common. This chapter furnishes guidelines regarding appropriate types of specimens to be submitted to the anaerobic bacteriology laboratory. Recommended techniques are presented for collection and transport of such specimens. Although selection, collection, and transport of human clinical specimens are described in this chapter, many of the principles are equally applicable to veterinary specimens.

Selection and Collection of Clinical Specimens for Anaerobic Bacteriology

Once a physician suspects that he or she is dealing with an anaerobe-associated infectious process, the physician must properly select and collect a clinical specimen and arrange for its proper and rapid transport to the laboratory. These steps are extremely important in the successful outcome of an anaerobic culture.

Selection, collection, and transport of clinical specimens occur **before** the specimen arrives in the laboratory. This does **not** mean, however, that these important events are beyond the control of clinical laboratory professionals. On the contrary, clinical laboratory professionals can play a major role in selecting, collecting, and transporting specimens for anaerobic bacteriology by taking the following steps:

- **Education** is an excellent first step. Through no fault of their own, physicians and nurses often receive very little information about anaerobes during their educational programs. Thus, clinical laboratory professionals should make every effort to provide them with information as to which specimens are appropriate for anaerobic culture and help them understand why other specimens are inappropriate. This type of training could be presented at in-services and rounds.

- The laboratory can disseminate **written guidelines** for the proper selection, collection, and transport of specimens. (Additional information regarding laboratory directives can be found in Chapter 11.)

- Laboratory personnel can ensure that the **proper collection and transport devices** are on hand.

- Finally, **criteria for rejection of inappropriate specimens** can be developed, serving to strengthen written directives that emanate from the laboratory. Such criteria, which are in the best interest of the patients, must always be developed with the knowledge, cooperation, and consent of the hospital's physicians. Whenever a specimen is rejected, the microbiologist must explain to the requesting physician the reasons for rejection and the consequences of working up improper specimens. A specimen must never be discarded without first talking with the physician. Some specimens, especially those taken at the time of surgery, would be impossible to replace.

Acceptable Specimens

Many types of specimens are acceptable for anaerobic culture; these are listed in Table 5-1. The best specimens for anaerobic bacteriology are those collected by needle and syringe, thus minimizing contamination of the specimen with endogenous anaerobes. Percutaneous transtracheal aspiration, protected brush, and suprapubic bladder aspiration procedures are illustrated in Figures 5-1, 5-2, and 5-3, respectively.

Table 5-1. Acceptable Specimens for Anaerobic Bacteriology

Anatomic Source	Specimens and Recommended Methods of Collection
Central nervous system	Cerebrospinal fluid, carefully aspirated abscess material, tissue from biopsy or autopsy
Dental/ENT specimens	Carefully aspirated material from abscesses, biopsied tissue
Localized abscesses	Needle and syringe aspiration of closed abscesses
Decubitus ulcers	Aspiration of pus from beneath skin flaps or from deep pockets, following thorough cleansing of the area with antiseptic
Sinus tracts or draining wounds	Aspiration by syringe through a small plastic catheter introduced as deeply as possible through a decontaminated skin orifice
Deep tissue or bone	Specimens obtained during surgery from depths of wound or underlying bone lesion
Pulmonary	Percutaneous transtracheal aspiration; aspirate obtained by direct lung puncture; pleural fluid obtained by thoracentesis; biopsied tissue: "sulfur granules" from draining fistula; if properly used, a double catheter with bronchial brush may be suitable for bronchoscopic specimens
Intraabdominal	Aspirate from abscess, ascitic fluid, biopsied tissue
Urinary tract	Suprapubic bladder aspiration
Female genital tract	Aspirate from loculated abscess; culdocentesis specimen, preferably following decontamination of the vagina using povidone-iodine; a double catheter with bronchial brush or a sterile swab may be used for uterine cavity specimens
Other	Blood, bone marrow, aspirated synovial fluid, biopsied tissue from any normally sterile site

Source: Based on references 1 and 5.

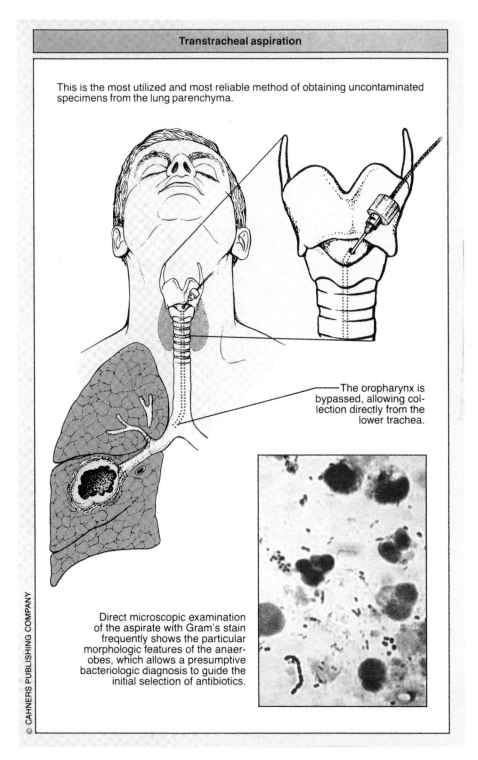

Transtracheal aspiration

This is the most utilized and most reliable method of obtaining uncontaminated specimens from the lung parenchyma.

The oropharynx is bypassed, allowing collection directly from the lower trachea.

Direct microscopic examination of the aspirate with Gram's stain frequently shows the particular morphologic features of the anaerobes, which allows a presumptive bacteriologic diagnosis to guide the initial selection of antibiotics.

© CAHNERS PUBLISHING COMPANY

Figure 5-1. Percutaneous transtracheal aspiration procedure (from reference 3, p. 42, © 1990, reproduced from *Hospital Medicine,* May 1990, with permission of Cahners Publishing Company)

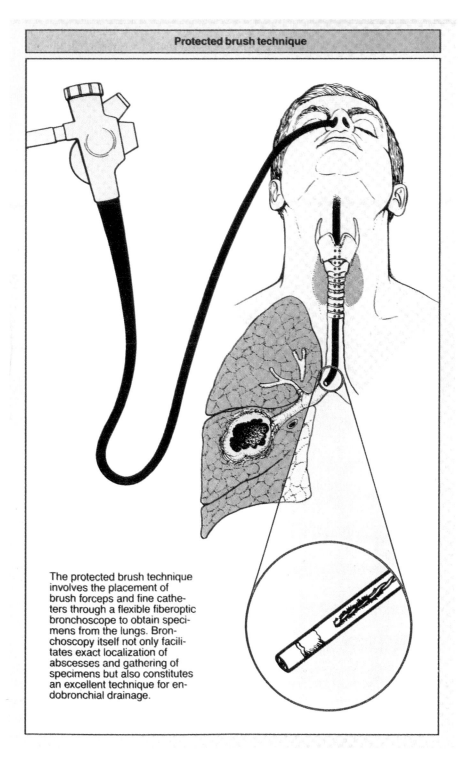

The protected brush technique involves the placement of brush forceps and fine catheters through a flexible fiberoptic bronchoscope to obtain specimens from the lungs. Bronchoscopy itself not only facilitates exact localization of abscesses and gathering of specimens but also constitutes an excellent technique for endobronchial drainage.

Figure 5-2. Protected brush technique (from reference 3, p. 43, © 1990, reproduced from *Hospital Medicine,* May 1990, with permission of Cahners Publishing Company)

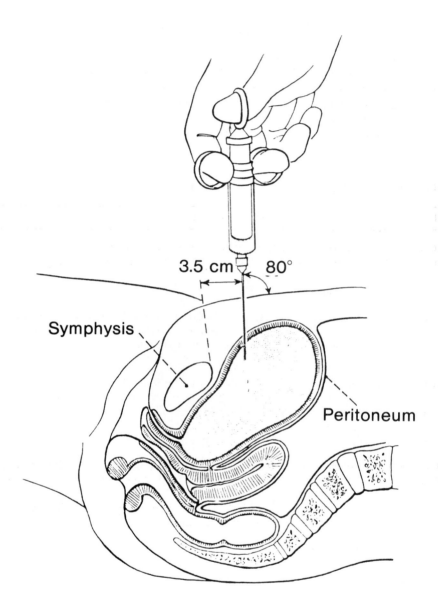

Figure 5-3. Suprapubic bladder aspiration procedure (from reference 2, p. 362, with permission of Charles C. Thomas, Publisher, Springfield, IL). Urine collected by suprapubic bladder aspiration is an acceptable specimen for diagnosis of anaerobic bacteriuria. "The patient is asked to report with a full bladder. The lower abdomen is prepared with alcohol and, at a point approximately two finger-breadths above the symphysis pubis, the skin is infiltrated with a local anesthetic. A long needle is then inserted directly into the bladder, at an angle of approximately 10° off the vertical."[2]

Unacceptable Specimens

Table 5-2 contains a list of specimens not recommended for anaerobic culture. Most of them have been collected by swab. In general, specimens collected by swabbing mucosal surfaces or skin are unacceptable for anaerobic bacteriology because they are contaminated with indigenous microflora (including endogenous anaerobes). Likewise, expectorated sputum, feces, and voided urine would all be expected to contain endogenous anaerobes.

Table 5-2. Unacceptable Specimens for Anaerobic Bacteriology

Clinical Specimen	Brief Explanation
Exudates and other material collected by swabs from superficial wounds, abscesses, burns, cysts, or ulcers	Such specimens would be contaminated with skin flora anaerobes (e.g., *Propionibacterium acnes* and anaerobic, gram-positive cocci).
Vaginal, cervical, or urethral swabs	Such specimens would be contaminated with a variety of vaginal flora anaerobes or those that colonize the distal urethra (e.g., *Bacteroides* spp. and *Fusobacterium* spp.).
Respiratory tract specimens collected by swab, nasotracheal or orotracheal suction, or bronchoscope; expectorated sputum	Such specimens would be contaminated with oral flora anaerobes (e.g., *Porphyromonas* spp., *Fusobacterium*, spp., *Veillonella* spp.).
Stool specimens, rectal swabs	Such specimens would be contaminated with a variety of gastrointestinal tract flora anaerobes (e.g., *Bacteroides* spp., *Clostridium* spp., *Eubacterium* spp.) Fecal specimens may be cultured when specific pathogens (e.g., *C. difficile* or *C. botulinum*) are being sought.
Voided or catheterized urine	Such specimens would be contaminated with endogenous anaerobes that colonize the distal urethra (e.g., *Bacteroides* and *Fusobacterium* spp.).

Source: Based on reference 5.

The Accu-CulShure® collection instrument (Technology for Medicine, Inc., Pleasantville, NY), illustrated in Figure 5-4, appears to be an exception to general statements concerning the unacceptability of swab specimens. The instrument is designed in a manner that exposes the swab only at the collection site (e.g., the cervix) and protects it from indigenous microflora contamination both before and after collecting the specimen.

Workup of inappropriate specimens is a waste of time, effort, and money. Culture results must be clinically relevant and must reveal something about the patient's infection—not about his or her mucosal or skin flora.

EASY TO USE; ONE-PIECE CONSTRUCTION:
No plastic bags, no ampules to crush, easy to unseal

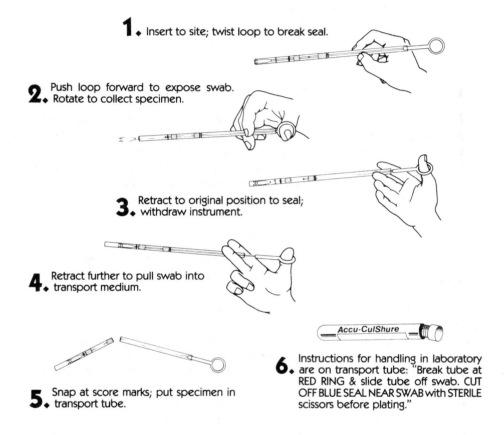

1. Insert to site; twist loop to break seal.

2. Push loop forward to expose swab. Rotate to collect specimen.

3. Retract to original position to seal; withdraw instrument.

4. Retract further to pull swab into transport medium.

5. Snap at score marks; put specimen in transport tube.

6. Instructions for handling in laboratory are on transport tube: "Break tube at RED RING & slide tube off swab. CUT OFF BLUE SEAL NEAR SWAB with STERILE scissors before plating."

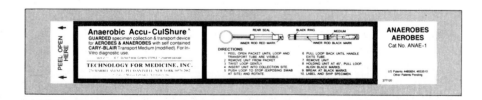

The Accu-CulShure Collection Instrument is available in 10" and 13" lengths. In addition to the anaerobic/aerobic model, the product line includes chlamydial, viral and parasitic instruments and an anaerobic/aerobic endometrial unit.

Figure 5-4. Accu-Culshure® collection instrument (courtesy of Technology for Medicine, Inc. Pleasantville, NY)

Clinical laboratory professionals must assume the responsibility for educating those who select, collect, and transport specimens for anaerobic bacteriology. They must always take time to explain the consequences of working up improper specimens and those that have been incorrectly collected or transported.

The specimens listed in Table 5-2 would be expected to contain indigenous microflora contaminants. Neither the microbiologist nor the physician would know whether the anaerobes isolated from these specimens represent contaminants or organisms actually involved in the infectious process (i.e., true pathogens).

Proper Transport of Specimens

Regardless of the type of specimen being submitted for anaerobic bacteriology, it must be transported as rapidly as possible and with minimum exposure to oxygen. Specimens are usually collected from a warm, moist environment that is low in oxygen. Thus, it is important to avoid "shocking" the anaerobes by exposing them to oxygen or permitting them to dry out. In addition, the amount of time they remain at room temperature should be minimized.

Aspirates

As previously mentioned, specimens collected by needle and syringe are better for anaerobic bacteriology than those collected by swab. Following aspiration of the specimen, any air present in the syringe and needle should rapidly be expelled. To prevent production of a potentially infectious aerosol, place an alcohol-soaked gauze pad over the needle while cautiously expelling the air.

In the past, it was acceptable to place a rubber stopper over the point of the needle and then rapidly transport the needle and syringe assembly to the laboratory. However, in view of current mandatory infectious disease safety precautions, this type of specimen handling and transport is now considered unsafe because the individual transporting the specimen may stick him or herself with the needle.

Ideally, the aspirate will be injected into some type of oxygen-free transport tube or vial, preferably one containing prereduced, anaerobically sterilized (PRAS) transport medium. PRAS media are prepared by boiling (to remove dissolved oxygen), autoclaving (to sterilize the mixture), and replacing any air with an oxygen-free gas mixture.[1a]

The transport container depicted in Figure 5-5 is manufactured by Anaerobe Systems (San Jose, CA). The plastic screw cap can be removed for insertion of a swab or small piece of tissue, or an aspirate may be carefully injected through a rubber diaphragm in the cap. Thus, the tube can be used to transport an aspirate, a small piece of tissue, or a swab specimen. Whenever such a transport system is used, care must be taken not to tip the container while the cap is removed (e.g., when inserting a swab or small tissue specimen). Because the oxygen-free gas mixture is heavier than room air, it would spill from the tipped container and be displaced by room air. This would, of course, defeat the primary purpose of using such a transport system. Other types of oxygen-free transport containers are manufactured by Scott Laboratories, Becton Dickinson, and other companies (see Appendix B and Figures 5-6 and 5-7).

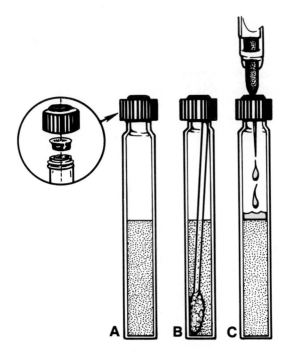

Figure 5-5. Anaerobic transport system for transporting either aspirates or swabs. **(A)** An anaerobic transport system (tube) fitted with a Hungate stopper, which consists of a rubber diaphragm and a plastic screw cap. **(B)** With the plastic cap removed and the tube held in an upright position (to prevent loss of the anaerobic gas mixture), a swab specimen is inserted into the transport system. The tube is then tightly capped. **(C)** Following careful removal of air from the syringe and needle and decontamination of the rubber diaphragm, an aspirate is injected into the transport system through the diaphragm. Extreme caution should be exercised to avoid a needle-stick injury.

Swabs

In those rare instances where swabs are deemed necessary, use commercially available, oxygen-free swabs. To prevent drying of the material during transit to the laboratory, place the swab in a tube containing a PRAS medium as well as an oxygen-free environment. Such containers are available from a number of commercial sources (see Appendix B).

Tissue

As previously mentioned, tissue specimens collected by biopsy or at autopsy from usually sterile sites are acceptable specimens for anaerobic culture. Small pieces of tissue can be placed in oxygen-free transport tubes or vials containing a PRAS medium to keep the tissue moist. To ensure that larger tissue specimens do not dry out enroute to the laboratory, place them in sterile petri dishes containing a small amount of sterile saline.

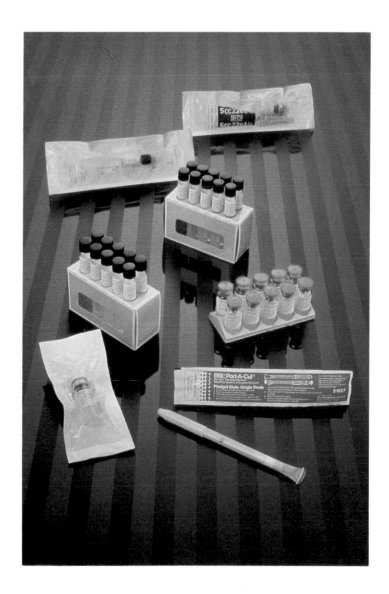

Figure 5-6. Anaerobic transport containers (courtesy of Becton Dickinson Microbiology Systems, Cockeysville, MD). These anaerobic transport containers (Port-A-Cul® tubes and vials available from Becton Dickinson) contain prereduced transport medium with reducing agents and a redox indicator. According to the manufacturer, viability of a wide variety of aerobes, facultative anaerobes, and obligate anaerobes is maintained for up to 72 hours at 20 to 25°C.

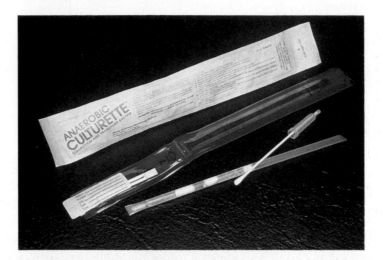

Figure 5-7. Anaerobic swab collection system (courtesy of Becton Dickinson Microbiology Systems, Cockeysville, MD). After activation, the Anaerobic Culturette® (Becton Dickinson) provides a reduced oxygen environment within ten minutes and maintains that environment for up to 72 hours. According to the manufacturer, even the strictest anaerobes will survive in this system.

To ensure that the specimen is not exposed to oxygen enroute to the laboratory, place the petri dish in a bag or pouch containing an oxygen-free atmosphere. Such bags/pouches, which are available commercially from Becton Dickinson (Bio-Bags and GasPak Pouches) and Difco (Anaerobe Pouches), are described in more detail in Chapter 6.

Blood

A discussion of the variety of commercially available blood culture systems is beyond the scope of this book. It is important to note, however, that blood for culture must be very carefully collected so as to minimize contamination with skin flora. This is usually accomplished by meticulous preparation of the venipuncture site with a bactericidal agent, such as tincture of iodine or an iodophor. Common contaminants of blood cultures include various gram-positive bacteria that inhabit the skin and, less frequently, gram-negative bacteria or yeast. Some of the more commonly encountered gram-positive contaminants are *Propionibacterium acnes, P. avidum, Staphylococcus epidermidis, S. aureus, S. saprophyticus,* and *S. saccharolyticus.*

Blood must be cultured in such a manner as to recover any and all bacteria or yeasts that may be present. This usually requires aseptic inoculation of both an anaerobic (unvented) bottle and an aerobic (vented) bottle. Investigations have shown that single-bottle systems are less efficient in isolating certain organisms (including anaerobes) than two-bottle systems.[4] Once inoculated, blood culture bottles should be rapidly transported to the laboratory, where they are incubated at 35 to 37°C.

Chapter in Review

- Proper selection, collection, and transport of specimens are essential for good anaerobic bacteriology. Much of the success of anaerobic bacteriology depends on things that happen to a specimen before it arrives in the laboratory. Was it **properly selected**? Was it **collected properly,** so as to avoid contamination with indigenous microflora? Was it protected from contact with room air during transport? Was it **rapidly transported** to the laboratory for processing?

- To ensure that the information generated in the anaerobic bacteriology laboratory is clinically relevant, the only specimens that should be processed are those that (1) are appropriate, (2) have been collected in such a manner as to minimize contamination by indigenous microflora, and (3) are transported in such a manner as to minimize exposure to air.

- Clinical laboratory professionals have a responsibility to assist in the education of those involved in selecting, collecting, and transporting specimens for anaerobic bacteriology. Further, they must ensure the availability of anaerobic transport containers.

- Contaminated specimens lead to wasted laboratory effort and money. Perhaps more important, an extensive workup of a contaminated specimen provides useless information that could result in mismanagement of patients by physicians. The quality of anaerobic bacteriology performed in any laboratory can only be as good as the quality of specimens received.

Self-Assessment Exercises

1. The following specimens have been received in the laboratory for anaerobic culture. Indicate which are acceptable (A) and which are unacceptable (U) for routine anaerobic bacteriology:

    _____ a. Voided urine

    _____ b. Blood

    _____ c. Cerebrospinal fluid

    _____ d. Feces

    _____ e. Catheterized urine

    _____ f. Synovial fluid

    _____ g. Swab containing material from oral abscess

    _____ h. Swab containing pus from skin lesion

    _____ i. Bone marrow

    _____ j. Tissue biopsy specimen

2. In general, the best types of clinical specimens to submit to the anaerobic bacteriology laboratory are those that have been collected with a_____ and _____.

   Why? _____

   _____

   _____

3. In general, the worst types of clinical specimens to submit to the anaerobic bacteriology laboratory are those that have been collected with a_____.

   Why? _____

   _____

   _____

4. An elderly, alcoholic patient is suspected of having aspiration pneumonia. Which one of the following specimens is recommended for isolation of any anaerobic bacteria that might be contributing to the infectious process?

   a. Bronchial washings
   b. Expectorated (coughed) sputum
   c. Percutaneous transtracheal aspirate
   d. Specimen collected by nasopharyngeal tube
   e. Sputum induced by nebulization

5. A female patient is suspected of having endometritis. Which of the following specimens is recommended for isolation of any anaerobic bacteria that might be contributing to the infectious process?

   a. Vaginal swab
   b. Cervical swab
   c. Specimen collected by culdocentesis

6. You have been asked to present a short in-service to nursing personnel regarding important considerations in the **transport** of clinical specimens for anaerobic bacteriology. List three of the important considerations you will include in your presentation.

   a. _____

   b. _____

   c. _____

7. The clinical microbiology section of a laboratory continues to receive specimens for anaerobic bacteriology that are unacceptable. Describe three ways clinical laboratory professionals could improve the situation.

   a. _____

   b. _____

   c. _____

8. What is the meaning of the term "PRAS media"?

   _____

   _____

References

1. Finegold, S. M., W. L. George, and M. E. Mulligan. 1986. Anaerobic infections. Year Book Medical Publishers, Chicago.
1a. Holdeman, L. V., E. P. Cato, and W. E. C. Moore (eds.). 1977 (with 1987 update). VPI anaerobe laboratory manual, 4th ed. Virginia Polytechnic Institute and State University, Blacksburg.
2. Martin, J. M., and J. W. Segura. 1974. Urinary tract infections due to anaerobic bacteria, p. 359–367. *In* A. Balows, R. M. DeHaan, V. R. Dowell, Jr., and L. B. Guze (eds.), Anaerobic bacteria: role in disease. Charles C. Thomas, Springfield, IL.
3. Mostow, S. R. 1990. Anaerobic infections of the lung. Hosp. Med. 33–48.
4. Reller, L. B., P. R. Murray, and J. D. MacLowry. 1982. Cumitech 1A: blood cultures II. American Society for Microbiology, Washington, D. C.
5. Sutter, V. L., D. M. Citron, M. A. C. Edelstein, and S. M. Finegold. 1985. Wadsworth anaerobic bacteriology manual, 4th ed. Star Publishing, Belmont, CA.

CHAPTER 6

SPECIMEN PROCESSING

At the conclusion of this chapter, you will be able to:

1. Describe the steps in macroscopic and microscopic examinations of specimens submitted to the clinical microbiology laboratory for isolation of anaerobes.
2. Describe the types of holding systems that could be used to prevent undue exposure of anaerobes (particularly obligate and microaerotolerant anaerobes) to oxygen during examination of clinical materials or cultures.
3. Explain how various types of clinical specimens (e.g., aspirates, tissue, and swabs) are processed in the laboratory prior to inoculation of media.
4. Discuss general criteria in the selection of media for use in anaerobic bacteriology.
5. Name four specific types of plated media recommended for use in the primary isolation setup of anaerobic bacteriology cultures.
6. State the purpose of each of the four types of plated media used in the primary isolation setup.
7. List the three anaerobic systems used most often in the clinical laboratory for anaerobic incubation of inoculated plates.
8. Compare and contrast the three most commonly used methods for anaerobic incubation of inoculated plates.

Definitions

Lecithin: One of a group of phosphoglycerides found in various mammalian tissues, including egg yolk, nerve tissue, semen, and cell membranes

Lecithinase: An enzyme that catalyzes the breakdown of lecithin; also called phospholipase C. Lecithinase production is a useful criterion for identification of various gram-positive and gram-negative bacteria that produce it

Lipase: Any of a group of enzymes that catalyzes the breakdown of triglycerides and phospholipids. Lipases occur in milk, the pancreas, adipose tissue, the stomach, and various other tissues.

This chapter describes actions to be taken once a properly selected, collected, and transported specimen arrives in the anaerobic bacteriology laboratory. There are many ways to process a specimen for anaerobic bacteriology, but regardless of the method selected, emphasis should be placed on **accuracy** and **speed.** Any anaerobes present within the specimen have been at room temperature during transit to the laboratory and may have been exposed to oxygen during collection. They should not have been exposed to additional oxygen during transport, and the specimen should not have dried out. Although this chapter describes processing procedures for human clinical specimens, many of the principles and procedures are equally applicable to veterinary specimens.

Ideally, once a specimen arrives in the laboratory, it can be immediately passed into an anaerobic chamber to prevent further exposure of clinical materials to oxygen. Anaerobic chambers are described in detail later in this chapter. For now, you need only understand that an anaerobic chamber allows **all steps** in the processing of a specimen to be performed in an oxygen-free environment. In those laboratories not equipped with anaerobic chambers, holding systems may be used (also described in detail later in this chapter). To comply with mandatory infectious disease safety policies, follow appropriate safety precautions. Wear disposable latex gloves when handling clinical specimens containing potentially infectious agents, and use a laminar flow safety cabinet whenever deemed appropriate to do so.

The initial processing of a clinical specimen involves the following steps, each of which is discussed in detail:

- **Macroscopic examination** of the specimen

- **Microscopic examination** of the specimen (to include examination of a Gram-stained smear)

- **Inoculation of appropriate plated and tubed media,** including media specifically designed for culturing anaerobes

- **Anaerobic incubation** of inoculated media

Macroscopic Examination of Specimens

Each specimen received in the anaerobic bacteriology section should first be examined macroscopically, and pertinent observations should be recorded on a worksheet (see Figure 7-2 in Chapter 7). Some of the features/characteristics to note during the macroscopic examination are listed in Table 6-1.

Table 6-1. Features/Characteristics To Note During the Macroscopic Examination of a Specimen

Questions To Ask	Comments
Is it an appropriate specimen?	Inappropriate specimens should be rejected (see Chapter 11).
Was it submitted in an appropriate transport container?	Improperly transported specimens should be rejected (see Chapter 11).
How old is the specimen? Are the date and time of collection recorded on the accompanying request slip?	Specimens that are too old may be cause for rejection (see Chapter 11).
Is there evidence that the specimen has dried out during transit?	Specimens that have dried out during transport should be rejected (see Chapter 11).
Does the specimen have a foul odor?	Many anaerobes, especially fusobacteria and pigmented *Porphyromonas* spp., have foul-smelling metabolic end products. However, the lack of a foul odor does not exclude anaerobes.
Does the specimen fluoresce brick-red when exposed to a Wood's lamp? (A Wood's lamp emits long-wave [366 nm] ultraviolet light.)	Pigmented species of *Porphyromonas* and *Prevotella* produce substances that fluoresce under long-wave UV light prior to becoming darkly pigmented. Although a brick-red fluorescence is presumptive evidence of these organisms, some members of this group fluoresce colors other than brick-red.

Table 6-1. Features/Characteristics To Note During the Macroscopic Examination of a Specimen (*continued*)

Questions To Ask	Comments
Is the necrotic tissue or exudate black?	Such discoloration may be due to the pigment produced by pigmented species of *Porphyromonas* and *Prevotella*.
Does the specimen contain sulfur granules?	Such granules are associated with actinomycosis, a condition caused by *Actinomyces* spp., *Propionibacterium propionicus*, and closely related organisms, such as *Propionibacterium acnes*.
Is the specimen bloody?	Such information should be included in a preliminary report to the requesting physician.
Is the specimen purulent?	Such information should be included in a preliminary report to the attending physician.

Microscopic Examination of Specimens

Gram-Stained Preparations

Examination of a thin, Gram-stained preparation is one of the most important diagnostic procedures performed in anaerobic bacteriology laboratories. Such a smear should be made of all appropriate clinical materials that are received.

In some laboratories, smears are stained before inoculation of media, and microscopic observations serve as a guide in selecting the media to be inoculated. In other laboratories, a routine battery of media is inoculated before staining and examining the smear, thus eliminating any further delay in media inoculation. The latter approach is recommended.

The recommended Gram-staining procedure for anaerobes is described in Appendix D. Direct smears should be methanol-fixed rather than heat-fixed.[17] Studies have shown that methanol fixation preserves the morphology of red and white blood cells, as well as the bacteria.

Perhaps the best way for microbiologists to demonstrate for themselves the dramatic differences between methanol fixation and heat fixation is in the staining of smears of positive blood cultures. Methanol fixation will greatly reduce the amount of red-staining amorphous debris seen whenever such smears are heat-fixed.[13]

Gram-negative anaerobes frequently stain a very pale pink when safranin is used as the counterstain and are thus easily overlooked in Gram-stained smears of clinical specimens and blood cultures. To enhance the red color of gram-negative anaerobes, Sutter et al. recommend use of a modified Gram-staining procedure, with basic fuchsin as the counterstain.[17] Alternatively, counterstaining with safranin can be increased to three to five minutes to make results consistent with the basic fuchsin technique.

Examination of the Gram-stained smear is useful for several reasons. First, the Gram stain reveals the various types of microorganisms present, as well as their relative numbers. This information should be included in a preliminary report to the requesting physician. Remember that infectious processes involving anaerobes are frequently **polymicrobic** and complex in composition, including aerobes (obligate and microaerophilic) and anaerobes (facultative, aerotolerant, microaerotolerant, and obligate). The presence of multiple distinct morphologic forms is strong presumptive evidence that the infectious process involves obligate anaerobes or a mixture of anaerobes and facultative anaerobes. The most important and fundamental information needed by the clinician is whether or not anaerobic bacteria are contributing to the infectious process.

Second, the Gram stain will often reveal the presence of leukocytes. Infectious processes involving anaerobes are usually pyogenic but may be necrotic. The presence of leukocytes in a preparation is an important observation that should also be included in the preliminary report. It is important to remember, however, that leukocytes may **not** be observed in certain anaerobic infectious processes. In cases of clostridial myonecrosis, for example, few intact neutrophils are seen in exudates, and those that are present are usually distorted by the action of clostridial toxins, such as *C. perfringens* alpha toxin.[6] Thus, the **absence** of leukocytes does not rule out the involvement of anaerobes, and the absence of leukocytes in a Gram-stained smear should never be used as a criterion for rejecting a wound specimen.

Third, it is **sometimes** possible to make a **presumptive** identification of organisms based upon their appearance in Gram-stained smears. For example, large, gram-positive bacilli **may** be clostridia. Thin, gram-negative bacilli with tapered ends **may** be fusobacteria. Extremely pleomorphic, gram-negative bacilli with bizarre shapes are **suggestive of** *F. mortiferum* or *F. necrophorum*. Tiny, round to oval, gram-negative cocci with a tendency to stain gram-variable are **suggestive of** *Veillonella* species. Gram-negative coccobacilli **may** be *Bacteroides*, *Porphyromonas*, or *Prevotella* spp.

As previously mentioned, the presence of certain morphotypes can serve as a guide to media selection. For example, if large, gram-positive bacilli, with or without spores, are seen in the smear (suggestive of *Clostridium* spp.), an egg yolk agar (EYA) plate might be included in the primary isolation setup, and/or a direct Nagler test could be performed. An EYA plate is used to detect lecithinase, lipase, and proteolytic activities of clostridia (described in Chapter 8). The direct Nagler test (described later in this chapter), once considered of value in presumptive identification of *Clostridium perfringens*, is nonspecific and infrequently used.

Finally, the Gram stain can serve as a quality control technique. Failure to isolate certain of the organisms observed in the Gram-stained smear might indicate that problems exist with the anaerobic technique being used. However, failure to recover certain morphotypes could also indicate that those particular organisms were dead or that the patient was receiving antimicrobial agents that inhibited growth of the organisms on the plated media.

In addition to Gram staining, some laboratories routinely examine wet mounts of clinical materials using regular transmitted light, phase contrast microscopy, or darkfield illumination. These procedures aid in detection of motile organisms and refractile spores. In laboratories where fluorescent antibody (FA) procedures are performed, direct and indirect FA techniques have been used successfully to identify various anaerobes (e.g., *Clostridium, Bacteroides, Fusobacterium,* and *Propionibacterium* spp.).

Specimen Processing Techniques

Ideally, specimen processing should take place in an anaerobic chamber. In all laboratories, whether or not equipped with anaerobic chambers, the emphasis should be placed on speed. The following specimen processing techniques are recommended by Sutter et al.[17]

Aspirates

Transport containers should be vortexed to ensure even distribution of the aspirate, especially when the material is grossly purulent. Using a sterile Pasteur pipette, add one drop of purulent material or two to three drops of nonpurulent material to each plate, and streak it in such a manner as to obtain well-isolated colonies. Inoculate 0.5–1 milliliter of the specimen into the bottom of a tube of enriched thioglycollate broth. Spread one drop evenly over an alcohol-cleaned glass slide for Gram staining.

Tissue and Bone Fragments

One milliliter of sterile thioglycollate broth is first added to a sterile tissue grinder. The piece of tissue or bone fragment is then homogenized until a thick suspension is obtained. Ideally, this procedure can be performed within an anaerobic chamber. If a chamber is not available, the grinding must be accomplished as quickly as possible at the workbench. The suspension is then inoculated as described previously for an aspirate.

Swabs

The swab should be inserted into a tube containing about 1 milliliter of sterile thioglycollate broth, then stirred vigorously to remove all material from the swab, and finally, pressed firmly against the inner wall of the tube to remove as much of the liquid from the swab as possible. The resulting liquid suspension is then inoculated as described previously for an aspirate.

Inoculation of Appropriate Plated and Tubed Media

In laboratories **not** equipped with an anaerobic chamber, inoculation of plated and tubed media could be performed in conjunction with a suitable nitrogen gas holding system (see Figure 6-1). A holding system, which may employ a jar, box, or other small chamber, allows uninoculated plates to be held under anaerobic conditions until needed and inoculated plates to be held under near-anaerobic conditions until placed into an anaerobic chamber, jar, or bags. Care should be taken to ensure that inoculated plates do not remain in the holding jar at room temperature for extended periods of time (preferably not longer than an hour). Also, because the holding jar should remain as anaerobic as possible, care should be taken to minimize convection currents whenever freshly inoculated plates are added to the jar.

Some microbiologists believe that it is better to batch process specimens rather than process each specimen as it arrives in the laboratory; this relieves their concern that freshly inoculated plates will remain at room temperature in a holding system

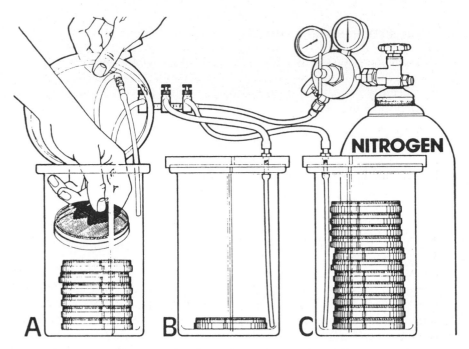

Figure 6-1. Holding jar system for anaerobic bacteriology (from reference 10, p. 412, courtesy of J.B. Lippincott, Philadelphia, PA). This diagram illustrates the type of holding jar system that could be used at the specimen processing station in laboratories not equipped with anaerobic chambers. (**A & B**) Uninoculated plates are held under anaerobic conditions until needed. (**C**) Inoculated plates are held under near-anaerobic conditions until sufficient plates have accumulated to set up and incubate a Gas-Pak® jar. A similar holding jar system should **always** be located at the anaerobe work-station so that culture and subculture plates may be held under anaerobic conditions while they are not physically being worked on.

that may contain some oxygen. Batch processing is certainly an acceptable alternative for aspirates and specimens received in proper transport containers (i.e., those being kept moist under anaerobic conditions). It would be **unacceptable,** however, to batch process improperly submitted specimens (e.g., dry swabs) or clinical materials apt to contain rapidly growing bacteria. The delay would further expose anaerobes to molecular oxygen, increase the likelihood of specimens drying out, and decrease the probability of recovering anaerobes.

Culture Media Recommendations

The choice of media for use in the anaerobic bacteriology laboratory is an extremely important aspect of successful anaerobic bacteriology. Media exposed to air for extended periods of time may contain toxic substances, produced as the result of the reduction of molecular oxygen (described in Chapter 1). Such media may also have oxidation-reduction (redox) potentials above that required for anaerobes to initiate growth.

The ideal media for use in the culture of anaerobes are those **never** exposed to oxygen or exposed for only a brief period of time. Such media include freshly prepared media and those stored under anaerobic conditions from the time they were made. It is, of course, not practical to prepare media fresh each time there is a need for it. However, freshly prepared media can be stored within an anaerobic chamber or holding system until used.

An alternative is to use commercial media that have been prepared, packaged, shipped, and stored under anaerobic conditions (i.e., not exposed to oxygen until they are inoculated or not exposed at all when plates are inoculated within an anaerobic chamber). Such media (shown in Figure 6-2), available from Anaerobe Systems (San Jose, CA), are prereduced and anaerobically sterilized (PRAS).[8] Growth is initiated quickly on PRAS media, and many anaerobes produce sufficient growth to work with after only 24 hours of incubation. Studies have demonstrated that these media perform better than fresh media or certain other commercially available media.[5,12]

Reducing agents (such as palladium chloride) are sometimes added to media prior to autoclaving in an attempt to "prereduce" them. Care must be taken in the selection of reducing agents, however, because some are toxic to certain anaerobes.[8,16] Media containing reducing agents are available from a variety of commercial sources. Storing them in gas-permeable cellophane sleeves and/or at refrigerator temperatures may decrease their shelf life. In all likelihood, there is a finite limit to the length of time that reducing agents can maintain sufficiently low oxidation-reduction potentials when they are in constant contact with oxygen. Additional studies are needed to determine the extent to which the toxic substances mentioned in Chapter 1 (e.g., superoxide anions and hydrogen peroxide) can actually be removed from media by incubating such plates anaerobically prior to inoculation. Readers wanting additional information about reducing agents and culture media should refer to the introductory chapters of Smith and Williams.[16]

Some microbiologists mistakenly believe that their media are performing well because they are routinely isolating *Bacteroides fragilis* and *Clostridium perfringens* from clinical specimens and/or because their quality control strains of these particular anaerobes are growing well. They seem to forget that some anaerobes are aerotolerant and will grow on media **unable** to support the growth of obligate anaerobes. Routine recovery of anaerobic, gram-positive cocci, fusobacteria, and pigmented *Porphyromonas* spp. is a far better test of the suitability of media than recovery of the more aerotolerant *B. fragilis* and *C. perfringens*.

Media for Use in the Primary Isolation Setup

Recommendations of different authorities in the area of anaerobic bacteriology vary slightly with regard to specific media to be included in the primary isolation setup of anaerobe cultures. Table 6-2 contains the names and uses of many of the enriched, nonselective, and selective media that have been recommended for isolation of anaerobes from clinical specimens. Described in detail in the following text are those recommended by the group at the Wadsworth Veterans Administration Medical Center in Los Angeles.[17] Regardless of which media are selected, they must be inoculated using sterile loops made of plastic or nonoxidized metal (e.g., platinum, platinum-iridium, or stainless steel). Nichrome loops are unacceptable.

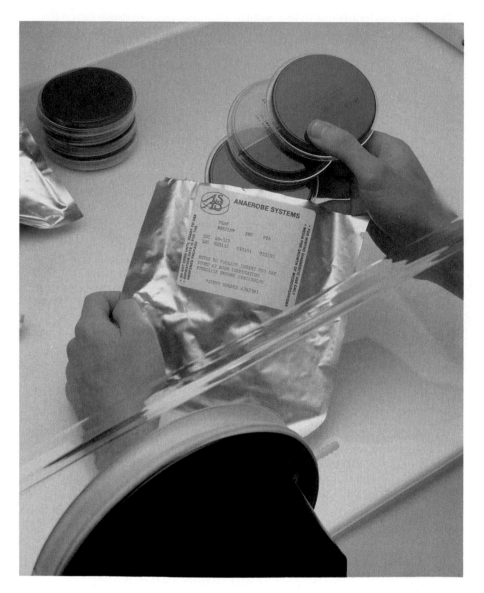

Figure 6-2. Prereduced, anaerobically sterilized (PRAS), plated media (courtesy of Anaerobe Systems, San Jose, CA). Prereduced, anaerobically sterilized (PRAS), plated media are manufactured, packaged, shipped, and stored under anaerobic conditions. Media are packaged as single plates, several plates of one particular type of medium (e.g., BRU/BA plates), or sets of media (e.g., the variety of media used in the primary setup of one specimen). Initial exposure of these media to oxygen occurs when the gas-impermeable foil pouches are opened at the bench at the time of specimen inoculation. Exposure to oxygen is totally avoided when the pouches are opened within an anaerobic chamber (as shown here).

Table 6-2. Media Recommended for Recovery of Anaerobic Bacteria

Media	Purpose
Primary isolation setup	
Brucella blood agar (BRU) (acceptable alternatives: CDC anaerobe blood agar, enriched brain heart infusion blood agar, Schaedler blood agar)	Nonselective; for isolation of any anaerobes present in the specimen
Bacteroides bile esculin (BBE) agar	Selective and differential; for isolation of members of the *Bacteroides fragilis* group
Kanamycin-vancomycin-laked blood (KVLB) agar (acceptable alternatives: kanamycin-vancomycin blood agar [KVA], paromomycin-vancomycin blood agar [PVA])	Selective for most nonsporeforming, gram-negative bacilli
Phenylethyl alcohol (PEA) agar (acceptable alternative: colistin-naladixic acid blood agar)	Permits growth of both gram-positive and gram-negative anaerobes; inhibits swarming *Proteus* spp. and other *Enterobacteriaceae*
Thioglycollate broth (THIO)	For emergency backup only
For special situations	
Egg yolk agar (EYA)	When clostridia are suspected
Direct Nagler Test	When *Clostridium perfringens* is suspected
Cycloserine-cefoxitin-fructose agar (CCFA) (or cycloserine mannitol agar [CMA] or cycloserine mannitol blood agar [CMBA])	When *Clostridium difficile* is suspected
Crystal violet-erythromycin (CVE) blood agar	For isolation of *Fusobacterium nucleatum* and *Leptotrichia buccalis*
Bacteroides gingivalis agar (BGA)	For isolation of *Porphyromonas gingivalis* (formerly, *Bacteroides gingivalis*)
Clindamycin blood agar (CBA)	For isolation of *Eikenella corrodens* and *Wolinella* spp.

Note: This is not intended to be an all-inclusive list of special media.

Brucella Blood Agar (BRU). This is an enriched medium that, when incubated anaerobically, allows the growth of both obligate and facultative anaerobes. It contains 5% (v/v) sheep blood for enrichment and detection of hemolysis, vitamin K_1 (required by some *Porphyromonas* spp.), and hemin (which enhances growth of members of the *B. fragilis* group and certain other *Bacteroides* spp). BRU contains no inhibitory substances. Similar excellent media (e.g., CDC anaerobe blood agar [BA], enriched brain heart infusion blood agar, and Schaedler blood agar) are also available for recovering anaerobes from clinical specimens. The abbreviation BRU/BA will be used to represent any of these acceptable anaerobic blood agars. BRU has been reported to support growth of anaerobic, gram-negative bacilli better than either CDC or Schaedler base formulations, and BA has been reported to support growth of anaerobic, gram-positive cocci better than BRU.[14] However, such differences were detected using commercial, non-PRAS media, and results varied from one manufacturer to another.

Bacteroides Bile Esculin (BBE) Agar. This medium is included in the primary isolation setup solely for the rapid isolation and presumptive identification of members of the *B. fragilis* group. Frequently, such a presumptive identification can be made after only 24 hours of incubation. BBE agar contains gentamicin, which inhibits most aerobic organisms, and 20% bile, which inhibits most anaerobes. This medium will support the growth of bile-tolerant *Bacteroides* species but not bile-sensitive species. Some strains of *Fusobacterium mortiferum*, *Klebsiella pneumoniae*, enterococci, and yeast may grow to a limited extent on this medium, producing colonies much smaller than those of *Bacteroides*. Members of the *B. fragilis* group will turn the light yellow medium to brown because they are capable of hydrolyzing esculin.

Kanamycin-Vancomycin-Laked Blood (KVLB) Agar. This medium is highly selective for obligately anaerobic and microaerotolerant, gram-negative bacilli, especially *Bacteriodes* and *Prevotella* spp. The presence of laked blood allows earlier detection of pigmented colonies of *Prevotella* spp. than BRU/BA. The kanamycin inhibits most facultative, gram-negative bacilli, but yeasts and kanamycin-resistant organisms will grow on KVLB agar. The vancomycin inhibits most gram-positive organisms and *Porphyromonas* spp. A similar medium, which substitutes paromomycin for kanamycin, will inhibit any kanamycin-resistant, facultative, gram-negative bacilli that might be present in the specimen.

Phenylethyl Alcohol (PEA) Agar. The primary purpose of this medium (also called phenylethanol agar or phenethylalcohol agar) is to inhibit facultative, gram-negative bacilli (e.g., the *Enterobacteriaceae*). Thus, a swarming *Proteus* species that would swarm over the surface of the BRU plate would be inhibited on this medium. Most obligate anaerobes (both gram-positive and gram-negative) will grow on PEA agar, as will gram-positive, facultative anaerobes.

Thioglycollate Broth (THIO). Thioglycollate broth will support the growth of most anaerobes. Those present in the broth can, however, be overgrown by more rapidly growing facultative organisms, which thrive in THIO. Furthermore, the anaerobes could be killed by toxic metabolic by-products produced by the facultative organisms. Thus, broth cultures should **never** be relied upon exclusively for isolating anaerobes from clinical specimens. The sole purpose for inoculating a broth medium is to provide a **backup** source of culture material. In the event of an anaerobic jar failure, for example, the broth culture could perhaps serve as a backup. Or, in the event of inhibition of growth on plated media due to antimicrobial agents present in the specimen, the broth culture could perhaps serve as a backup; the antimicrobial agents would be diluted in the broth to the point that they are no longer effective.

To improve recovery of anaerobes, some manufacturers (e.g., Anaerobe Systems, BBL, and Remel) produce THIO that contains supplements such as vitamin K_1 and hemin. On the day of use, non-PRAS THIO tubes should be placed in a boiling water bath to drive off oxygen and then cooled to 25 to 35°C prior to inoculation. Chopped meat carbohydrate broth could be used in place of thioglycollate broth.

Whenever clostridia are suspected, either clinically or as a result of Gram stain observations (i.e., the presence of large, gram-positive bacilli, with or without spores), you can inoculate an EYA plate and/or perform the direct Nagler test.

Egg Yolk Agar (EYA). EYA is useful for detecting enzymes produced by some clostridia (i.e., lecithinase, lipase, and proteolytic enzymes).

Direct Nagler Test. If *C. perfringens* is suspected (gram-positive, boxcar-shaped bacilli seen on a Gram stain), a direct Nagler test can be performed as part of the primary isolation setup. This test employs an egg yolk agar plate and a reagent known as *C. perfringens* type A antitoxin. This reagent will inhibit the lecithinase reaction produced by *C. perfringens* and three other species of clostridia. The test is sometimes used in the presumptive identification of *C. perfringens*, although a positive test result is **not specific** for this organism. The Nagler test and its interpretation are described in Appendix D.

Nonselective and selective media recommended by the CDC Anaerobe Reference Laboratory for isolating various aerobic and anaerobic bacteria from clinical specimens include CDC anaerobe blood agar, chocolate agar, MacConkey agar, modified kanamycin-vancomycin blood agar, modified phenylethyl alcohol blood agar, modified thioglycollate medium, chopped meat glucose medium, and *Bacteroides gingivalis* blood agar.[1, 3, 7, 9, 10, 11]

Cycloserine-Cefoxitin-Fructose Agar (CCFA)

Whenever a patient is suspected of having antibiotic-associated diarrhea (AAD) or pseudomembranous colitis (PMC), inoculate a CCFA plate with a fecal specimen from that patient. CCFA is a selective and differential medium for recovery and presumptive identification of *Clostridium difficile*, a common cause of AAD and PMC. On CCFA, *C. difficile* will produce yellow, ground-glass colonies, and the originally pink agar will turn yellow in the vicinity of the colonies.[7]

Plates To Be Incubated Aerobically

In addition to the thioglycollate broth and plated media to be incubated anaerobically, a variety of plated media is inoculated and incubated aerobically in a CO_2 incubator. The specific media to be incubated aerobically vary somewhat from one specimen type to another and from one laboratory to another, but they usually include a blood agar plate, a MacConkey agar plate, and a chocolate agar plate as a minimum. Some laboratories routinely include a PEA plate or colistin-naladixic acid (CNA) plate to inhibit facultative, gram-negative bacilli and select for gram-positive organisms.

The results of culturing a hypothetical "wound" specimen containing mixed anaerobic and facultative organisms are shown in Figure 6-3.

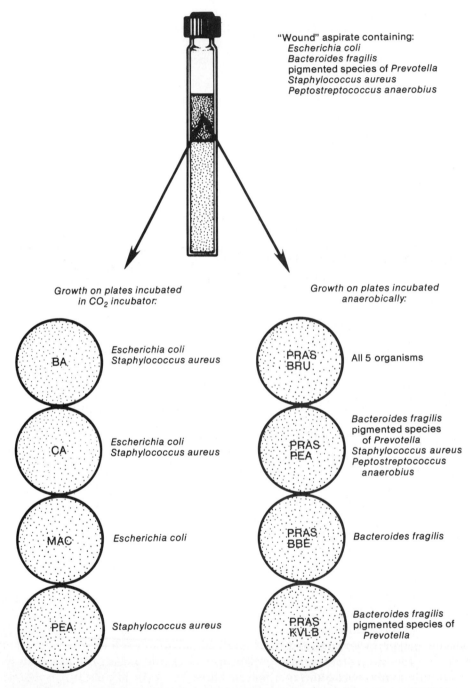

Figure 6-3. Culture results that might be obtained from the primary isolation setup of a hypothetical "wound" specimen. This diagram illustrates the media and atmospheric conditions that would support growth of various organisms contained in a hypothetical "wound" specimen.

Anaerobic Incubation of Inoculated Media

After specimens are rapidly processed and inoculated onto/into appropriate media, the inoculated plates must be incubated anaerobically at 35 to 37°C. The most common and practical choices of anaerobic incubation systems for clinical laboratories are:

- Anaerobic chambers

- Anaerobic jars

- Anaerobic bags or pouches

The choice of system will be influenced by a number of factors, including financial considerations, the number of anaerobic cultures performed, and space limitations.

Anaerobic Chambers

The ideal anaerobic incubation system is an anaerobic chamber, which provides an environment free of oxygen for inoculating media and incubating cultures. Identification and susceptibility test systems can also be inoculated within the chamber, if necessary.

A variety of anaerobic chambers is available commercially. Some models (called "glove boxes") are fitted with airtight rubber gloves (see Figure 6-4). The microbiologist inserts his or her arms into the gloves and manipulates specimens, plates, and tubes inside the chamber. Some users find the gloves bulky and difficult to work with. They are often too large for their hands, there is a loss of manual dexterity, and the gloves frequently cause the user's hands to perspire. Some, but not all, of these problems can be overcome with experience. Glove boxes have traditionally been constructed of flexible vinyl. They are available from companies such as Coy Corporation, Forma Scientific, and Lab-Line Instruments.

"Gloveless" anaerobic chambers are also available commercially (Figure 6-5; Anaerobe Systems, San Jose, CA). Airtight rubber sleeves that fit snugly against the user's bare forearms are used in place of gloves, enabling the microbiologist to work within an anaerobic environment with his or her bare hands. However, to comply with mandatory infectious disease safety precautions, it is recommended that thin, disposable, latex gloves be worn if clinical specimens are being processed within gloveless chambers. Subsequent operations may be performed with bare hands. The newer gloveless models have a dissecting microscope mounted on the front of the rigid plexiglass chamber. This enables the user to observe colony morphology within the chamber, eliminating the need to remove plates from the chamber and eliminating exposure of the colonies to oxygen.

All anaerobic chambers contain a catalyst (usually palladium-coated allumina pellets) that removes residual oxygen from the atmosphere within the chamber. With time, the catalyst pellets become inactivated by gaseous metabolic end products produced by anaerobes growing within the chamber. A product called Anatox (Don Whitley Scientific, Shipley, West Yorkshire, England) has been shown to absorb these metabolites and prolong catalyst life[2] A similar product is available from Anaerobe Systems (San Jose, CA).

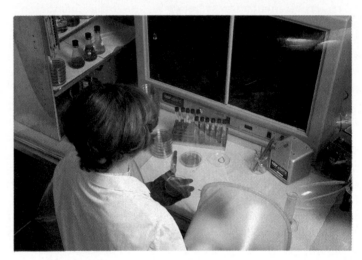

Figure 6-4. "Glove box" type of anaerobic chamber (courtesy of Coy Corporation, Ann Arbor, MI). These photographs illustrate the "glove box" type of anaerobic chamber manufactured by Coy Laboratory Products, Inc. The flexible, clear vinyl chambers are available in three sizes (lengths of 36", 59", and 78"), including one fitted with two pairs of gloves so that two microbiologists can use the chamber simultaneously. Chambers are 32" deep and 40" high. Specimens, plates, and other items are introduced into and removed from the chamber by way of an automatic airlock (located on the right.) Inoculated plates and tubes are stored within an incubator (located in the rear). The chamber also contains shelving and a catalyst box (out of view). Glove ports can be fitted with various size gloves to accommodate the difference in hand sizes from one user to another.

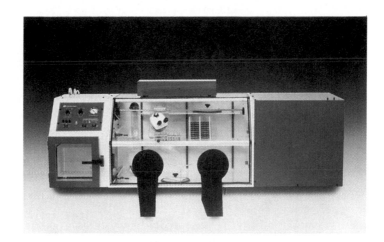

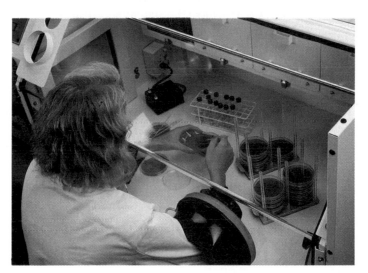

Figure 6-5. "Gloveless" type of anaerobic chamber with dissecting microscope attachment (courtesy of Anaerobe Systems, San Jose, CA and Sheldon Manufacturing, Inc., Cornelius, OR). These photographs illustrate the stainless steel and plexiglass, "gloveless" type of anaerobic chamber manufactured by Anaerobe Systems. An automatic airlock is located on the left side of the chamber, and an incubator is located on the right side. In place of rubber gloves are rubber sleeves, the ends of which fit snugly around the user's bare forearms. After the sleeves are flushed with an oxygen-free gas mixture (using foot pedals), the round "porthole-type" entry doors are removed and placed inside the chamber. Instruments constantly monitor and adjust temperature and anaerobic conditions within the chamber. A high level of relative humidity is constantly maintained within the chamber. The BACTRON I, BACTRON II, AND BACTRON IV (shown here) are 48", 62", and 88.5" wide, respectively; all three models are 31" deep and 27" high. The incubator in the BACTRON IV has a 600-plate capacity.

Anaerobic Jars

For small laboratories, where the volume of anaerobic cultures may not justify the purchase of anaerobic chambers, alternative systems are available. These do not provide all the features or advantages of anaerobic chambers, and cost analyses may reveal that, over time, a chamber is actually more cost-effective.

One such alternative is the GasPak® jar, manufactured by Becton Dickinson (Figure 6-6). The jars have been used in clinical laboratories for many years, enabling even small laboratories to perform satisfactory anaerobic bacteriology. Newer models accommodate a greater number of plates than the smaller, older model jars and will accommodate microtiter susceptibility trays and anaerobe identification strips and trays. Anaerobic jars are also available from Scott Laboratories, Oxoid, Difco Laboratories, Remel, and other companies. The components of the Difco anaerobic jar are shown in Figure 6-7.

The major disadvantage of any of the anaerobic jar systems is that the plates have to be removed from the jars to be examined. This, of course, exposes the colonies to oxygen, which is especially hazardous to the anaerobes during their first 48 hours of growth. A suitable holding system should always be used in conjunction with anaerobic jars. Plates can be removed from the anaerobic jar, placed in an oxygen-

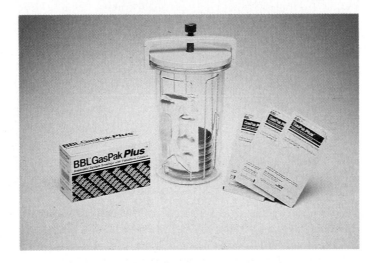

Figure 6-6. Anaerobic jar (courtesy of Becton Dickinson Microbiology Systems, Cockeysville, MD). This photograph depicts one of the many different types of commercially available anaerobic jars. Inoculated plates, gas-generating envelopes, and an indicator strip are placed within the jar, which contains fresh or rejuvenated catalyst pellets. Many of the commercially available, gas-generating packets (e.g., GasPak Plus™) contain built-in catalyst and indicator. Water is added to the packet, the jar is sealed, and carbon dioxide and hydrogen are released from the packet. If the catalyst is performing properly, the hydrogen combines with the oxygen to form water vapor. The indicator system lets the user know that anaerobic conditions were achieved.

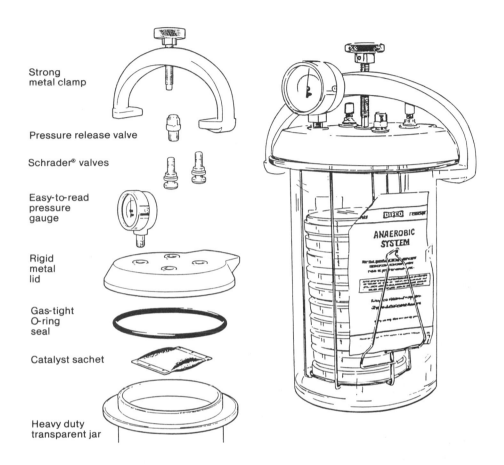

Strong
metal clamp

Pressure release valve

Schrader® valves

Easy-to-read
pressure
gauge

Rigid
metal
lid

Gas-tight
O-ring
seal

Catalyst sachet

Heavy duty
transparent jar

ANAEROBIC
SYSTEM

Figure 6-7. Components of the Difco anaerobic jar (courtesy of Difco Laboratories, Detroit, MI).

free holding system, removed one by one for rapid microscopic examination of colonies, and then quickly returned to the holding system. Plates should never remain in room air on the open bench.

When water is added to the GasPak® envelope, two gases are generated: carbon dioxide and hydrogen (see Equations 6-1 and 6-2). The carbon dioxide is important because some anaerobes require an increased concentration of it for growth (i.e., they are **capnophiles**). In the presence of a catalyst, the hydrogen combines with oxygen in the jar to form water vapor (see Equation 6-3). Hydrogen is explosive, and, should the catalyst not be functioning properly, hydrogen gas will be present in the jar. Thus, never uncover the jar in the vicinity of an open flame.

A methylene blue indicator strip is always added to the jar to verify that an anaerobic atmosphere was truly achieved. If the catalyst performed correctly, water vapor will be present on the inside of the jar when the jar is removed from the incubator, and the indicator strip will be colorless. Some of the newer gas-generating packets contain a built-in indicator. Failure to achieve anaerobic conditions could be due to "poisoned" catalyst pellets or a crack in the jar, lid, or "O-ring."

Equation 6-1. Production of Carbon Dioxide

$$C_6H_8O_7 + 3\ NaHCO_3 \longrightarrow Na_3(C_6H_5O_7) + 3\ H_2O + 3\ CO_2$$

Equation 6-2. Production of Hydrogen

$$NaBH_4 + 2\ H_2O \longrightarrow NaBO_2 + 4\ H_2$$

Equation 6-3. Removal of Oxygen

$$2\ H_2 + O_2 \xrightarrow{\text{CATALYST}} 2\ H_2O$$

Some jars contain reusable catalyst pellets. Because the catalyst becomes poisoned by gases (particularly H_2S) produced by anaerobes, it is recommended that the catalyst be rejuvenated **after every use.** [17] This is accomplished by heating the catalyst in a 160°C oven for a minimum of two hours. Some of the newer gas-generating packets contain a built-in supply of fresh catalyst, thus eliminating the need to maintain a container of rejuvenated catalyst in the lid of the jar.

Anaerobic Bags or Pouches

Other alternatives to an anaerobic chamber are the anaerobic bags or pouches (Figure 6-8). Such products are manufactured by Becton Dickinson (Type A Bio-Bag® and GasPak Pouch®) and Difco Laboratories (Anaerobic Pouch®). These products will be collectively referred to as "bags." One or two inoculated plates are placed into a bag, an oxygen-removal system is activated, and the bag is sealed and incubated. Theoretically, the plates can be examined for growth without removing them from the bags—thus without exposing the colonies to oxygen. However, with some of the products, a water vapor film on the inner surface of the bag or the lid of the plate can sometimes obscure vision. [15] In such cases, the plates must be removed from the bag for observation of growth, and a new bag and oxygen-removal system must be used whenever additional incubation is required. As with the anaerobic jar, plates must be removed from the bags in order to work with the colonies at the bench. Thus, an anaerobic holding system should always be used in conjunction with any of the anaerobic bags.

A comparison of anaerobic bags and anaerobic chambers revealed that more anaerobes were recovered from clinical specimens when chambers were used. [4] However, the catalyst-free Anaerobic Pouch System® (Difco Laboratories, Detroit, MI) performed well enough for use in laboratories not equipped with anaerobic chambers.

Any of these bags are useful as transport devices. For example, a biopsy specimen could first be placed in a sterile petri dish containing sterile saline. The petri dish could then be placed in one of these bags, the oxygen-removal system activated, and the specimen carefully transported to the laboratory within the bag. Hold the petri dish horizontally enroute to the laboratory to avoid spillage of its contents.

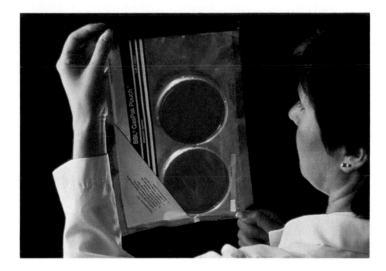

Figure 6-8. Anaerobic pouch (courtesy of Becton Dickinson Microbiology Systems, Cockeysville, MD). This photograph depicts the GasPak® Pouch, one of the commercially available anaerobic bag or pouch systems. After inoculated plates are added to the pouch, an oxygen removal system is initiated. The pouch is then tightly sealed and placed into a nonanaerobic incubator. In some cases, the bags or pouches permit observation of growth directly through the plastic, thus eliminating the need to remove plates from the bag or pouch to make such observations. Identification and susceptibility testing systems requiring anaerobic incubation can be incubated within some of the bags or pouches. Bags or pouches are also useful for transporting specimens anaerobically to the laboratory.

Chapter in Review

- Processing of a specimen for anaerobic bacteriology includes a macroscopic examination of the specimen, a microscopic examination of the specimen, inoculation of appropriate plated and tubed media, and anaerobic incubation of inoculated media. Each of these steps is critical to the successful recovery of clinically relevant anaerobes from specimens.

- Important characteristics to note during the macroscopic examination of a specimen include its odor, color, turbidity, and fluorescence, in addition to the presence of blood, leukocytes, or "sulfur granules."

- Careful preparation and thorough examination of a Gram-stained smear is the major component of the microscopic examination. Furthermore, it is one of the most important procedures performed in the anaerobic bacteriology laboratory. Information gained from the macroscopic examination and the Gram stain often provides a physician with his or her first indication that anaerobes are involved in a given infectious process.

- Recommended modifications to the "conventional" Gram staining procedure include methanol fixation in place of heat fixation, counterstaining with safranin for three to five minutes, or substituting basic fuchsin for safranin as the counterstain. These modifications improve the morphology of organisms and host cells present in the specimen and cause gram-negative bacteria to stain a deeper red.

- The ideal media for use in the anaerobic bacteriology laboratory are (1) freshly prepared media, (2) media that have been stored under anaerobic conditions since their preparation, and (3) media that have been manufactured, packaged, shipped, and stored under anaerobic conditions ("PRAS media"). Studies have shown that use of such media improves the yield of anaerobes from clinical specimens and permits more rapid growth of any anaerobes present.

- Media recommended by the Wadsworth V.A. Medical Center group for use in the primary isolation setup of specimens for anaerobic bacteriology are Brucella blood agar (BRU), phenylethyl alcohol (PEA) blood agar, Bacteroides bile esculin (BBE) agar, and kanamycin-vancomycin-laked blood (KVLB) agar or acceptable alternatives. In addition, specimens should be inoculated to media (e.g., blood agar, chocolate agar, MacConkey agar, phenylethyl alcohol agar) that will be incubated in a CO_2 incubator or in air.

- Nonselective and selective media recommended by the CDC Anaerobe Reference Laboratory for isolating various aerobic and anaerobic bacteria from clinical specimens include CDC anaerobe blood agar, chocolate agar, MacConkey agar, modified kanamycin-vancomycin blood agar, modified phenylethyl alcohol blood agar, modified thioglycollate medium, chopped meat glucose medium, and *Bacteroides gingivalis* blood agar.

- Anaerobic incubation of inoculated plates can be accomplished using an anaerobic chamber, anaerobic jars, or anaerobic bags or pouches. Although anaerobic chambers are recommended for use in laboratories doing a high volume of anaerobic bacteriology, jars and bags have proven to be acceptable alternatives for smaller laboratories.

Self-Assessment Exercises

1. List four types of solid culture media recommended in this chapter for primary isolation of anaerobes from clinical specimens.

   a. _____

   b. _____

   c. _____

   d. _____

2. While examining a Gram-stained smear of material aspirated from a festering leg wound, a microbiologist observes numerous neutrophils, large, gram-positive bacilli, gram-negative coccobacilli, and gram-positive cocci. In addition to the four plates routinely used for primary isolation of anaerobes, what additional plated medium or test should be included in the primary isolation setup?

    Why? _____

    _____

    _____

3. List the three most commonly used methods employed by clinical microbiology laboratories to incubate inoculated plates anaerobically.

    a. _____

    b. _____

    c. _____

4. Match the media used in the primary isolation setup for anaerobes with their intended purpose:

    a. Bacteroides bile esculin (BBE) agar

    _____

    b. Phenylethyl alcohol (PEA) agar

    _____

    c. Kanamycin-vancomycin-blood agar (KVA) or kanamycin-vancomycin-laked blood (KVLB) agar

    _____

    1. To enhance the brown-black color of colonies of pigmented *Prevotella* spp.

    2. For rapid presumptive identification of organisms in the *Bacteroides fragilis* group

    3. To suppress the growth of any *Enterobacteriaceae*, including swarming *Proteus* species

5. A properly selected, collected, and transported specimen was received in the laboratory for anaerobic culture. Appropriate plated and tubed media were inoculated and incubated within an anaerobic jar. No anaerobes were recovered from any of the plated media, but a *Bacteroides* species was recovered from the thioglycollate broth. Give two possible explanations for these findings.

    a. _____

    _____

    b. _____

    _____

6. What selective and differential media could be used when attempting to recover *Clostridium difficile* from fecal specimens?

   a. _____

   b. _____

   c. _____

7. What solid medium is used to determine an organism's ability to produce the enzymes lecithinase and lipase?

   _____

8. What do the initials "CCFA" stand for?

   _____

9. TRUE or FALSE: Because most infectious processes involving anaerobes are pyogenic, it would be impractical to culture a wound specimen for anaerobes when no leukocytes are observed on the Gram stain of the specimen.

10. To obtain anaerobic incubation conditions, a laboratory is using anaerobic jars containing reusable palladium-coated alumina catalyst pellets. Indicate whether the following statements are TRUE or FALSE.

    a. Anaerobic bacteria produce gases (including H_2S) that can "poison" the catalyst pellets, rendering them ineffective. _____

    b. Following incubation, the presence of a water vapor condensate on the inner walls of the anaerobe jar is an indication that the pellets are "poisoned" and not operating effectively. _____

    c. To "rejuvenate" catalyst pellets, keep them in a 160°C oven for a minimum of two hours. _____

    d. It is recommended that catalyst pellets should be "rejuvenated" after every use. _____

    e. If catalyst pellets are properly "rejuvenated," it is not necessary to include an anaerobic indicator strip (e.g., a methylene blue strip) each time a jar is set up. _____

11. A wound aspirate containing five organisms (*Escherichia coli, Bacteroides fragilis,* a pigmented species of *Prevotella, Staphylococcus aureus,* and *Peptostreptococcus anaerobius*) is inoculated onto plates to be incubated in a CO_2 incubator (blood agar, CA, MAC, and PEA) and plates to be incubated anaerobically (PRAS BRU/BA, PRAS PEA, PRAS BBE, and PRAS KVLB). State which of the five organisms would be expected to grow on each of the plates.

Plates incubated in the CO_2 incubator:

Blood agar _____

CA _____

MAC _____

PEA _____

Plates incubated anaerobically:

PRAS BRU/BA _____

PRAS PEA _____

PRAS BBE _____

PRAS KVLB _____

References

1. Allen, S. D., J. A. Siders, and L. M. Marler. 1985. Anaerobic bacteria, p. 413–433. *In* E. H. Lennette, A. Balows, W. J. Hausler, Jr., and H. J. Shadomy (eds.), Manual of clinical microbiology, 4th ed. American Society for Microbiology, Washington, D.C.

2. Brazier, J. S. 1982. Appraisal of Anotox, a new anaerobic atmospheric detoxifying agent for use in anaerobic cabinets. J. Clin. Pathol. 35:233–238.

3. Dowell, V. R., Jr., G. L. Lombard, F. S. Thompson, and A. Y. Armfield. 1977. Media for isolation, characterization, and identification of obligately anaerobic bacteria. Centers for Disease Control, Atlanta.

4. Downes, J., J. I. Mangels, J. Holden, M. J. Farraro, and E. J. Baron. 1990. Evaluation of two single-plate incubation systems and the anaerobic chamber for the cultivation of anaerobic bacteria. J. Clin. Microbiol. 28:246–248.

5. Ferguson, D. A., Jr., O. M. Momani, K. P. Ferguson, and D. W. Lambe, Jr. 1987. Comparison of commercially prepared anaerobic agar media with freshly prepared reduced media for cultivation of *Bacteroides.* Abstract C-11, Annual Meeting of the American Society for Microbiology, Atlanta.

6. Finegold, S. M., W. L. George, and M. E. Mulligan. 1986. Anaerobic infections. Year Book Medical Publishers, Chicago.

7. George, W. L., V. L. Sutter, D. Citron, and S. M. Finegold. 1979. Selective and differential medium for isolation of *Clostridium difficile.* J. Clin. Microbiol. 9:214–219.

8. Holdeman, L. V., E. P. Cato, and W. E. C. Moore (eds.). 1977 (with 1987 update). VPI Anaerobe Laboratory Manual, 4th ed. Virginia Polytechnic Institute and State University, Blacksburg.

9. Hunt, D. E., J. V. Jones, and V. R. Dowell, Jr. 1986. Selective medium for isolation of *Bacteroides gingivalis.* J. Clin. Microbiol. 23:441–445.

10. Koneman, E. W., S. D. Allen, V. R. Dowell, Jr., W. M. Janda, H. M. Sommers, and W. C. Winn, Jr. 1988. Color atlas and textbook of diagnostic microbiology, 3rd ed. J. B. Lippincott, Philadelphia.

11. Livingston, S. J., S. D. Kominos, and R. B. Yee. 1978. New medium for selection and presumptive identification of the *Bacteroides fragilis* group. J. Clin. Microbiol. 7:448–453.

12. Mangels, J. I., and B. P. Douglas. 1989. Comparison of four commercial Brucella agar media for growth of anaerobic organisms. J. Clin. Microbiol. 27:2268–2271.

13. Mangels, J. I., M. E. Cox, and L. H. Lindberg. 1984. Methanol fixation: an alternative to heat fixation of smears before staining. Diag. Microbiol. Infect. Dis. 2:129–137.

14. Sheppard, A., C. Cammarata, and D. H. Martin. 1990. Comparison of different medium bases for the semiquantitative isolation of anaerobes from vaginal secretions. J. Clin. Microbiol. 28:455–457.

15. Slifkin, M., and C. Engwall. 1988. Evaluation of two compact microenvironment systems for the isolation of anaerobic bacteria. Clin. Lab. Sci. 1:114–117.

16. Smith, L. D. S., and B. L. Williams. 1984. The pathogenic anaerobic bacteria, 3rd ed. Charles C. Thomas, Springfield, IL.

17. Sutter, V. L., D. M. Citron, M. A. C. Edelstein, and S. M. Finegold. 1985. Wadsworth anaerobic bacteriology manual, 4th ed. Star Publishing, Belmont, CA.

CHAPTER 7

PROCESSING ANAEROBIC ISOLATES

At the conclusion of this chapter, you will be able to:

1. Identify four indications of the presence of anaerobes that might be noted during examination of the anaerobically incubated primary plates.
2. Explain the rationale for identifying anaerobic isolates.
3. List the types of organisms that would be expected to grow on or in each of the media used in the primary isolation setup (BRU/BA, PEA, BBE, KVLB, EYA, and THIO).
4. Describe the appearance of lecithinase and lipase reactions on egg yolk agar (EYA).
5. Name the types of disks that could be added to the pure culture/subculture plate and the circumstances under which specific disks would be used.
6. Identify features/characteristics on primary plates that would permit presumptive identification of *Clostridium perfringens*, *Clostridium difficile*, members of the *Bacteroides fragilis* group, and pigmented species of *Porphyromonas* and *Prevotella*.

Definitions

Actinomyces-like: A term used throughout this book to describe relatively thin, branching, anaerobic, gram-positive bacilli. However, *Actinomyces* spp. do not always have this appearance and in fact may at times be diphtheroidal in appearance.

Aerotolerance testing: A procedure used to determine the atmospheric requirements of an isolate suspected of being an anaerobe

Clostridium-like: A term used throughout this book to describe relatively large, rectangular-shaped, anaerobic, gram-positive bacilli that may or may not be sporulating. However, some species of *Clostridium* do not fit this description. These species would, therefore, not be *Clostridium*-like in appearance.

Diphtheroid: A term used to describe gram-positive bacilli that morphologically resemble *Corynebacterium diphtheriae*. *C. diphtheriae* and diphtheroids often appear as relatively short rods that tend to form aggregates of cells, sometimes resembling Chinese letters or picket fences. *Propionibacterium* spp. are frequently (but not always) diphtheroidal in appearance.

Morphotype: A term used in reference to a specific morphological appearance; can be used in reference to colony appearance (colony morphotype) or Gram stain appearance (cellular morphotype). A given organism (e.g., *Fusobacterium nucleatum*) could have more than one colony morphotype.

Nagler test: A test once widely used to presumptively identify *Clostridium perfringens*. It is used less often in today's laboratories because it is not specific for *C. perfringens*. The Nagler test employs an antitoxin that will neutralize the lecithinase produced by *C. perfringens* (and three additional *Clostridium* spp.).

Earlier chapters described proper selection, collection, and transport of clinical specimens for anaerobic bacteriology. Also presented were recommendations regarding the processing of such specimens, including appropriate plated and tubed media to be inoculated and options for anaerobic incubation of cultures. In this chapter, techniques for processing isolates suspected of being anaerobes are described. Although this chapter discusses isolates that may be obtained from human clinical specimens, the principles and procedures are equally applicable to isolates from veterinary specimens.

Examining Primary Plates

When To Examine Primary Plates

The first major decision in the examination of primary plates is **when** to examine them. Although plates being incubated in an anaerobic chamber can be safely examined at any time, opening an anaerobic jar, bag, or pouch exposes colonies to the potentially damaging effects of oxygen.

Should the jars, bags, and pouches be opened after 24 or 48 hours of incubation? Some of the slower growing anaerobes will be too small to work with at 24 hours. However, many of the more rapidly growing, pathogenic anaerobes, such as some *Clostridium* and *Bacteroides* spp., could be worked up at 24 hours, especially when appropriate PRAS media are used for the primary isolation setup. Thus, waiting until 48 hours to open jars or bags could delay the identification of clinically important anaerobes by a day.

If a holding system (described in Chapter 6) is available at the anaerobe workstation and is used in conjunction with jars, bags, or pouches, exposure to oxygen will be minimized, and plates can routinely be examined at 24 hours. Each plate is removed from the jar or bag/pouch, rapidly examined for growth, and placed immediately into a nitrogen gas holding system. Plates are then removed one by one from the holding system, colonies are examined quickly using a dissecting microscope, and plates are promptly returned to the holding system following examination. As soon as all plates are examined, those requiring additional incubation are returned to an anaerobic system and incubated for an appropriate period of time before reexamination. Cultures should be held five to seven days to allow growth of particularly slow growing anaerobes.

Indications of the Presence of Anaerobes

The following are some of the findings/observations that should alert the microbiologist that anaerobes may be present on the primary plates:[6]

- A foul odor upon opening an anaerobic jar or bag/pouch suggests that anaerobes are present in the cultures. Some anaerobes, especially *Clostridium difficile*, fusobacteria, and pigmented species of *Porphyromonas*, produce foul-smelling metabolic end products that are readily apparent when the jars, bags, or pouches are opened.

- Colony morphotypes present on the anaerobically incubated blood agar plates but not on the CO_2-incubated blood or chocolate agar plates are probably anaerobes.

- Good growth (gray colonies > one millimeter in diameter) on the BBE plate is indicative of anaerobes. (Such a finding is characteristic of members of the *B. fragilis* group.)

- Colonies on either the KVLB agar or BRU/BA plate that fluoresce brick-red under UV light or are brown to black in ordinary light indicate the presence of anaerobes. (Such a finding is characteristic of pigmented *Porphyromonas* and *Prevotella* spp., although the former will not grow on KVLB.)

Examining Specific Plates in the Primary Isolation Setup

Brucella Blood Agar (BRU) or CDC Anaerobe Blood Agar (BA) Plates.   When incubated anaerobically, BRU and BA plates will support the growth of facultative anaerobes as well as anaerobes. Thus, not all colonies appearing on the plates will be anaerobes. **Aerotolerance testing** (described later in this chapter) must be performed on each colony type to determine if it is truly an anaerobe. Figure 7-1 presents an overview of procedures to be performed on each colony morphotype present on the BRU/BA plate. These procedures are described in detail in the following sections.

Each colony morphotype should be enumerated, described, and Gram stained, and all observations should be recorded on some type of worksheet. An example of such a worksheet is shown in Figure 7-2. Growth of a particular morphotype can be semi-quantitated using the terms "light," "moderate," or "heavy" or by using some type of coding system (e.g., 1+, 2+, 3+, 4+). Descriptions of each colony morphotype should include features depicted in Figure 7-3.

The BRU/BA plate should be examined for evidence of pitting, swarming, hemolysis, and discoloration of the medium (greening, browning, etc.). Pitting, anaerobic, gram-negative bacilli include *Bacteroides ureolyticus, B. gracilis,* and *Wolinella* spp. Pitting is depicted in Figure 7-4. Swarming, anaerobic, gram-positive bacilli include *Clostridium septicum* and *C. tetani.*

A double zone of hemolysis (an inner zone of complete beta hemolysis and an outer zone of partial hemolysis) is characteristic of *C. perfringens;* this double zone of hemolysis is shown in Figure 7-5.

Characteristic colonies and morphologic features in Gram-stained preparations of selected anaerobes are described in Tables 7-1 and 7-2. A BRU/BA subculture plate is inoculated for the aerotolerance test, and appropriate disks (described later in this chapter) are added to this subculture plate.

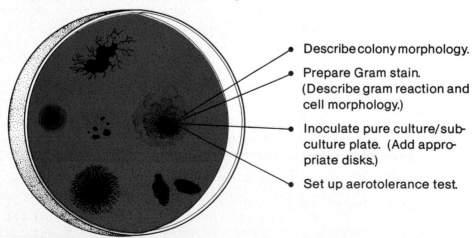

Figure 7-1. Procedures to be performed on each colony morphotype

DATE	SPECIMEN NO.	PATIENT NAME AND/OR HOSPITAL ID NO.

SPECIMEN TYPE	DIRECT GRAM STAIN OBSERVATIONS

GROWTH ON AEROBICALLY INCUBATED PLATES

COLONY DESIGNATION	GRAM STAIN RESULTS	BRUCELLA/BA PLATE
A.		
B.		
C.		
D.		
E.		

FLUORESCENCE	PIGMENT	HEMOLYSIS	PITTING	AEROTOL.TEST
A.				
B.				
C.				
D.				
E.				

BBE PLATE	KVLB PLATE	PEA PLATE	EYA PLATE
A.			
B.			
C.			
D.			
E.			

VA-5	CS-10	K-1000	SPS	NIT	CAT	IND	ESC	LIP	LEC	UREASE
A.										
B.										
C.										
D.										
E.										

PRESUMPTIVE IDENTIFICATION
A.
B.
C.
D.
E.

DEFINITIVE IDENTIFICATION
A.
B.
C.
D.
E.

SUSCEPTIBILITY TESTING INFORMATION
A.
B.
C.
D.
E.

Figure 7-2. Example of a worksheet for use in the anaerobic bacteriology laboratory. This worksheet provides areas to record information about the specimen, direct Gram stain, and growth on plates incubated aerobically. There is also sufficient space to record information about five different colony morphotypes (A-E) suspected of being anaerobes. The following abbreviations are used: vancomycin VA, colistin CS, kanamycin K, sodium polyanethol sulfonate SPS, nitrate NIT, catalase CAT, indole IND, esculin ESC, lipase LIP, lecithinase LEC.

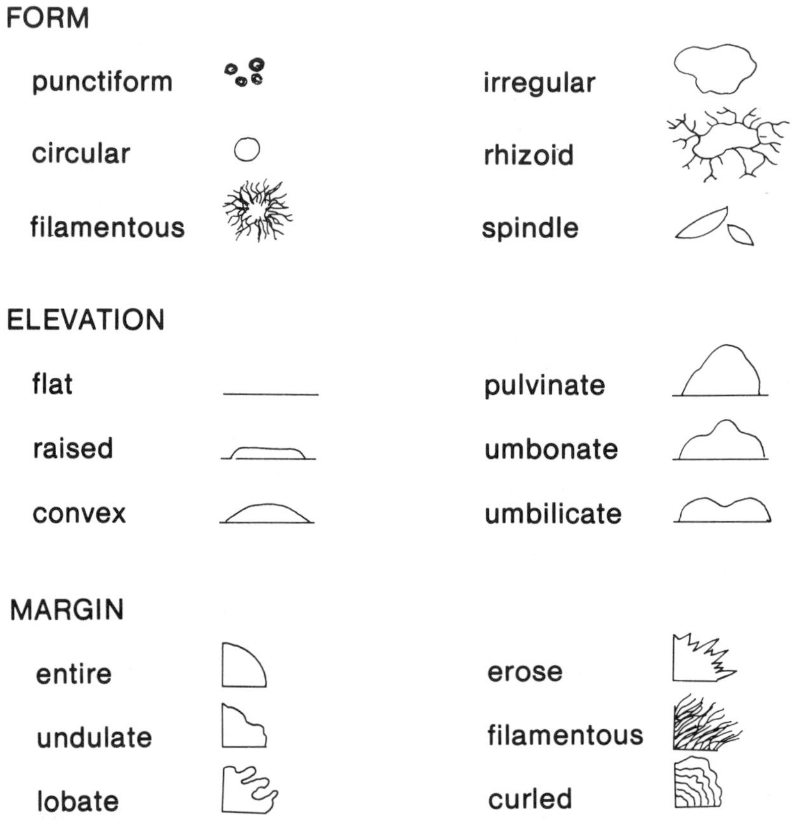

FORM

punctiform

circular

filamentous

irregular

rhizoid

spindle

ELEVATION

flat

raised

convex

pulvinate

umbonate

umbilicate

MARGIN

entire

undulate

lobate

erose

filamentous

curled

Figure 7-3. Features/characteristics to be included in colony descriptions (based on reference 3)

Figure 7-4. Pitting of the agar surface by *Bacteroides ureolyticus,* 48-hour-old culture on BRU agar

Table 7-1. Features/Characteristics To Note When Examining Colonies of Anaerobic Bacteria

Feature/Characteristic	Comments
Color/pigment	Pigmented species of *Porphyromonas/Prevotella* produce brown-black colonies. *Actinomyces* colonies may be red, pink, tan, or yellow.
Surface	For example, glistening or dull.
Density	For example, opaque, translucent, transparent; members of the *B. ureolyticus* group typically have translucent to transparent colonies.
Consistency	For example, butyrous (butterlike), viscous, membranous, brittle.
Form, elevation, margin	See Figure 7-3.
Fluorescence under UV light	Pigmented species of *Porphyromonas* and *Prevotella* often fluoresce brick-red. *F. nucleatum* and *C. difficile* fluoresce chartreuse. Some species of *Veillonella* fluoresce red.
Pitting of the agar	Caused by *B. ureolyticus, B. gracilis,* and *Wolinella* spp.
Double zone of hemolysis	Most strains of *C. perfringens* produce a small inner zone of complete beta hemolysis and a wider outer zone of incomplete hemolysis.
Swarming	*C. tetani* and *C. septicum* frequently swarm over the entire surface of the plate.
Odor	Some anaerobes have a characteristic odor (e.g., *C. difficile* has a horse stable odor, and *P. anaerobius* has a sweet, unpleasant odor).
Spider colonies	Young *Actinomyces israelii* and *Propionibacterium propionicus* colonies often appear as thin masses of wooly filaments originating at a single point.
Molar tooth colonies	Older colonies of *A. israelii* and *P. propionicus* often resemble a molar tooth.
Breadcrumb colonies	Some strains of *F. nucleatum* produce white colonies that resemble breadcrumbs.
Ground-glass colonies	Other strains of *F. nucleatum* produce colonies having a ground-glass appearance when viewed under a dissecting microscope (referred to by some authors as "speckled opalescence" or "internal flecking").
Fried egg colonies	Colonies of *F. mortiferum* and *F. varium* typically have a fried egg appearance. This characteristic is not specific for these organisms, however.

Table 7-2. Features/Characteristics To Note When Examining Gram-Stained Smears of Anaerobic Isolates

Feature/Characteristic	Comments
Gram reaction	Is the organism gram-positive, gram-negative, or gram-variable? Some gram-positive organisms (e.g., *Clostridium ramosum* and *C. clostridiiforme*) routinely stain gram-negative.
Cell morphology	
Cocci	Are the cocci arranged singly, in pairs (diplococci), in tetrads, in clusters, or in chains? Are they especially large (like *Peptostreptococcus magnus*)? Are they especially small (like *Peptostreptococcus micros* and *Veillonella* spp.)?
Bacillus	If gram-positive, are the cells branched (*Actinomyces*-like)? Are they diphtheroid in appearance (like *Propionibacterium* spp.)? Are they large (*Clostridium*-like)? Not all clostridia are large. Are the ends of the cells bifurcated (forked) like some *Actinomyces* and *Bifidobacterium* spp.? Are the cells beaded? Is there evidence of spores? If gram-negative, are they cocco-bacilli? Are they curved? Are they pleomorphic? Are there bizarre forms (swellings and balloon forms)? Are they filamentous? Are they safety pin–like (bipolar staining)? Are they long, thin, and tapered at the ends ("fusiform," like *Fusobacterium nucleatum*)?

Bacteroides Bile Esculin (BBE) Agar Plates.   The sole purpose for including the BBE agar plate in the primary isolation setup is the rapid (often 24-hour) presumptive identification of the *B. fragilis* group. If members of this group are present, the gray colonies will be a minimum of one millimeter in diameter, and the originally light-yellow medium will usually be brown in the area around the colonies. A presumptive identification of *B. fragilis* group can be made at this point. Good growth is the result of bile resistance, and browning of the medium is due to esculin hydrolysis. A dark precipitate (stippling) in the medium around the areas of heavy growth is suggestive of the species *B. fragilis*.[4] The appearance of *B. fragilis* group organisms on BBE agar is depicted in Figure 7-6. Although *B. vulgatus* is a member of the *B. fragilis* group, it does not hydrolyze esculin and thus does not produce a brown discoloration of the medium. Although *B. splanchnicus* and *B. eggerthii* are not members of the *B. fragilis* group, they are bile resistant and esculin hydrolysis positive; thus, colonies of these organisms will have the same appearance on BBE agar as members of the *B. fragilis* group. Depending upon commercial source, age, and storage conditions of the medium, other organisms (e.g., *Fusobacterium mortiferum, Klebsiella pneumoniae, Enterococcus* spp., and yeast) may also grow on BBE agar, but their colony size will be less than one millimeter in diameter. Gram staining and aerotolerance testing will aid in recognition of these other organisms.

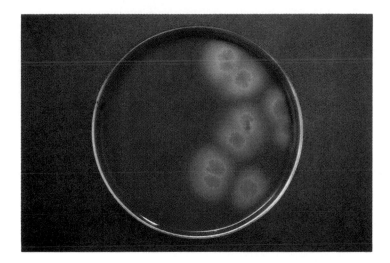

Figure 7-5. Double zone of hemolysis produced by *Clostridium perfringens,* 48-hour-old culture on PEA agar

Figure 7-6. Appearance of *Bacteroides thetaiotaomicron* (a member of the *B. fragilis* group) on KVLB agar/BBE agar biplate, 48-hour-old culture (The originally yellow BBE agar has turned brown in the vicinity of colonies due to esculin hydrolysis.)

Kanamycin-Vancomycin-Laked Blood (KVLB) Agar Plates. Although most *Bacteroides* and *Prevotella* spp. will grow on KVLB agar, this medium is included in the primary isolation setup to accelerate production of pigment by certain species of *Prevotella* (formerly classified as pigmented species of *Bacteroides*). The characteristic appearance of such pigmented organisms is shown in Figure 7-7. Most *Porphyromonas* strains are inhibited by the vancomycin in KVLB agar.

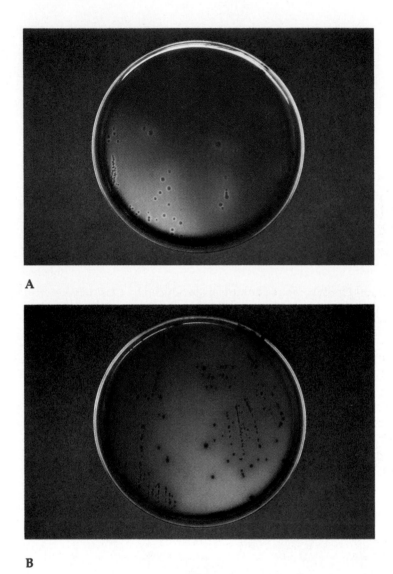

A

B

Figure 7-7. *Prevotella melaninogenica,* 72-hour-old cultures. (A) Organism on
BRU agar (centers of colonies becoming pigmented). (B) Same
organism on laked blood agar (entire colonies have turned brown-
black).

Darkening of the colonies may not always occur by 48 hours, and some strains
may require up to three weeks or more before pigmentation is observed. For this
reason, always subject nonpigmented colonies on KVLB agar (or BRU) to long-wave
ultraviolet (UV) light (such as a Wood's lamp). Although some species of pigmented
Porphyromonas and *Prevotella* fluoresce colors other than brick-red (e.g., brilliant red,
yellow, orange, and pink-orange), and some do not fluoresce at all (e.g., *Porphyro-*

monas gingivalis), a brick-red (red-brown to red-orange) fluorescence allows presumptive identification of the pigmented *Porphyromonas/Prevotella* group.[5] The brick-red fluorescence of such pigmented species under UV light is depicted in Figure 7-8. Organisms other than *Bacteroides* and *Prevotella* spp. will sometimes grow on KVLB agar (including some yeasts), so Gram staining and aerotolerance testing of colonies are absolutely necessary.

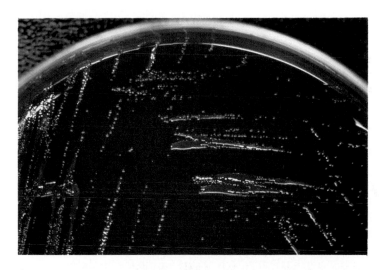

A

B

Figure 7-8. *Porphyromonas asaccharolytica,* 72-hour-old culture. (A) Colonies on BRU agar in ordinary light. (B) Brick-red fluorescence of same colonies under long-wave UV light.

Phenylethyl Alcohol (PEA) Agar Plates.   Essentially all anaerobes (gram-positive as well as gram-negative) will grow on anaerobically incubated PEA agar. In most cases, their colony appearance will be identical to that on BRU/BA plates. Gram-positive, facultative anaerobes will also grow on PEA agar. The sole reason for including a PEA agar plate in the primary isolation setup is to inhibit growth of *Enterobacteriaceae*, especially swarming *Proteus* spp. When *Proteus* is present in the specimen, it often swarms over the entire surface of the BRU/BA plate, thus making it impossible to work with other colonies on the plate. PEA agar will inhibit swarming by *Proteus* species, permitting the microbiologist to work on other colony morphotypes that are present.

Egg Yolk Agar (EYA) Plates.   If an EYA plate was included as part of the primary isolation setup, examine it for lecithinase, lipase, and proteolytic reactions. Certain clostridia produce enzymes that cause these reactions, but so do a few organisms other than *Clostridium* species.

A colony of a lecithinase-positive organism will be surrounded by a wide zone of opacity. This opacity is actually in the medium and is not a surface phenomenon. The appearance of the **lecithinase reaction** is shown in Figure 7-9. Lecithinase-positive clostridia include *C. bifermentans, C. sordellii, C. perfringens, C. novyi* type A, and *C. barati.*

A colony of a lipase-positive organism will be covered with an iridescent, multicolored sheen, sometimes described as resembling the appearance of oil on water or mother of pearl. This multicolored sheen may also appear on the surface of the agar in a narrow zone around the colony. The **lipase reaction** occurs only at the agar surface, unlike the lecithinase reaction, which occurs within the medium. The appearance of the lipase reaction is depicted in Figure 7-10. Examples of lipase-positive clostridia are *C. botulinum, C. novyi* type A, and *C. sporogenes.* Other anaerobes that can be lipase-positive include *Fusobacterium necrophorum, Prevotella intermedia,* and some isolates of *P. loescheii.*

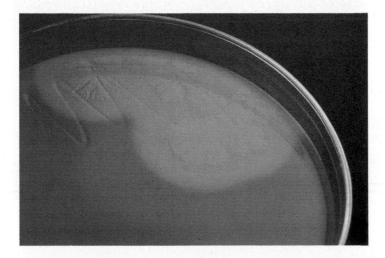

Figure 7-9. Lecithinase reaction of *Clostridium perfringens* on EYA, 48-hour-old culture

A

B

Figure 7-10.   Lipase reaction. (A) *Fusobacterium necrophorum* on EYA, 48-hour-old culture. (B) Close-up view of a lipase reaction, showing the oil on water/mother of pearl appearance. (from reference 6; courtesy of Sydney M. Finegold, M.D. and Star Publishing, Belmont, CA)

   Organisms that produce proteolytic enzymes will have a completely clear zone (often quite narrow) around colonies. **Proteolysis** is best observed by holding the plate up to a strong light source. It is reminiscent of the complete clearing seen with beta-hemolytic organisms on BRU/BA plates.

Direct Nagler Test.   Although not commonly employed, a direct Nagler test can be set up whenever *C. perfringens* is suspected. Such a suspicion is usually based upon observation of boxcar-shaped, gram-positive bacilli in a Gram-stained preparation of the specimen. The direct Nagler test, which is described in Appendix D, was initially developed as a means of presumptively identifying *C. perfringens*. It is now known that three other species of clostridia (*C. barati, C. bifermentans,* and *C. sordellii*) are also Nagler test–positive. Thus, a positive Nagler test result can be caused by any of four species of *Clostridium.*

Cycloserine-Cefoxitin-Fructose Agar (CCFA) Plates.   If a patient was suspected of having antibiotic-associated diarrhea (AAD) or pseudomembranous colitis (PMC), a CCFA plate might have been inoculated with a fecal specimen from that patient. CCFA is a selective and differential medium for recovery and presumptive identification of *Clostridium difficile*. *C. difficile* colonies on CCFA are yellow and ground glass in appearance, and the originally pink agar is yellow in the vicinity of the colonies. The appearance of *C. difficile* on CCFA is shown in Figure 7-11. Although other organisms may grow on CCFA, their colonies will be smaller than and will not resemble the characteristic colonies of *C. difficile*. Also, *C. difficile* has a characteristic horse stable odor. Colonies on BRU/BA will fluoresce chartreuse under UV light. Isolation of this organism from fecal specimens does not always indicate AAD or PMC; *C. difficile* can be part of a patient's GI flora. There are more accurate ways to diagnose AAD or PMC caused by *C. difficile,* including cytotoxin assays and latex agglutination procedures (see Chapter 9). *C. difficile* is not the only etiologic agent of these conditions.

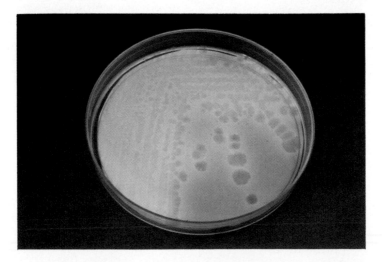

Figure 7-11.   The appearance of *Clostridium difficile* on CCFA, 48-hour-old culture (large, ground-glass colonies and yellowing of the originally pink medium in the vicinity of colonies)

Processing Colonies Suspected of Being Anaerobes

As previously mentioned, colonies growing on the anaerobically incubated primary plates are only suspected of being anaerobes. They could, in fact, be facultative anaerobes or even the **same** organisms growing on the plates that were incubated in the CO_2 incubator. Thus, it is necessary to perform an **aerotolerance test** to prove that they are anaerobes. In addition to aerotolerance testing, colonies must be Gram stained, and a subculture plate (usually a BRU or BA plate) must be inoculated. Table 7-3 summarizes the workup of each colony morphotype.

Table 7-3. Logical, Stepwise Approach to the Workup of Isolates Suspected of Being Anaerobes

Step	Action(s) To Be Taken
1	Record colony morphology and other features/characteristics on the worksheet, using the criteria in Figure 7-3 and Table 7-1. Include growth/appearance on selective and differential media, such as BBE and KVLB agars.
2	Prepare and examine a Gram-stained smear of a portion of the colony. Record the gram reaction and cell morphology on the worksheet, using the criteria in Table 7-2.
3	Using portions of the **same** colony, perform an aerotolerance test by inoculating a BRU/BA subculture plate (streak for isolation) and a portion of a chocolate agar plate.
4	Based upon the gram reaction (positive or negative) and cell morphology, place the appropriate disks on the BRU/BA subculture plate (refer to Figure 7-12 and Table 7-4).
5	Incubate the plates appropriately (the BRU/BA plate anaerobically, the chocolate agar plate in a CO_2 incubator.).

Colony Morphology

Carefully examine each colony morphotype. A dissecting microscope is recommended for this purpose so that fine details (e.g., a ground-glass appearance to the colony) can be seen. Record colony morphology and other features/characteristics on a worksheet (as shown in Figure 7-2) using the criteria outlined in Figure 7-3 and Table 7-1. Also record, in semiquantitative terms, the number of colonies of each morphotype present on the primary plates.

Gram Reaction and Cell Morphology

A portion of the colony is used to prepare a smear for Gram staining. Record the gram reaction and cell morphology on the worksheet using the criteria outlined in Table 7-2. The gram reaction and morphologic appearance of the organism will help you select the disks to add to the pure culture/subculture plate, discussed in the following section.

Inoculating a Pure Culture/Subculture (PC/SC) Plate

Inoculate a portion of the colony to a BRU/BA plate in such a manner as to obtain well-isolated colonies. Following anaerobic incubation, this plate will serve as a pure culture of the organism. Thus, it is referred to as the pure culture/subculture (PC/SC) plate.

Based upon the gram reaction and morphology of the organism, you will add certain disks to this plate before anaerobic incubation. Place the disks on the heavily inoculated first third of the PC/SC plate, as shown in Figure 7-12. The specific disks to be added, shown in Table 7-4, are discussed in the next section. In many cases, disk results will enable presumptive identification of the organisms.

Disks To Add to the PC/SC Plate

Three special potency antimicrobial disks (vancomycin, kanamycin, and colistin) are added to the PC/SC plate whenever the Gram stain reveals a gram-negative bacillus. As is described in Chapter 8, these disks aid in determining the **true** gram reaction of a pink-staining bacillus. You may recall from Chapter 2 that certain clostridia routinely stain pink although they are technically considered to be gram-positive. If the organism is truly gram-negative, the special potency antimicrobial disks also aid in differentiating the genus *Fusobacterium* from other genera of anaerobic, gram-negative bacilli; details can be found in Chapter 8. Other disks that can be added to the PC/SC plate of a gram-negative bacillus are a bile disk and a nitrate disk, uses of which are explained in Chapter 8.

Figure 7-12. Placement of the disks on the PC/SC plate. Inoculate the BRU/BA subculture plate in such a manner as to ultimately obtain well-isolated colonies. Then add the disks to the heavily inoculated area of the plate (usually the first quarter or first third of the plate, depending upon the particular manner of streaking). Then incubate the plate anaerobically. The specific disks to add are shown in Table 7-4.

Table 7-4. Disks to Add to the PC/SC Plate

Gram Stain Appearance	Disk(s) To Add
Gram-positive	
Cocci	Sodium polyanethol sulfonate (SPS) and nitrate
Bacilli	
Clostridium-like*	None
Actinomyces-like	Nitrate
Diphtheroidal	Nitrate
Other	Nitrate
Gram-negative	
Cocci	Nitrate
Bacilli	Vancomycin (5 μg), colistin (10 μg), kanamycin (1 mg), nitrate, and bile

* Large, unbranched, gram-positive rods, with or without spores (Not all clostridia are large, and not all clostridia will stain as gram-positive rods.)

Source: Based on reference 6.

Whenever the Gram stain reveals the isolate to be a gram-positive coccus, the most important disk to add to the PC/SC plate is a sodium polyanethol sulfonate (SPS) disk. This disk is useful in the presumptive identification of *Peptostreptococcus anaerobius,* the only anaerobic, gram-positive coccus susceptible to this substance. A nitrate disk may also be added to the PC/SC plate of a gram-positive coccus.

Aerotolerance Testing

The PC/SC plate serves as one of the plates used in the aerotolerance test. One or two additional plates are also needed:

- A chocolate agar (CA) plate, which will be incubated in a CO_2 incubator. By dividing the CA plate up into sections, you can use one plate to test several different isolates.

- Some authorities recommend also inoculating a blood agar plate for incubation in a non-CO_2 incubator. This plate will enable differentiation between aerobic and capnophilic organisms. By dividing the blood agar plate up into sections, you can use one plate to test several different isolates.

All aerotolerance test plates are then incubated appropriately for 24–48 hours. Following incubation, the aerotolerance test is interpreted in the following manner:

- Theoretically, an anaerobe should grow only on the anaerobically incubated BRU/BA plate. But some aerotolerant anaerobes (e.g., certain *Clostridium*, *Actinomyces*, *Propionibacterium*, *Bifidobacterium*, and *Lactobacillus* spp.) are capable of growing on the CO_2-incubated CA plate. Anaerobes usually grow **better**, however, on the anaerobically incubated plate. The catalase test is useful for distinguishing between *Bacillus* spp. (aerobic, sporeforming, gram-positive bacilli) and aerotolerant species of *Clostridium*. Most *Bacillus* spp. are catalase-positive, whereas most *Clostridium* spp. are catalase-negative. In addition, *Bacillus* species sporulate aerobically, whereas *Clostridium* species form spores only under anaerobic conditions.[6]

- A noncapnophilic, facultative organism will grow on all three plates.

- A capnophilic organism (capnophile) will grow on the CO_2-incubated CA plate but not on the aerobically incubated blood agar plate. It may or may not grow on the anaerobically incubated BRU/BA plate. *Haemophilus influenzae*, for example, will grow on the BRU/BA and CO_2-incubated CA plates but not on the aerobically incubated blood agar plate.

Review of Processing Procedures

Whenever a colony is suspected of being an anaerobe, the colony morphology should first be described on the worksheet. A portion of the colony should next be Gram stained, and observations should be recorded. A portion of the colony must be streaked for isolation on a PC/SC plate. Based upon the gram reaction and morphology of the organism, appropriate disks are added to the PC/SC plate. Additional plates are inoculated as part of the aerotolerance test, and these are incubated appropriately. Lastly, if clostridia are suspected (e.g., based upon Gram stain appearance and colony morphology), an EYA plate should be streaked, and a direct Nagler test can be performed. Remember that some aerotolerant anaerobes are capable of growing on the CO_2-incubated CA plate, and these can be particularly troublesome. However, they grow better anaerobically than they do in the CO_2 atmosphere.

At this point, we'll assume that the isolate suspected of being an anaerobe did not grow on the CO_2-incubated CA plate, that it is deemed clinically significant, and that we have a pure culture of the anaerobe on the PC/SC plate. Is there a need to further identify the isolate?

That question is perhaps best answered by a statement made by Dr. Stephen Allen: "Isolation, identification, and susceptibility tests are required for anaerobes for the same reasons they are required for nonanaerobic microorganisms that are involved in infectious diseases."[1] Listed below are three reasons further identification of anaerobes is warranted:

- It seems logical to assume that empirical therapy is apt to be more effective if the physician's choice of antimicrobial agent(s) is based upon knowledge of the infecting organism(s).

- Identification of an anaerobic isolate can often indicate to the physician the probable **source** of the infectious process (e.g., certain endogenous anaerobes are known to inhabit sites above the waist, whereas others inhabit sites below the waist). *Clostridium septicum* bacteremia, for example, suggests colon cancer or other diseases of the colon.[2] Regardless of the site of isolation, recovery of pigmented *Porphyromonas* species usually suggests an oral origin, whereas isolation of members of the *B. fragilis* group suggests an intestinal origin.

- Over time, the identification of anaerobes provides a data base of information regarding the role of certain anaerobic bacteria in infectious processes, in much the same way that susceptibility testing results provide a data base of information that helps physicians select antimicrobial agents for empirical therapy.

Having agreed that anaerobes **should** be identified, we must find ways to do so. Chapter 8 describes procedures that can be used to make **presumptive identifications** of anaerobic isolates. Chapter 9 provides information on **definitive identifications**. Appendix B offers cost-containment guidelines, including how far to go in identifying anaerobic isolates.

Chapter in Review

- When examining the anaerobically incubated primary plates, certain findings/characteristics should alert the microbiologist that anaerobes may be present. These include (1) a foul odor when the anaerobic jar is opened, (2) the presence on the anaerobically incubated plates of certain colony morphotypes that are absent on the CO_2-incubated plates, (3) good growth on the BBE agar plate, and (4) colonies on the BRU/BA or KVLB agar plate that are brown-black or fluoresce brick-red.

- Good growth on the BBE agar plate and browning of the medium provide presumptive evidence of the *Bacteroides fragilis* group. Brown-black pigmented colonies on BRU/BA or KVLB agar or colonies that fluoresce brick-red under UV light are presumptive evidence of pigmented *Porphyromonas/Prevotella* spp.

- *Clostridium*-like, anaerobic, gram-positive bacilli that produce a lecithinase reaction on EYA might be *C. bifermentans, C. sordellii, C. perfringens, C. novyi* type A, or *C. barati.*

- Anaerobic, gram-positive bacilli that produce a lipase reaction on EYA might be *C. botulinum, C. novyi* type A, or *C. sporogenes.* Some anaerobic, gram-negative bacilli are also lipase-positive, including *Fusobacterium necrophorum, Prevotella intermedia,* and some isolates of *P. loescheii.*

- Processing of a colony suspected to be an anaerobe includes (1) describing the colony, (2) preparing and examining a Gram stain from it, (3) inoculating a pure culture/subculture plate, and (4) performing an aerotolerance test.

- Specific disks are placed on a PC/SC plate, depending upon the gram reaction and morphology of the isolate. For example, three special potency antimicrobial disks will be added to the PC/SC plate if the organism is a gram-negative bacillus. An SPS disk will be added to the PC/SC plate if the organism is a gram-positive coccus.

- In the aerotolerance test, obligate anaerobes will grow on the anaerobically incubated BRU/BA plate, but not on the CO_2-incubated CA plate or the aerobically incubated blood agar plate. Although some aerotolerant anaerobes can grow on a CO_2-incubated CA plate, they will grow better on an anaerobically incubated BRU/BA plate.

- Anaerobic isolates are identified for the same reasons that nonanaerobic bacteria are identified: to provide physicians with the names of potential pathogens that have been isolated from their patient's infectious processes, to enable physicians to more accurately prescribe appropriate empirical therapy, to indicate the probable source of organisms involved in infectious processes, and to add to the data base of information regarding the role of particular anaerobes in infectious processes.

Self-Assessment Exercises

1. Name the anaerobe (genus and species) that produces a characteristic double zone of hemolysis on BRU/BA.

2. Which two genera of anaerobes produce colonies on BRU/BA and/or KVLB agar that fluoresce brick-red under long-wave (366 nm) ultraviolet light?

3. The main reason for including PRAS PEA agar in the primary isolation setup is to inhibit the growth of what specific facultative anaerobe?

4. TRUE or FALSE: The Nagler test is useful for presumptive identification of *Clostridium perfringens* because *C. perfringens* is the only organism that produces a positive result with this test.

5. What characteristic odor is associated with colonies of *Clostridium difficile*?

6. TRUE or FALSE: No anaerobes are capable of growing on a chocolate agar plate incubated at 35–37°C in a CO_2 incubator.

7. Which disk is especially important to add to the pure culture/subculture plate of an anaerobic, gram-positive coccus?

   _____

8. What rapid test is useful in differentiating a *Bacillus* species from an aerotolerant species of *Clostridium*? Why?

   _____

   _____

9. Cite three reasons why identifying anaerobic isolates is important.

   a. _____

   _____

   b. _____

   _____

   c. _____

   _____

10. Which of the bacteria in the following list would be able to grow on or in the following anaerobically incubated media?   (Place the letters of the correct answers on the lines following the types of media.)

   a. Virtually all gram-positive anaerobes

   b. Virtually all gram-negative anaerobes

   c. Facultative, gram-positive organisms

   d. Facultative, gram-negative organisms

   e. Virtually all *Bacteroides* and *Prevotella* spp.

   f. Members of the *Bacteroides fragilis* group

BRU/BA   _____

PEA agar   _____

EYA   _____

BBE agar   _____

KVLB agar   _____

THIO broth   _____

References

1. Allen, S. D. 1984. Current relevance of anaerobic bacteriology. Clin. Microbiol. Newsl. 6:147–149.
2. Bryan, C. S. 1989. Clinical implications of positive blood cultures. Clin. Microbiol. Rev. 2:329–353.
3. Dowell, V. R., Jr., and T. M. Hawkins. 1974. Laboratory methods in anaerobic bacteriology: CDC laboratory manual. Centers for Disease Control, Atlanta.
4. Moore, H. B. 1981. Rapid methods in microbiology: IV. Presumptive and rapid methods in anaerobic bacteriology. Amer. J. Med. Technol. 47:705–712.
5. Slots, J., and H. S. Reynolds. 1982. Long-wave UV light fluorescence for identification of black-pigmented *Bacteroides* spp. J. Clin. Microbiol. 16:1148–1151.
6. Sutter, V. L., D. M. Citron, M. A. C. Edelstein, and S. M. Finegold. 1985. Wadsworth Anaerobic Bacteriology Manual, 4th ed. Star Publishing, Belmont, CA.

CHAPTER 8

PRESUMPTIVE IDENTIFICATIONS

> Simple Observations and Inexpensive Tests for
>   Presumptive Identification (PID) of Anaerobic
>   Isolates
> Logical, Stepwise Approach for PID of Anaerobic
>   Isolates
> PID Flow Charts
> Presumpto Plates

At the conclusion of this chapter, you will be able to:

1. Differentiate between the terms "presumptive identification" and "definitive identification."
2. List five colony observations/characteristics of value in making presumptive identifications of anaerobes.
3. List ten simple test results or reactions of value in making presumptive identifications of anaerobes.
4. State the minimal information required to identify the following anaerobes: a member of the *Bacteroides fragilis* group, pigmented species of *Porphyromonas* and *Prevotella*, *Clostridium difficile*, *C. perfringens*, *Fusobacterium mortiferum*, *F. necrophorum*, *F. nucleatum*, *Peptostreptococcus anaerobius*, *P. asaccharolyticus*, *Propionibacterium acnes*, and *Veillonella* spp.
5. Use logical, stepwise guidelines and flow charts to identify a variety of clinically encountered anaerobes.

Definitions

Definitive identification: A term used throughout this book to mean a valid identification of a microorganism to genus and species, regardless of the method(s) used to make the identification

Indole: A compound produced by the decomposition of the amino acid tryptophan; microorganisms capable of producing tryptophanase are indole-positive.

Presumptive identification: A term used throughout this book to mean an identification of a microorganism based upon simple colony and Gram stain observations and the results of relatively simple and inexpensive tests. Although a presumptive identification is believed to be valid, it may not be.

Urease: An enzyme that catalyzes the decomposition of urea into carbon dioxide and ammonia. Relatively few anaerobes produce urease, notable exceptions being *Bacteroides ureolyticus, Clostridium sordellii,* and *Peptostreptococcus tetradius,* which are urease-positive, and two species of *Actinomyces* (*naeslundii* and *viscosus*), which are variable for urease production.

Presumptive identifications of microorganisms have become more popular in recent years, due primarily to increased emphasis on speed and cost reduction. Aerobic and facultatively anaerobic bacteria commonly identified in the aerobic bacteriology laboratory using presumptive identification (hereafter referred to as PID) criteria include *Escherichia coli, Pseudomonas aeruginosa,* swarming species of *Proteus,* and Group A beta-hemolytic streptococci (see Table 8-1). Although many laboratories base their identifications of these organisms on PID criteria, others perform the additional procedures necessary for definitive identification.

Table 8-1. Examples of Aerobe and Facultative Anaerobe PIDs

Organism	PID Criteria
Escherichia coli	Relatively dry, often flat, bright pink (lactose-positive) colonies on MacConkey agar; indole-positive
Pseudomonas aeruginosa	Beta-hemolytic colonies on blood agar; non–lactose fermenter (colorless colonies) on MacConkey; green pigment; grapelike (or tortilla-like) odor; oxidase-positive
Swarming *Proteus* spp.	Gram-negative bacillus; swarming on blood agar; non–lactose fermenter on MacConkey; *P. vulgaris* if indole-positive; *P. mirabilis* if indole-negative
Streptococcus pyogenes	Beta-hemolytic, catalase-negative, gram-positive coccus; susceptible to bacitracin (A-disk-positive)

PIDs of many anaerobes are also possible. As with the previously mentioned examples from the aerobic bacteriology laboratory, it is an individual laboratory decision whether such PIDs will suffice or whether definitive identifications are necessary. By combining readily observable colony and Gram stain features with results of simple test procedures, even small laboratories are capable of making PIDs of many commonly isolated and clinically important anaerobes. However, many anaerobes **cannot** be identified using PID criteria. Definitive identification procedures are described in Chapter 9.

Because of the excellent "tools" available for use in today's laboratories (e.g., anaerobic chambers, jars, bags/pouches; special enriched, selective, and differential media for anaerobic bacteriology; a variety of commercially available products of value in identification of anaerobes), even small laboratories should be able to isolate anaerobes from clinical specimens, use special media, perform Gram staining, inoculate pure culture/subculture (PC/SC) plates, use disks (Chapter 7), and perform simple tests such as catalase and spot indole. As discussed in Chapter 10, it is even possible for small laboratories to perform susceptibility testing of important anaerobic isolates. Table 8-2 highlights a few of the major events that have led to our current understanding of, and ability to isolate and identify, clinically significant anaerobes.

Table 8-2. Milestones in the Development of Techniques for Cultivation and Identification of Obligate Anaerobes

Date	Event
1861–1915	Use of various deep broths and agar shake cultures that were boiled and cooled before inoculation
1861–1916	Removal of oxygen from tubes and jars using vacuum pumps, inert gases, or illuminating gas
1916	Introduction of cooked-meat medium
1916–1966	Utilization of various anaerobic jars employing a catalyst to remove oxygen by oxidation of hydrogen (e.g., introduction of McIntosh-Fildes jar in 1916; introduction of GasPak® jar in 1966)
1940	Introduction of thioglycollate combined with agar in liquid media
1950	Development of the roll tube technique by Hungate
1964–1969	Introduction of various anaerobic chambers ("glove boxes")
1965	Roll tubes, prereduced anaerobically sterilized (PRAS) media, chromatographic techniques, and VPI inoculator described by Holdeman and Moore; development of antibiotic-containing selective media for isolation and identification of anaerobes
Early 1970s	Introduction of biochemical-based minisystems for identification of anaerobes

Table 8-2. Milestones in the Development of Techniques for Cultivation and Identification of Obligate Anaerobes (*continued*)

Date	Event
1975	Development of a gloveless anaerobic chamber with incubator by Cox
1978	Introduction of PRAS plated media by Cox
1980s	Development of rapid techniques (disks, chromogenic substrates, preexisting enzyme-based minisystems) for identification of anaerobes (Wadsworth group and others)

Source: Based on reference 7.

Simple Observations and Inexpensive Tests for Presumptive Identification (PID) of Anaerobic Isolates

The following are simple observations and inexpensive tests that can be used to make PIDs of many of the anaerobes most commonly encountered in clinical materials. Only tests that provide results on the **same day** the PC/SC becomes available are included here. Rather than use overnight tests that merely provide PIDs, microbiologists might prefer to use one of the commercial identification systems that provides **same day definitive** identifications (Chapter 9). Here the requirement for speed must be weighed against cost. The decision to use the more expensive commercial identification systems must be made independently by each laboratory. Many of the simple PID procedures described in this chapter were developed by Citron, Edelstein, Finegold, Sutter, and others at the Wadsworth Veterans Administration Medical Center in Los Angeles, CA.[12]

Here are some simple observations for making PIDs:

- Growth/no growth on selective media such as BBE agar, KVLB agar, and CCFA (described in Chapter 7)

- Colony morphology, including color/pigment, surface, density, consistency, form, elevation, margin, pitting, hemolysis, swarming, and odor (described in Chapter 7)

- Gram reaction and cell morphology (described in Chapter 7)

Here are some inexpensive tests for making PIDs. Appendix D contains detailed descriptions of these tests, as well as specific references.

- Results of aerotolerance testing (described in Chapter 7).

- Fluorescence under long-wave (366 nm), ultraviolet (UV) light. Many pigmented species of *Porphyromonas* and *Prevotella* will fluoresce brick-red under UV light, a characteristic that can be used to presumptively identify this group of organisms. Some strains fluoresce colors other than brick-red (e.g., brilliant red, yellow, orange, and pink-orange).[10] Some anaerobes other than pigmented *Porphyromonas* and *Prevotella* species will also fluoresce under UV light (e.g., *Fusobacterium nucleatum* and *Clostridium difficile* will fluoresce chartreuse, and *Veillonella* species, red). The red fluorescence of *Veillonella* is culture medium–dependent, weaker than the fluorescence produced by pigmented species of *Porphyromonas* and *Prevotella*, and fades rapidly upon exposure of colonies to air (fades completely after five to ten minutes).[3]

- **Special potency antimicrobial disks.** As shown in Figure 7-13, the three special potency antimicrobial disks are added to the BRU/BA subculture plate whenever the Gram stain reveals the isolate to be a gram-negative bacillus (i.e., the organism stains pink). The primary purpose of the disks is to verify that the organism is **truly** a gram-negative bacillus, as opposed to a *Mobiluncus* species, for example, or one of the clostridia (e.g., *C. ramosum* or *C. clostridioforme*) that routinely stain gram-negative. In one study, 97.5% of 281 gram-positive anaerobes were susceptible to a 5 µg vancomycin disk, whereas 98.3% of 650 gram-negative anaerobes were resistant.[2] In addition to determining the true gram reaction, the disks can often reveal whether the gram-negative bacillus is a *Fusobacterium* species. It is critical to use disks of the prescribed potency: a 5 µg vancomycin disk, a 10 µg colistin disk, and a 1 mg (1000 µg) kanamycin disk. Disks must be pressed firmly to the surface of the PC/SC plate to ensure uniform diffusion of the agent into the medium. Disk results are analyzed in a stepwise manner, as shown later in this chapter, and give no information whatsoever about antimicrobial agents that might be used for therapy. Typical disk results are shown in Figure 8-1. Interpretation of disk results is described in Appendix D.

- **SPS disk.** As shown in Figure 7-13, a sodium polyanethol sulfonate (SPS) disk is added to the BRU/BA subculture plate whenever the Gram stain reveals the isolate to be a gram-positive coccus. The SPS disk is especially valuable for presumptive identification of *Peptostreptococcus anaerobius*, which is susceptible to SPS (Figure 8-2).

- **Nitrate disk.** The nitrate disk is used to determine an organism's ability to reduce nitrate. The test is merely a "miniaturized" version of the conventional nitrate reduction test used in the aerobic bacteriology laboratory (Figure 8-3).

- **Bile disk.** A bile disk can be used to determine an organism's ability to grow in the presence of relatively high concentrations of bile. As shown in Figure 7-13, a bile disk can be added to the BRU/BA subculture plate whenever the Gram stain reveals the isolate to be a gram-negative bacillus. Other indicators of bile resistance include good growth on the BBE plate and growth in 20% bile broth. An anaerobic, gram-negative bacillus that is bile tolerant and catalase-positive can be presumptively identified as a member of the *Bacteroides fragilis* group.[14] (However, *Bilophila wadsworthia* also fits this description.)

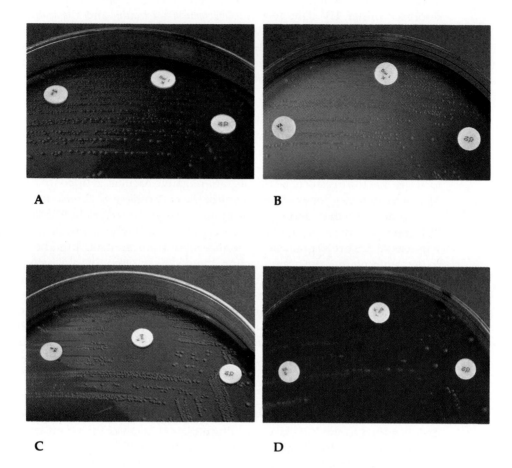

Figure 8-1. Typical special potency antimicrobial disk results: (A) *Bacteroides thetaiotaomicron,* 48-hour-old culture on BRU, showing the typical pattern of members of the *B. fragilis* group (resistant to all three agents). (B) *Prevotella buccae,* 48-hour-old culture on BRU, resistant to vancomycin and kanamycin but susceptible to colistin. (C) *Clostridium ramosum,* 48-hour-old culture on BRU, resistant to colistin but susceptible to vancomycin and kanamycin. (D) *Fusobacterium necrophorum,* 48-hour-old culture on BRU, showing the typical *Fusobacterium* pattern (resistant to vancomycin but susceptible to kanamycin and colistin). Reading from left to right, the disks in all four photographs are vancomycin, kanamycin, and colistin.

Figure 8-2. *Peptostreptococcus anaerobius,* 48-hour-old culture on BRU, showing susceptibility to sodium polyanethol sulfonate (SPS)

Figure 8-3. *Propionibacterium acnes,* 48-hour-old culture on BRU, showing a positive nitrate disk result. The red color indicates the presence of nitrite, resulting from reduction of nitrate.

- **Catalase test.** Using a plastic, disposable inoculating loop or a wooden applicator stick, remove some of the growth from the PC/SC plate, and rub it onto a small area of a glass microscope slide. Be careful not to transfer any of the agar because sheep red blood cells contain catalase, which might cause a false positive test result. Add a drop of 3% hydrogen peroxide, and watch for the production of bubbles of oxygen gas. The presence of gas bubbles indicates a positive test, whereas the absence of gas bubbles constitutes a negative test. Along with other uses, the catalase test is of value in differentiating aerotolerant strains of *Clostridium* (catalase-negative) from *Bacillus* spp. (catalase-positive).

- **Spot indole test.** Saturate a small piece of filter paper with the spot indole reagent (p-dimethylaminocinnamaldehyde; DMCA). Using an inoculating loop, remove some of the growth from the PC/SC plate, and rub it onto the saturated area. Rapid development of a blue to green color indicates a positive test (i.e., production of indole from the amino acid tryptophan), whereas a pink to orange color indicates a negative test. The spot indole test can also be performed directly on a pure culture plate, as described in Appendix D and shown in Figure 8-4.

  When DMCA was compared to Kovac and Ehrlich reagents for detection of indole production by anaerobes, it proved to be the most sensitive, and Kovac reagent the least sensitive, of the three reagents.[8] In addition, DMCA detects certain indole derivatives (skatole, 3-indolepropionic acid, and 3-indolebutyric acid) that are produced by some *Clostridium* spp. (e.g., *C. botulinum* and *C. sporogenes*), in which case a deep violet/lavender color is produced.

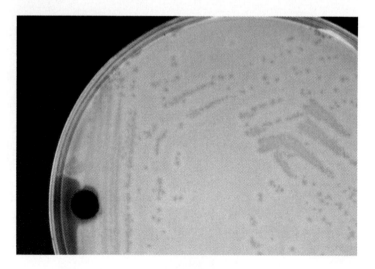

Figure 8-4. *Fusobacterium necrophorum,* 48-hour-old culture on EYA, showing a positive spot indole test (blue color following addition of p-dimethylaminocinnamaldehyde reagent)

- **Motility test.** Motility may be determined using either very young (4–6 hour) broth cultures or 24–48 hour colonies on BRU/BA. Both techniques are described in Appendix D. Motile, gram-negatively staining anaerobes include *Wolinella* spp., *Mobiluncus* spp., and *Campylobacter concisus*, among many others. Remember that *Mobiluncus* spp. are actually gram-positive organisms, even though they stain pink with the Gram stain technique.

- **Lecithinase** and **lipase** reactions. An egg yolk agar plate is used to determine the activities of these enzymes. These reactions are of value in identifying many species of *Clostridium*, as well as lipase-positive strains of *Prevotella intermedia, P. loescheii,* and *Fusobacterium necrophorum.* The opaque lecithinase reaction occurs within the medium in a relatively wide zone around colonies, whereas the lipase reaction occurs on the surfaces of colonies and on the agar surface in a relatively narrow zone around the colonies. Additional information can be found in Chapter 7 and Appendix D.

- **Urease test.** The ability of anaerobes to hydrolyze urea can be determined in a variety of ways, including rapid tube, disk, and spot tests. Perhaps the easiest way is by using Urea Differentiation Disks® (Difco Laboratories). First suspend a loopful of organisms from the PC/SC plate in a small volume of sterile water. Following addition of the urea disk, incubate the tube aerobically for up to four hours. A color change from pale yellow to dark pink represents a positive test. The disk method is accurate and does not require preparation of reagents.[9] Urease production can also be detected by anaerobic incubation of urea agar slants. The urease test is especially important in distinguishing *B. ureolyticus* (urease-positive) from other pitting, anaerobic, gram-negative bacilli (urease-negative). It is also a useful test in identifying *Clostridium sordellii* and *Peptostreptococcus tetradius,* which are urease-positive, and *Actinomyces naeslundii* and *A. viscosus,* some strains of which are urease-positive.

- **Oxidase test.** Kovac oxidase reagent can be used to detect cytochrome oxidase, which is produced by a small number of anaerobes. Oxidase-positive anaerobes include *Bacteroides ureolyticus, Campylobacter concisus,* and most strains of *Wolinella curva.* Approximately 50% of *Anaerobiospirillum* isolates are also oxidase-positive.[1]

Some companies have combined several of the rapid PID tests. The Microring® devices, manufactured by Medical Wire & Equipment Co. USA (Victory Gardens, NJ), are examples of such combination tests. The Microring® AN, which contains six antimicrobial agent disks (erythromycin, rifampicin, colistin, penicillin G, kanamycin, and vancomycin), is of some value in presumptively identifying anaerobic, gram-negative bacilli and distinguishing between gram-positive and gram-negative anaerobic cocci. The Microring® AC, which combines metronidazole, SPS, and novobiocin, is of further value in identifying anaerobic, gram-positive cocci.

Flow Chart 8-1. PID of anaerobic, gram-positive cocci

Starting point: The organism is a gram-positive coccus. Read the SPS disk results on the BRU/BA PC/SC plate.

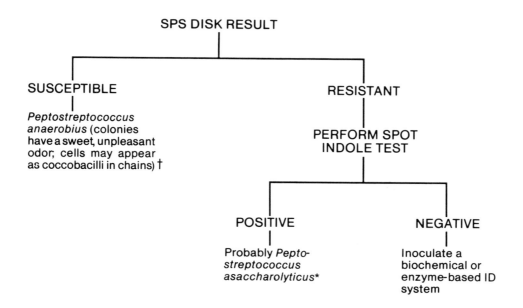

† Although some strains of *P. micros* may also be SPS-sensitive, the cells will usually be much smaller than *P. anaerobius*.

* *P. indolicus* is also indole-positive but is rarely encountered in human specimens. Perform either a nitrate test or a coagulase test on any SPS-resistant, indole-positive, anaerobic, gram-positive cocci recovered from veterinary specimens. This would enable differentiation of *P. indolicus* (nitrate- and coagulase-positive) from *P. asaccharolyticus* and *P. hydrogenalis* (nitrate- and coagulase-negative). *P. hydrogenalis* can be differentiated from *P. asaccharolyticus* either by glucose fermentation or alkaline phosphatase activity. *P. hydrogenalis* is positive for both, whereas *P. asaccharolyticus* is negative for both.

Flow Chart 8-2. PID of anaerobic, gram-positive bacilli/coccobacilli

Starting point: The organism is a gram-positive bacillus. Assign it to one of the following four categories:

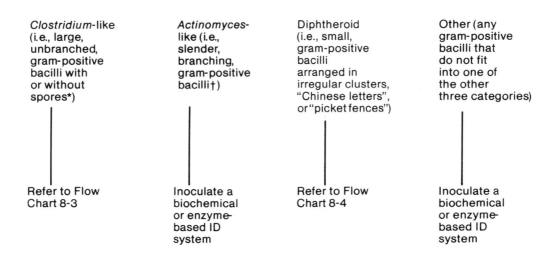

Clostridium-like (i.e., large, unbranched, gram-positive bacilli with or without spores*)	*Actinomyces*-like (i.e., slender, branching, gram-positive bacilli†)	Diphtheroid (i.e., small, gram-positive bacilli arranged in irregular clusters, "Chinese letters", or "picket fences")	Other (any gram-positive bacilli that do not fit into one of the other three categories)
Refer to Flow Chart 8-3	Inoculate a biochemical- or enzyme-based ID system	Refer to Flow Chart 8-4	Inoculate a biochemical- or enzyme-based ID system

* Not all clostridia are large, and not all clostridia will stain gram-positive. Those species that are not large will fall into the "other" category and will require inoculation of some type of biochemical- or enzyme-based ID system. It is hoped that those species that stain gram-negative will be identified correctly following inoculation of a biochemical- or enzyme-based ID system.

† Anaerobic, gram-positive bacilli other than *Actinomyces* spp. may also have this morphology, and *Actinomyces* spp. will not always have this morphology.

Note: The terms "*Clostridium*-like" and "*Actinomyces*-like" are intended solely for intralaboratory use. They are of value in preliminary groupings of isolates. Because they are merely descriptive terms, based on Gram stain characteristics alone, they should **never** be used in laboratory reports.

Flow Chart 8-3. PID of *Clostridium*-like,* anaerobic, gram-positive bacilli

Starting point: The organism is a *Clostridium*-like gram-positive bacillus. Check
the BRU/BA PC/SC plate for beta hemolysis.

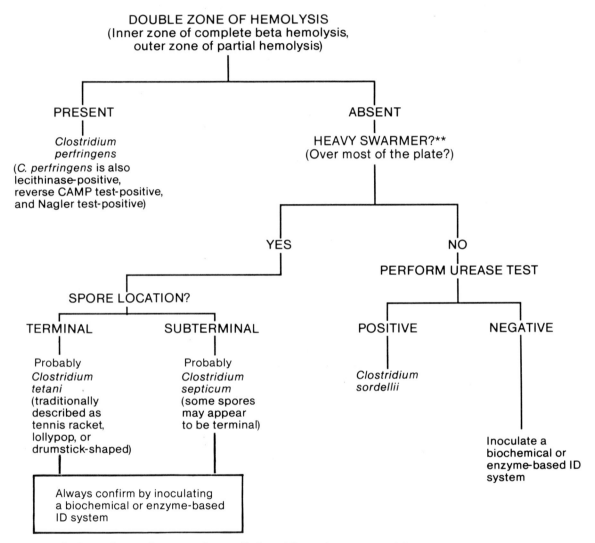

* Not all clostridia are large, and not all clostridia stain gram-positive.

** The extent of swarming is influenced by the moisture content of the medium.

Flow Chart 8-4. PID of anaerobic, gram-positive diphtheroids

Starting point: The organism is a gram-positive diphtheroid. Perform a catalase test. (If removing growth from the BRU/BA PC/SC plate, avoid transferring any of the red cells in the medium because these contain catalase, which could cause a false-positive reaction.)

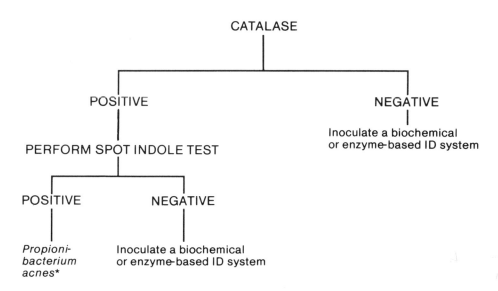

CATALASE

POSITIVE — PERFORM SPOT INDOLE TEST

    POSITIVE — *Propioni-bacterium acnes**

    NEGATIVE — Inoculate a biochemical or enzyme-based ID system

NEGATIVE — Inoculate a biochemical or enzyme-based ID system

* Some strains are catalase-negative and some strains are spot indole-negative. Most strains are nitrate-positive.

Flow Chart 8-5. PID of anaerobic, gram-negative cocci

Starting point: The organism is a gram-negative coccus. Perform a nitrate reduction test by adding appropriate reagents to the nitrate disk on the BRU/BA PC/SC plate (or by performing an alternative nitrate reduction procedure).

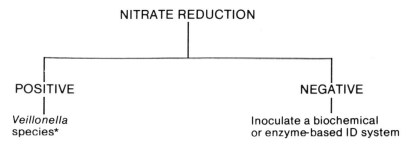

NITRATE REDUCTION

POSITIVE — *Veillonella* species*

NEGATIVE — Inoculate a biochemical or enzyme-based ID system

* Some strains of *Veillonella* spp. are nitrate-negative. These must be identified using a biochemical- or enzyme-based identification system.

Flow Chart 8-6. PID of anaerobic, gram-negative bacilli/coccobacilli

Starting point: The organism is a gram-negative bacillus or coccobacillus. Check the results of the vancomycin disk on the BRU/BA PC/SC plate.

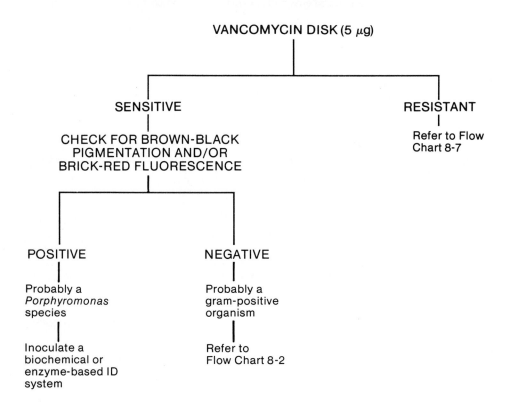

Note: Spot tests of value in rapid presumptive identification of black-pigmented, anaerobic, gram-negative bacilli are described in an article by Moncla et at., J. Clin. Microbiol. 29: 1955-1958, 1991.

Flow Chart 8-7. PID of vancomycin-resistant, anaerobic, gram-negative bacilli/coccobacilli

Starting point: The organism (a gram-negative bacillus or coccobacillus) is vancomycin-resistant. Read the results of the colistin and kanamycin disks on the BRU/BA PC/SC plate.

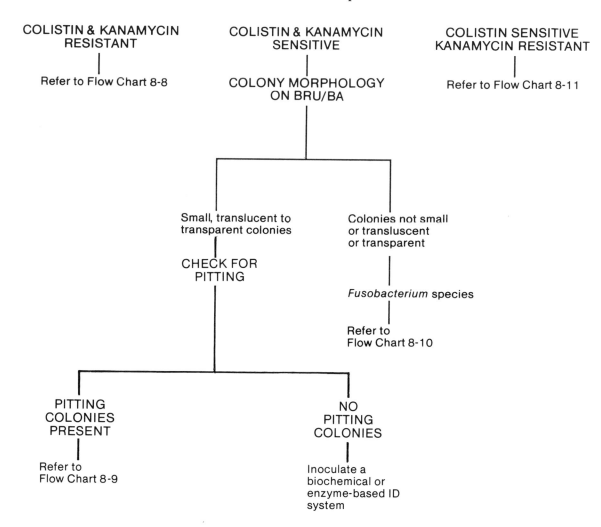

COLISTIN & KANAMYCIN
RESISTANT

Refer to Flow Chart 8-8

COLISTIN & KANAMYCIN
SENSITIVE

COLONY MORPHOLOGY
ON BRU/BA

COLISTIN SENSITIVE
KANAMYCIN RESISTANT

Refer to Flow Chart 8-11

Small, translucent to
transparent colonies

CHECK FOR
PITTING

Colonies not small
or transluscent
or transparent

Fusobacterium species

Refer to
Flow Chart 8-10

PITTING
COLONIES
PRESENT

Refer to
Flow Chart 8-9

NO
PITTING
COLONIES

Inoculate a
biochemical or
enzyme-based ID
system

Flow Chart 8-10. PID of *Fusobacterium* spp.

Starting point: The organism (a gram-negative bacillus) is vancomycin-resistant but susceptible to *both* colistin and kanamycin. Its colonies are *not* small or translucent to transparent. Check the cellular morphology.

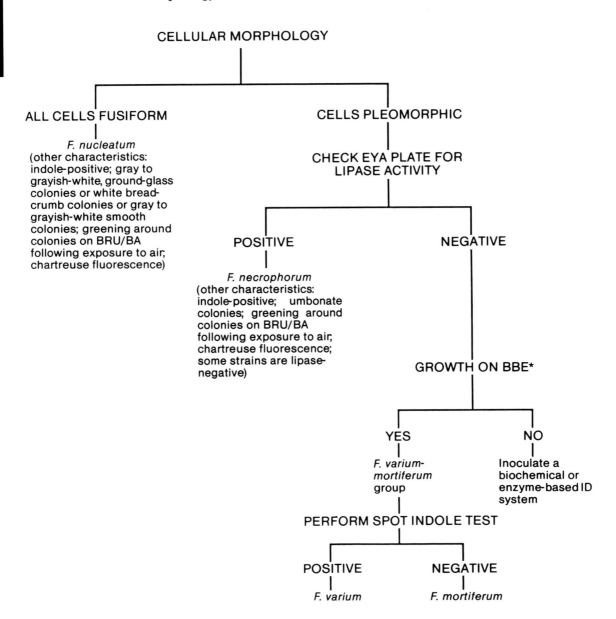

CELLULAR MORPHOLOGY

ALL CELLS FUSIFORM

F. nucleatum
(other characteristics: indole-positive; gray to grayish-white, ground-glass colonies or white bread-crumb colonies or gray to grayish-white smooth colonies; greening around colonies on BRU/BA following exposure to air; chartreuse fluorescence)

CELLS PLEOMORPHIC

CHECK EYA PLATE FOR LIPASE ACTIVITY

POSITIVE

F. necrophorum
(other characteristics: indole-positive; umbonate colonies; greening around colonies on BRU/BA following exposure to air; chartreuse fluorescence; some strains are lipase-negative)

NEGATIVE

GROWTH ON BBE*

YES

F. varium-mortiferum group

NO

Inoculate a biochemical or enzyme-based ID system

PERFORM SPOT INDOLE TEST

POSITIVE

F. varium

NEGATIVE

F. mortiferum

* Did the organism grow on the BBE plate included in the primary isolation setup?

Note: Dzink et al. (Int. J. Syst. Bacteriol. 40:74-78, 1990) described three subspecies of *F. nucleatum* and Shinjo et al. (Int. J. Syst. Bacteriol. 41:395-397, 1991) described two subspecies of *F. necrophorum*.

Flow Chart 8-11. PID of anaerobic, gram-negative bacilli/coccobacilli resistant to vancomycin and kanamycin but susceptible to colistin

Starting point: The organism (a gram-negative bacillus or coccobacillus) is resistant to vancomycin and kanamycin but susceptible to colistin. Check carefully for brown-black pigmentation of colonies on KVLB or BRU/BA.

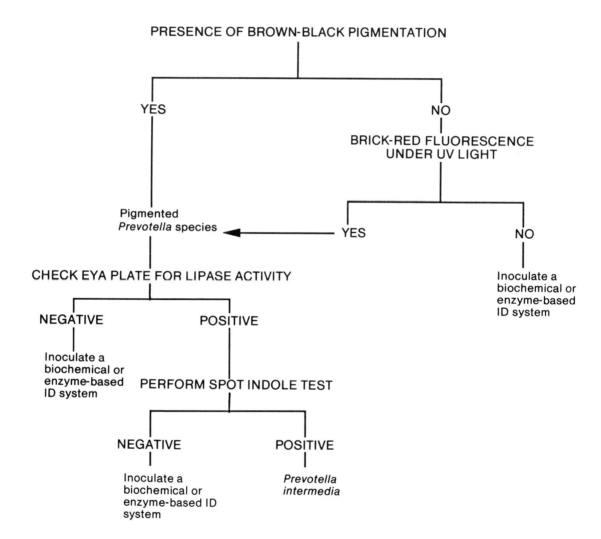

PRESENCE OF BROWN-BLACK PIGMENTATION

YES — NO

BRICK-RED FLUORESCENCE UNDER UV LIGHT

Pigmented *Prevotella* species ← YES — NO → Inoculate a biochemical or enzyme-based ID system

CHECK EYA PLATE FOR LIPASE ACTIVITY

NEGATIVE — POSITIVE

Inoculate a biochemical or enzyme-based ID system

PERFORM SPOT INDOLE TEST

NEGATIVE — POSITIVE

Inoculate a biochemical or enzyme-based ID system

Prevotella intermedia

Note: Spot tests of value in rapid presumptive identification of black-pigmented, anaerobic, gram-negative bacilli are described in an article by Moncla et al., J. Clin. Microbiol. 29:1955-1958, 1991.

Presumpto Plates

Many of the previously described observations and simple tests are incorporated into specialized types of plated media, collectively referred to as **Presumpto plates** (or CDC Differential Agar Media). Three types of plates are available: Presumpto 1, Presumpto 2, and Presumpto 3. Each is a quad plate, thus providing a total of 12 separate quadrants when all three plates are used.

Presumpto 1 contains LD agar, LD-egg yolk agar, LD-esculin agar, and LD-bile agar.[5] A sterile, blank filter paper disk, later used to detect indole production, is added to the LD agar section after inoculation. Although originally used to presumptively identify gram-negative bacilli, Presumpto 1 is also of value in the PID of other anaerobes.

Presumpto 2, originally developed to aid in identifying clostridia, contains LD-DNA agar, LD-glucose agar, LD-milk agar, and LD-starch agar.[11] Presumpto 3 contains LD-gelatin agar and three key carbohydrate media (LD-mannitol agar, LD-lactose agar, and LD-rhamnose agar).[13] Formulations for all Presumpto plate media can be found in the CDC publication *Media for Isolation, Characterization, and Identification of Obligately Anaerobic Bacteria.*[6] Presumpto plates are available from several commercial sources, including Remel and Carr-Scarborough Microbiologicals, Inc.

Presumpto plates were developed before the advent of 4-hour commercial identification systems, which are described in Chapter 9. Like the 4-hour systems, Presumpto plates require an inoculum prepared from a pure culture of the anaerobic isolate, but unlike the 4-hour systems, Presumpto plates require 48 hours of incubation.

As shown in Table 8-5, the three Presumpto plate system enables a user to determine 18 different characteristics of an anaerobe that has been inoculated onto the plates. Reactions of many clinically encountered anaerobes on Presumpto plates can be found in the previously referenced CDC publication and product inserts. Figure 8-5 illustrates various reactions on LD-egg yolk agar.

Figure 8-5. CDC slide showing various reactions on LD-egg yolk agar: (1) lipase reaction, (2) lecithinase reaction, (3) proteolysis reaction, (4) no reaction (courtesy of Suzette L. Bartley, James D. Howard, and Ray Simon, Centers for Disease Control, Atlanta)

Table 8-5. Presumpto Plates

Plate	Quadrant	Observation or Reaction
Presumpto 1	LD* agar	Indole production; growth on LD medium
	LD-esculin agar	Esculin hydrolysis; H$_2$S production; catalase production
	LD-egg yolk agar	Lecithinase; lipase; proteolysis
	LD-bile agar	Bile tolerance/susceptibility (determined by comparing the amount of growth here with the amount of growth on LD agar); precipitate formation
Presumpto 2	LD-DNA	DNAse activity
	LD-glucose agar	Glucose fermentation; stimulation of growth by a fermentable carbohydrate
	LD-milk agar	Casein hydrolysis
	LD-starch agar	Starch hydrolysis
Presumpto 3	LD-mannitol agar	Mannitol fermentation
	LD-lactose agar	Lactose fermentation
	LD-rhamnose agar	Rhamnose fermentation
	LD-gelatin	Gelatin hydrolysis

* LD refers to the names of the individuals who developed these media, George L. Lombard and V. R. Dowell, Jr.

Source: Based on reference 4.

Chapter in Review

- Presumptive identifications (PIDs) of microorganisms are becoming increasingly popular, due primarily to the increased emphasis on cost containment. Bacteria commonly identified in the aerobic bacteriology laboratory using PID criteria include *Pseudomonas aeruginosa, Escherichia coli,* swarming species of *Proteus,* and Group A beta-hemolytic streptococci. PIDs of many clinically encountered anaerobes are also possible. By combining readily observable colony and Gram stain features with results of simple test procedures, even small laboratories are capable of making PIDs of commonly isolated and clinically important anaerobes. However, many anaerobes **cannot** be identified using PID criteria.

- Among the simple observations and inexpensive tests used to presumptively identify anaerobic isolates are growth/no growth on selective media, colony morphology, gram reaction and cell morphology, aerotolerance testing, fluorescence, special potency antimicrobial disks, SPS disk, nitrate disk, bile disk, spot indole test, catalase test, urease test, motility test, and oxidase test.

- Microbiologists should use a logical, stepwise approach to PIDs, always starting with the gram reaction, then asking the next appropriate question, and the next, and so on until the organism can be presumptively identified. For example, if the Gram stain and aerotolerance test results reveal an anaerobic, gram-positive coccus, the next piece of information required is the SPS disk result. If the organism is susceptible to SPS, a PID can be made at that point. However, if the organism is SPS-resistant, the next logical step is to determine the spot indole result. The flow charts in this chapter present a logical, stepwise approach to PID in the format of dichotomous keys.

- Many of the simple observations and tests used to presumptively identify anaerobes are incorporated into specialized plated media, referred to as **Presumpto plates** (or CDC Differential Agar Media). The three Presumpto plates, called Presumpto 1, 2, and 3, collectively provide information on 18 separate characteristics.

Self-Assessment Exercises

1. Presumptively identify the following anaerobes using Table 8-3 and the appropriate flow charts. The first one is worked through for you as an example.

GRAM STAIN:	Somewhat pleomorphic, gram-negative bacilli (Table 8-3 states, "If the organism is a gram-negative bacillus or coccobacillus, start with Flow Chart 8-6.")
VANC DISK:	Resistant (Flow Chart 8-6 indicates that if the organism is resistant to vancomycin, refer to Flow Chart 8-7.)
COL DISK:	Resistant
KANA DISK:	Resistant (Flow Chart 8-7 indicates that if the organism is colistin- and kanamycin-resistant, refer to Flow Chart 8-8.)
BRU/BA:	Good growth; shiny, grayish, translucent colonies; slight beta-hemolysis; no evidence of brown-black colonies; no fluorescence
BBE:	Good growth; gray colonies; browning of medium; stippling in medium near heavy growth (Flow Chart 8-8 indicates that good growth on BBE agar, browning of the medium, and stippling in the medium is most likely to be *Bacteroides fragilis*, but could be *B. ovatus*.)
KVLB:	Good growth; shiny, grayish, translucent colonies; no evidence of brown-black colonies; no fluorescence

PEA:	Same as BRU/BA
EYA:	Good growth; no lecithinase or lipase reaction
CATALASE:	Positive (although some strains are catalase-negative)
SPOT INDOLE:	Negative (rules out *B. ovatus*; thus, PID is *B. fragilis*)

Presumptive identification: *Bacteroides fragilis.* The major clues are good growth on BBE agar and browning of the medium, which together provide presumptive evidence of a member of the *B. fragilis* group. The only members of this group that cause stippling (bile precipitation) in the medium are *B. fragilis* (indole-negative) and *B. ovatus* (indole-positive). However, bile precipitation is strain specific and a relatively rare event.

a.

GRAM STAIN:	Very pleomorphic, gram-negative bacilli
VANC DISK:	Resistant
COL DISK:	Sensitive
KANA DISK:	Sensitive
BRU/BA:	Good growth; circular, low convex, translucent colonies with a smooth consistency and erose edges; greening around colonies following exposure to air; chartreuse fluorescence
BBE:	Very small, cream-colored colonies; no browning of the medium
KVLB:	No growth
PEA:	Same as BRU/BA
EYA:	Good growth; positive lipase reaction

Presumptive identification: _____

b.

GRAM STAIN:	Short, rectangular-shaped, gram-positive bacilli with blunt ends; some pink-staining cells; no spores observed
VANC DISK:	N/A (used only for gram-negative bacilli)
COL DISK:	N/A (used only for gram-negative bacilli)

KANA DISK:	N/A (used only for gram-negative bacilli)
BRU/BA:	Good growth; low convex, semiopaque, glossy colonies with raised centers, flattened peripheries, and entire margins; double zone of hemolysis (inner zone clear, outer zone partial)
BBE:	No growth
KVLB:	No growth
PEA:	Same as BRU/BA
EYA:	Good growth; positive lecithinase reaction

Presumptive identification: _____

c.
GRAM STAIN:	Gram-positive diphtheroid; some metachromatic granules; no spores observed
VANC DISK:	N/A (used only for gram-negative bacilli)
COL DISK:	N/A (used only for gram-negative bacilli)
KANA DISK:	N/A (used only for gram-negative bacilli)
BRU/BA:	Good growth; small, glistening, white, opaque colonies
BBE:	No growth
KVLB:	No growth
PEA:	Same as BRU/BA
EYA:	Good growth; no lecithinase or lipase reaction (although some strains are lipase-positive)
CATALASE:	Positive
SPOT INDOLE:	Positive

Presumptive identification: _____

d.
GRAM STAIN:	Tiny, gram-negative cocci and diplococci in clusters and short chains
VANC DISK:	N/A (used only for gram-negative bacilli)

COL DISK: N/A (used only for gram-negative bacilli)

KANA DISK: N/A (used only for gram-negative bacilli)

BRU/BA: Good growth; grayish-white, opaque, butyrous
                    colonies; red fluorescence

BBE: No growth

KVLB: No growth

PEA: Same as BRU/BA

EYA: Was not inoculated

NITRATE: Positive

Presumptive identification: _____

e.   GRAM STAIN: Gram-negative coccobacilli

     VANC DISK: Resistant

     COL DISK: Sensitive

     KANA DISK: Resistant

     BRU/BA: Good growth; shiny, convex, gray colonies;
                         brick-red fluorescence; no hemolysis

     BBE: No growth

     KVLB: Same as BRU/BA except some colonies have light
                         brown edges; brick-red fluorescence

     PEA: Same as BRU/BA

     EYA: Good growth; positive lipase reaction

     SPOT INDOLE: Positive

     Presumptive identification: _____

f.   GRAM STAIN: Elongated, gram-positive cocci in chains

     VANC DISK: N/A (used only for gram-negative bacilli)

COL DISK:	N/A (used only for gram-negative bacilli)
KANA DISK:	N/A (used only for gram-negative bacilli)
SPS DISK:	Sensitive to SPS
BRU/BA:	Good growth; small, glistening, translucent, creamy white to gray colonies
BBE:	No growth
KVLB:	No growth
PEA:	Same as BRU/BA
EYA:	Was not inoculated

Presumptive identification: _____

g.
GRAM STAIN:	Long, thin, gram-negative bacilli that are tapered at the ends
VANC DISK:	Resistant
COL DISK:	Sensitive
KANA DISK:	Sensitive
BRU/BA:	Good growth; low convex, grayish, translucent, glossy, ground-glass colonies; greening around colonies following exposure to air; chartreuse fluorescence
BBE:	No growth
KVLB:	No growth
PEA:	Same as BRU/BA
EYA:	Good growth; no lecithinase or lipase reaction

Presumptive identification: _____

h.
GRAM STAIN:	Gram-positive cocci in chains
VANC DISK:	N/A (used only for gram-negative bacilli)

COL DISK:	N/A (used only for gram-negative bacilli)
KANA DISK:	N/A (used only for gram-negative bacilli)
SPS DISK:	Resistant to SPS
BRU/BA:	Good growth; small, circular, low convex, smooth, white to grayish colonies
BBE:	No growth
KVLB:	No growth
PEA:	Same as BRU/BA
EYA:	Was not inoculated
SPOT INDOLE:	Positive

Presumptive identification: _____

i.	GRAM STAIN:	Very pleomorphic, gram-negative bacilli, including large, spherical bodies
	VANC DISK:	Resistant
	COL DISK:	Sensitive
	KANA DISK:	Sensitive
	BRU/BA:	Good growth; low convex, grayish, translucent, glossy colonies
	BBE:	Very small, grayish colonies; no browning of the medium
	KVLB:	No growth
	PEA:	Same as BRU/BA
	EYA:	Good growth; no lecithinase or lipase reaction

Presumptive identification: _____

2. Differentiate between the terms "presumptive identification" and "definitive identification."

   _____

   _____

   _____

   _____

3. List five colony observations/characteristics that are of value in making presumptive identifications of anaerobes.

   a. _____

   b. _____

   c. _____

   d. _____

   e. _____

4. List ten simple test results or reactions that are of value in making presumptive identifications of anaerobes.

   a. _____

   b. _____

   c. _____

   d. _____

   e. _____

   f. _____

   g. _____

   h. _____

   i. _____

   j. _____

5. When used in conjunction with other characteristics, a long-wave (366 nm) ultraviolet light source (a Wood's lamp) is useful in the anaerobic bacteriology laboratory for presumptive identification of certain anaerobic isolates. Answer the following questions about fluorescence.

    a.   An anaerobic, gram-positive bacillus is producing large, ground-glass colonies on CCFA, is turning the usually pink medium a yellow color, and has a "horse stable" odor. What color fluorescence would you expect if a BRU/BA plate containing the same organism was held under UV light?

         _____

    b.   An anaerobic, gram-negative coccobacillus is growing well on KVLB agar, and some of the colonies have started to turn brown-black. What color fluorescence would you expect if the KVLB plate was held under UV light?

         _____

    c.   An anaerobe that produces grayish-white, bread crumb colonies on BRU/BA appears as fusiform, gram-negative bacilli on Gram stain. What color fluorescence would you expect if the BRU/BA plate was held under UV light?

         _____

    d.   A nitrate-positive anaerobe is growing well on BRU/BA and appears as tiny, gram-negative cocci and diplococci on Gram stain. What color fluorescence would you expect if the BRU/BA plate was held under UV light?

         _____

6. TRUE or FALSE: In some laboratories, it is not necessary to definitively identify every anaerobic isolate.

7. Coccobacillary, gram-negative bacilli were seen on the Gram-stained smear of a positive blood culture. The contents of the blood culture bottle were inoculated to four different plates. The organism did not grow on a CO_2-incubated chocolate agar plate. Although the organism grew well on an anaerobically incubated BRU/BA plate, it did not grow at all on the anaerobically incubated KVLB and BBE plates. A Gram stain of a BRU/BA plate colony confirmed that the organism was a gram-negative coccobacillus, and aerotolerance testing demonstrated that it was

truly an anaerobe. Young, nonpigmented colonies on the BRU/BA plate fluoresced brick-red under UV light; and, with age, the colonies became brown-black. This anaerobic, gram-negative coccobacillus was presumptively identified as a pigmented *Porphyromonas* sp. Indicate whether the following statements are TRUE or FALSE.

a.   There is insufficient information to presumptively identify this organism as a *Porphyromonas* species. _____

b.   The Gram stain results are inconsistent with a PID of a *Porphyromonas* species. _____

c.   The fluorescence and pigmentation results are consistent with a PID of a *Porphyromonas* species. _____

d.   The primary purpose for using a KVLB plate is to enhance the brown-black pigmentation of pigmented species of *Prevotella*.

_____

e.   A possible explanation for failure of the organism to grow on the KVLB plate is that it is a vancomycin-sensitive strain of *Porphyromonas*.

_____

8.   Presumptively identify (to genus and species) the following anaerobes **without** referring to the flow charts. Use the flow charts only if you are unable to identify the organisms without them.

a.   Anaerobic, gram-positive diphtheroids have been isolated from several blood culture bottles of a patient suspected of having subacute bacterial endocarditis. Simple test procedures have demonstrated that the organism is catalase-positive, spot indole–positive, and nitrate-positive.

Presumptive identification: _____

b.   Very pleomorphic, anaerobic, gram-negative bacilli have been recovered from an aspirate that was carefully collected from a liver abscess. Special potency antimicrobial disks have demonstrated that the organism is vancomycin-resistant, colistin-sensitive, and kanamycin-sensitive. Simple test procedures have shown that the organism is spot indole–positive, lecithinase-negative, and lipase-positive.

Presumptive identification: _____

c. Anaerobic, gram-positive cocci have been recovered from a polymicrobic infectious process that developed following GU surgery. Simple test procedures have demonstrated that the organism is catalase-negative, SPS-resistant, spot indole–positive, and nitrate-negative.

Presumptive identification: _____

d. Large, rectangular, anaerobic, gram-positive bacilli have been recovered from a wound resulting from puncture of the patient's foot by a rusty nail. No spores were observed in the Gram-stained preparations of either the specimen or the pure culture. A double zone of hemolysis was seen on the BRU/BA subculture plate. The organism was lecithinase-positive and lipase-negative.

Presumptive identification: _____

e. Somewhat pleomorphic, anaerobic, gram-negative bacilli were recovered from a brain abscess that developed subsequent to GI surgery. The organism grew well on BBE, turned the medium brown, and produced a stippling in the medium around areas of heavy growth. The organism grew well on KVLB, but the colonies were nonpigmented and did not fluoresce under UV light. The organism was catalase-positive and indole-negative.

Presumptive identification: _____

f. An anaerobic, gram-positive coccus is recovered in pure culture from material that was carefully aspirated from a cervical abscess. The organism is catalase-negative, SPS-sensitive, spot indole–negative, and nitrate-negative.

Presumptive identification: _____

References

1. Bartley, S. L. (Centers for Disease Control). 1990. Personal communication.
2. Bourgault, A-M. , and F. Lamothe. 1988. Evaluation of the KOH test and the antibiotic disk test in routine clinical anaerobic bacteriology. J. Clin. Microbiol. 26:2144–2146.
3. Brazier, J. S., and T. V. Riley. 1988. UV red fluorescence of *Veillonella* spp. J. Clin. Microbiol. 26:383–384.
4. Dowell, V. R., Jr., and G. L. Jones. 1981. Procedures for use of differential agar media in the identification of anaerobic bacteria. Centers for Disease Control, Atlanta.
5. Dowell, V. R., Jr., and G. L. Lombard. 1977. Presumptive identification of anaerobic nonsporeforming gram-negative bacilli. Centers for Disease Control, Atlanta.

6. Dowell, V. R., Jr., G. L. Lombard, F. S. Thompson, and A. Y. Armfield. 1977. Media for isolation, characterization, and identification of obligately anaerobic bacteria. Centers for Disease Control, Atlanta.

7. Lindberg, L. H. (San Jose State University). 1990. Personal communication.

8. Lombard, G. L., and V. R. Dowell, Jr. 1983. Comparison of three reagents for detecting indole production by anaerobic bacteria in microtest systems. J. Clin. Microbiol. 18:609–613.

9. Mills, C. K., B. Y. Grimes, and R. L. Gherna. 1987. Three rapid methods compared with a conventional method for detection of urease production in anaerobic bacteria. J. Clin. Microbiol. 25:2209–2210.

10. Slots, J., and H. S. Reynolds. 1982. Long-wave UV light fluorescence for identification of black-pigmented *Bacteroides* spp. J. Clin. Microbiol. 16:1148–1151.

11. Story, S., and V. R. Dowell, Jr. 1978. Development of a presumpto plate for identification of clostridia. Abstract C24, Annual Meeting of the American Society for Microbiology, Las Vegas.

12. Sutter, V. L., D. M. Citron, M. A. C. Edelstein, and S. M. Finegold. 1985. Wadsworth Anaerobic Bacteriology Manual, 4th ed. Star Publishing, Belmont, CA.

13. Wanderlinder, L., D. N. Whaley, and V. R. Dowell, Jr. 1980. Abstract, Annual Meeting of the American Society for Microbiology, Miami Beach.

14. Weinberg, L. G., L. L. Smith, and A. H. McTighe. 1983. Rapid identification of the *Bacteroides fragilis* group by bile disk and catalase tests. Lab. Med. 14:785–788.

DEFINITIVE IDENTIFICATIONS

At the conclusion of this chapter, you will be able to:

1. Explain what is meant by a "conventional" or "traditional" system for definitive identification of anaerobes.
2. Identify two advantages and one disadvantage of biochemical-based minisystems for anaerobe identification, when compared to a "conventional" identification system.
3. Describe four advantages and one disadvantage of preexisting enzyme–based minisystems for anaerobe identification, when compared to a "conventional" identification system.
4. Identify two advantages of preexisting enzyme–based minisystems for anaerobe identification, when compared to biochemical-based minisystems.
5. State three components of the "Hungate technique" or "VPI system" for isolation and identification of anaerobes.
6. Explain the principles of gas-liquid chromatography, including the meaning of "mobile phase" and "stationary phase."
7. Describe two ways in which gas-liquid chromatography can be used to identify anaerobes.
8. List five genera of anaerobes that contain flagellated species.
9. Name two desulfoviridin-producing genera of anaerobes.
10. Briefly discuss laboratory confirmation of the following clostridial diseases: botulism, tetanus, *C. perfringens* food poisoning, *C. difficile*–induced diarrhea.

Definitions

Carboxylic acid: An organic acid containing only carbon, hydrogen, and oxygen and consisting of a carboxyl group (-COOH) attached to either an alkyl group (R-) or an aryl group (Ar-)

Chromatography: A general term referring to a technique used to separate complex mixtures on the basis of different physical and/or chemical characteristics. Specific examples include thin-layer, paper, ion-exchange, liquid-solid (adsorption), liquid-liquid (partition), gas-solid, and gas-liquid chromatography.

Desulfoviridin: An enzyme (sulfite reductase) produced by certain sulfate-reducing organisms (e.g., *Desulfovibrio* spp., *Desulfomonas* spp., and *Bilophila wadsworthia*)

Fatty acid: Any of a group of carboxylic acids made up of an alkyl group attached to a carboxyl group and readily obtained by acid, alkaline, or enzyme-catalyzed hydrolysis of fats. Saturated fatty acids have the general formula $C_nH_{2n}O_2$. Unsaturated fatty acids may have one or more double bonds. The prefix "iso-" is used to designate branched-chain fatty acids having a single methyl group branch at the end of the molecule most distant from the carboxyl group (examples include isobutyric, isovaleric, and isocaproic acids).

Flagellum: A long, mobile, whiplike projection from the free surface of a cell that serves as a locomotor organelle; composed of tightly wound chains of strands containing a protein called flagellin

Gas-liquid chromatography: A type of gas chromatography in which a gas serves as the mobile phase and a nonvolatile liquid serves as the stationary phase

Risus sardonicus: A distorted, grinning expression observed in patients with tetanus, caused by sustained contraction of facial muscles
Trismus: Motor disturbance of the trigeminal nerve causing spasm of the masticatory (chewing) muscles and difficulty in opening the mouth (i.e., "lockjaw")

Chapter 8 described presumptive identifications (PIDs) of some of the anaerobic bacteria most commonly encountered in clinical specimens. Many anaerobic isolates cannot be identified using PID techniques. A wide variety of techniques exist for making definitive identifications. Examples include prereduced, anaerobically sterilized (PRAS) and non-PRAS tubed biochemical test media, biochemical-based minisystems, preexisting enzyme–based minisystems, gas-liquid chromatographic analysis of metabolic end products, and cellular fatty acid analysis by gas-liquid chromatography (GLC). These approaches and additional techniques useful in making definitive identifications (e.g., flagella staining and the desulfoviridin test) are described in this chapter.

None of the commercially available biochemical-based or prexisting enzyme–based minisystems will identify **all** the anaerobes that can potentially be isolated from clinical specimens. Many of the rapid minisystems are designed to identify those most frequently encountered in clinical specimens and those most apt to be contributing to the infectious process. Thus, when reviewing published evaluations of commercial systems, pay particular attention to their ability to identify those specific anaerobes. The anaerobes of primary importance are members of the *Bacteroides fragilis* group, pigmented species of *Porphyromonas* and *Prevotella*, *C. perfringens*, *F. nucleatum*, and the anaerobic cocci.[13] It is far more important for a system to identify **these** organisms than obscure anaerobes only rarely isolated from clinical specimens and/or involved in infectious processes.

Regardless of the rapid minisystem used, you are likely to encounter anaerobes that can be identified only by using highly sophisticated techniques employing many tubes of PRAS or non-PRAS media and/or GLC. The decision to employ time-consuming, labor-intensive procedures to identify anaerobes of questionable clinical significance is one that must be weighed very carefully. In view of the current emphasis on cost containment in clinical laboratories, PIDs might suffice—especially when susceptibility test results are also furnished. Rarely isolated organisms that are difficult and/or time-consuming to identify might best be sent to reference laboratories. Chapter 12 contains additional information regarding cost considerations in the use of presumptive versus definitive identification procedures.

The use of tubed biochemical systems and GLC procedures is often necessary at the reference laboratory level because these laboratories frequently receive isolates that are deemed clinically significant but cannot be identified at the hospital level. Veterinary laboratories may also require tubed media and GLC to identify unique veterinary isolates not contained in the data bases of the rapid minisystems.

Definitive Identification of Anaerobic Isolates

"Conventional" Tubed Biochemical Identification Systems

Much of our knowledge of endogenous anaerobes stems from pioneering studies of rumen contents and sewage sludge by R. E. Hungate in the 1950s and 1960s. Major components of the "Hungate technique" are oxygen-free, flame-sterilized gassing jets or cannulae (illustrated in Figure 9-1), agar-containing media on the inner surfaces of tightly stoppered glass tubes ("roll tubes"), and a variety of tightly stoppered test tubes containing PRAS biochemicals. Roll tubes are inoculated under flowing oxygen-free gas, and the tightly stoppered, inoculated tubes serve as their own anaerobic culture chambers. A modified Hungate technique (sometimes referred to as the "VPI system") has been taught for many years at the Virginia Polytechnic Institute in Blacksburg, and detailed procedures for preparation and use of roll tubes and other aspects of this technique can be found in the *VPI Anaerobe Laboratory Manual.*[21] Although the VPI system is ideal for certain types of research and has been adopted for use in some reference laboratories, it does not lend itself to the volume, pace, and cost-conscious environment of today's hospital microbiology laboratories. Some authorities have described the system as time-consuming, labor-intensive, and expensive.[9, 33] Others have indicated that "roll-tube techniques are not as practical for clinical work" as other available methods[2], perhaps in part due to "delayed reporting of results to physicians."[9] Nevertheless, the VPI system of PRAS tubed biochemical test media remains one of the "gold standard" techniques to which newer methodologies are compared whenever they are evaluated.

In general, "conventional" or "traditional" systems for the identification of anaerobic isolates employ several to many large test tubes containing PRAS or non-PRAS biochemical test media. Such tubed media can either be prepared in the laboratory or purchased from commercial sources (e.g., Carr-Scarborough, Scott, Remel). Tests include, but are not limited to, a variety of carbohydrate fermentations (e.g., arabinose, glucose, lactose, maltose, mannitol, rhamnose, salicin, sucrose, trehalose, xylose), 20% bile, catalase, esculin hydrolysis, gas production, gelatin hydrolysis, H_2S production, indole production, iron milk (proteolysis), motility, nitrate reduction, and urease. Detailed information about these systems can be found in manuals available from the Centers for Disease Control or the Virginia Polytechnic Institute.[11, 21] The PRAS system is said to be "expensive" and "labor-intensive," with problems that include "uninterpretable biochemical results," a need for repeat testing "about 10–20% of the time," and incubation periods "for as long as one week" to obtain sufficient growth.[12]

Examples of the types of charts used in conjunction with the non-PRAS tubed biochemical identification system employed at the Centers for Disease Control are depicted in Tables 9-1 through 9-8. Such charts should be relied upon **only** when the test media being used are prepared in exactly the same manner as the media used to obtain the data for the charts. For example, because of possible differences in formulations of test media, utilizing these charts in conjunction with some of the newer, rapid minisystems could lead to misidentifications.

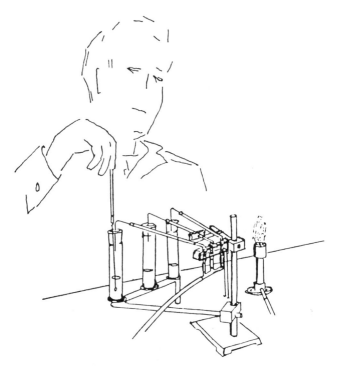

Figure 9-1. Three-place, swing-type tube and cannula holder of the VPI® system (from reference 21, used with the permission of Drs. L. V. H. and W. E. C. Moore, Department of Anaerobic Bacteriology, Virginia Polytechnic Institute). The permanently mounted cannulae deliver oxygen-free CO_2 or N_2 to the culture tubes. The cannulae are sterilized by being moved laterally into a bunsen burner flame using foot treadles, leaving the microbiologist's hands free to manipulate the cultures. A similar device and accessories are available from Bellco Glass, Inc. (Vineland, NJ).

Table 9-1

Table 9-1. Characteristics of Some Clinically Encountered Clostridia

Species	Relationship to Oxygen	Swarming on BRU/BA	Double Zone of Hemolysis	Chartreuse Fluoresence	Gram Reaction	Position of Spores	Motility	Flagella	Indole	Indole Derivatives	Esculin Hydrolysis	Lecithinase	Lipase	Proteolysis in Milk	Gelatin Hydrolysis	DNase	Glucose	Lactose	Mannitol	Rhamnose	Urease	Major Acids Produced in PYG	Species
C. bifermentans	MA, AN	−	−	−	+	ST	+	PF	+	−	−	+	−	+	−	−	+	−	−	−	−	A	C. bifermentans
C. clostridioforme	AN	−	−	−	−+	ST	−+	PF	−	−	+	−	−	−	−	−	+	+−	−	+−	−	A	C. clostridiiforme
C. difficile	AN	−	−	+	+	ST	+	SSF^{PF}	−	−	+−	−	−	−	−	−+	−+	+	−	+	−	A, IB, B, IV, V, IC	C. difficile
C. novyi type A	AN	−	−	−	+	ST	+−	PF^{NF}	−	−	−	+	+	−	+	NG	+−	−	−	−	−	A, P, B	C. novyi type A
C. perfringens	MA, AN	−	+	−	+	−ST	−		−	−	V	+	−	+	+	+	+	+	−	−	−	A, B	C. perfringens
C. ramosum	AN	−	−	−	−	(T)	−	−	−	−	+	−	−	−	−	−	+	+	+−V	−		A	C. ramosum
C. septicum	MA, AN	+	−	−	+	ST	+	PF	−	−	+	−	−	−	+	+	+	+	−	−	−	A, B	C. septicum
C. sordellii	MA, AN	−	−	−	+	ST	+	PF	+	−	−	+	−	+	−	−	+	−	−	−	+	A	C. sordellii
C. sphenoides	AN	−	−	−	−	TST	+	PF	+	−	−	−	−	V	V	+	+	+	+	+	−	A	C. sphenoides
C. sporogenes	AN	−	−	−	+	ST	+	PF	−	+	+	−+	+	+	−+	−+	+	−	−	−	−	A, IB, B, IV	C. sporogenes
C. tertium	FA, AT	−	−	−	+−	T	+	PF	−	−	+	−	−	−	−	+	+	+	+	−	−	A, B	C. tertium
C. tetani	AN	+	−	−	+	T	+	PF	+−	−	−	−	−	−	+	+	−	−	−	−	−	A, P, B	C. tetani

Key:

AN	obligate anaerobe	NF	no flagella	P	propionic acid
MA	microaerotolerant anaerobe	SSF	single subpolar flagellum	IB	isobutyric acid
AT	aerotolerant anaerobe	PF	peritrichous flagella	B	butyric acid
FA	facultative anaerobe	V or ( )	variable	IV	isovaleric acid
ST	subterminal	NG	no growth	V	valeric acid
T	terminal	A	acetic acid	IC	isocaproic acid

Source: Based on references 5 and 10.

Table 9-2. Characteristics of Some Clinically Encountered, Nonsporeforming, Anaerobic, Gram-Positive Bacilli

Species	Relationship to Oxygen	48-hr Colony <1mm Diam	Red Pigment	Beta-Hemolysis	Rough Colonies	Branched Rods	Catalase	Indole	Gelatin Hydrolysis	Proteolysis in Milk	Glucose	Lactose	Mannitol	Rhamnose	Major Acid Products in PYG	Species
Actinomyces spp.																*Actinomyces* spp.
A. israelii	FA, AT, MA, AN	+	−	−	+	+	−	−	−	−	+	+⁻	+⁻	+⁻	A, L, S	*A. israelii*
A. meyeri	FA, AT, MA, AN	−⁺	−	−	−	+	−	−	−	−	+	−⁺	−	−	A, L, S	*A. meyeri*
A. naeslundii	FA, AT, MA, AN	−⁺	−	−	−⁺	+	−	−	−	−	+	−⁺	−	−	A, L, S	*A. naeslundii*
A. odontolyticus	FA, AT, MA, AN	−	+	−	−	+	−	−	−	−	+	−⁺	−	−	A, L, S	*A. odontolyticus*
A. pyogenes	FA, AT	−	−	+	−	−	−	−	+	+	+	−	−	−	A, L, S	*A. pyogenes*
A. viscosus	FA, AT, MA, AN	−	−	−	−⁺	+	+	−	−	−	+	+⁻	−	−	A, L, S	*A. viscosus*
Bifidobacterium spp.																*Bifidobacterium* spp.
B. dentium	FA, AT, MA, AN	−	−	−	−	−	−	−	−	−	+	+	−	−	A, L	*B. dentium*
Eubacterium spp.																*Eubacterium* spp.
E. alactolyticum	AN	+	−	−	−	−	−	−	−	−	V	−	−	−	A, B, C	*E. alactolyticum*
E. lentum	AN	−	−	−	−	−	V	−	−	−	−	−	−	−	(A)	*E. lentum*
E. limosum	AN	−	−	−	−	−	−	−	−	−	+	−	+	−	A, B	*E. limosum*
Propionibacterium spp.																*Propionibacterium* spp.
P. acnes	FA, AT, MA, AN	−	−	−	−	−	+	+	+	+	+	−	−⁺	−	A, P, (L), (S)	*P. acnes*
P. propionicus	FA, AT, MA, AN	+	−	−	+	+	−	−	−	−	+	+⁻	−	−	A, P, L	*P. propionicus*

Key:
AN	obligate anaerobe	FA	facultative anaerobe	P	propionic acid	L	lactic acid
MA	microaerotolerant anaerobe	V or ( )	variable	B	butyric acid	S	succinic acid
AT	aerotolerant anaerobe	A	acetic acid	C	caproic acid		

Source: Based on references 5 and 10.

Table 9-2

Table 9-3

188 *Definitive Identifications*

Table 9-3. Characteristics of Some Clinically Encountered, Nonmotile, Anaerobic, Gram-Negative Bacilli*

Species	Brown-Black Pigmentation	Red Fluorescence	Pitted Colonies	Colonies ≤ 1mm Diam	Catalase	Oxidase	Indole	Inhibited by Bile	Esculin Hydrolysis	Lipase	DNase	Glucose	Lactose	Gelatin Hydrolysis	Proteolysis in Milk	Urease	Rifampin (15 µg)	Acids Produced in PYG Butyric	Succinic	Phenylacetic	Species
Bacteroides spp.																					*Bacteroides* spp.
B. gracilis	−	−	$-^{+}$	+	−	−	−	+	−	−	−	−	−	−	−	−	S	−	+	−	B. gracilis
B. ureolyticus	−	−	$+^{-}$	+	−	+	−	+	−	−	−	−	−	−	−	+	S	−	+	−	B. ureolyticus
Fusobacterium spp.																					*Fusobacterium* spp.
F. mortiferum	−	−	−	−	−	−	−	−	+	−	−	+	$+^{-}$	−	−	−	R	+	$-^{+}$	−	F. mortiferum
F. necrophorum	−	−	−	−	−	−	+	+	−	+	−	+	−	$-^{+}$	V	−	S	+	−	−	F. necrophorum
F. nucleatum	−	−	−	−	−	−	+	+	−	−	−	−	−	−	−	−	S	+	−	−	F. nucleatum
F. varium	−	−	−	−	−	−	$+^{-}$	−	−	−	−	+	−	−	−	−	R	+	−	−	F. varium
Porphyromonas spp.																					*Porphyromonas* spp.
P. asaccharolytica	+	+	−	−	−	−	+	+	−	−	−	−	−	−	+	+	−	S	+	+	−	P. asaccharolytica
P. endodontalis	+	$+^{-}$	−	−	−	−	+	+	−	−	−	−	−	−	+	+	−	S	+	+	−	P. endodontalis
P. gingivalis	+	−	−	−	−	−	+	+	−	−	−	−	−	−	+	+	−	S	+	+	+	P. gingivalis
Prevotella spp.																					*Prevotella* spp.
P. intermedia	+	+	−	−	−	−	+	+	−	+	+	+	−	+	+	−	S	−	+	−	P. intermedia

Key:
V   variable
R   resistant
S   sensitive

* *Bacteroides* (excluding the *B. fragilis* group), *Fusobacterium*, *Porphyromonas*, and *Prevotella* spp.

Source: Based on references 5 and 10.

Table 9-4. Characteristics of Some Clinically Encountered, Bile-tolerant, Saccharolytic *Bacteroides* Species (the *B. Fragilis* Group)

Species	Catalase	Indole	DNase	Penicillin (2 U Disk)	Rifampin (15 μg Disk)	Kanamycin (1 mg Disk)	Arabinose	Cellobiose	Mannitol	Rhamnose	Salicin	Trehalose	Butyric	Succinic	Species
B. caccae	−	−	+	R	S	R	+	+⁻	−	V	+	+	−	+	*B. caccae*
B. distasonis	+⁻	−	−	R	S	R	−	−	−	+	+	+	−	+	*B. distasonis*
B. fragilis	+⁻	−	−	R	S	R	−	−	−	−	−	−	−	+	*B. fragilis*
B. merdae	−	−	−	R	S	R	−	+	−	+	+	+	−	+	*B. merdae*
B. ovatus	+	+	+	R	S	R	+	+	+	+	+	+	−	+	*B. ovatus*
B. stercoris	−	+	+	R	S	R	−⁺	−	−	+	−	−	−	+	*B. stercoris*
B. thetaiotaomicron	+	+	+	R	S	R	+	+	−⁺	+	−⁺	+	−	+	*B. thetaiotaomicron*
B. uniformis	−	+	+	R	S	R	+	+	−	−⁺	+⁻	−	−	+	*B. uniformis*
B. vulgatus	−⁺	−	−	R	S	R	+	−	−	+	−	−	−	+	*B. vulgatus*

Key:
V   variable
R   resistant
S   sensitive

Source: Based on references 5 and 10.

Table 9-4

Table 9-5. Differentiation of Some Clinically Encountered, Nonmotile, Asaccharolytic, Anaerobic, Gram-Negative Bacilli

Species	Brown-Black Pigmentation	Pitted Colonies	Oxidase	Catalase	Indole	Desulfoviridin	Urease	Gelatin Hydrolysis	H$_2$S in TSI	Phenylacetic Acid in PYG	Species
Bacteroides spp.											*Bacteroides* spp.
B. gracilis	−	−	−	−	−	−	−	−	+	−	B. gracilis
B. putredinis	−	−	−	−	+	−	−	+	−	−	B. putredinis
B. ureolyticus	−	+⁻	+	−	−	−	+	−	+	−	B. ureolyticus
Bilophila spp.											*Bilophila* spp.
B. wadsworthia	−	−	−	+	−	+	+⁻	−	−	−	B. wadsworthia
Desulfomonas spp.											*Desulfomonas* spp.
D. pigra	−	−	−	+⁻	−	+	−	−	+	−	D. pigra
Porphyromonas spp.											*Porphyromonas* spp.
P. asaccharolytica	+	−	−	−	+	−	−	+	−	−	P. asaccharolytica
P. endodontalis	+	−	−	−	+	−	−	+	−	−	P. endodontalis
P. gingivalis	+	−	−	−	+	−	−	+	−	+	P. gingivalis

Source: Based on references 5 and 10.

Table 9-5

Table 9-6. Differentiation of Some Clinically Encountered, Motile, Asaccharo-lytic, Anaerobic, Gram-Negative Bacilli

Species	Relationship to Oxygen	Curved Rods	Catalase	Oxidase	Indole	Desulfoviridin	Penicillin (2 U Disks)	Kanamycin (1 mg Disk)	Species
Campylobacter spp.									*Campylobacter* spp.
C. *concisus*	MA, AN	+	−	+	−	−	R	R	C. *concisus*
C. *mucosalis*	AN	+	−	+	−	−	R	V	C. *mucosalis*
C. *sputorum* subsp. *sputorum*	MA	+	−	+	−	−	S	S	C. *sputorum* subsp. *sputorum*
Desulfovibrio spp.									*Desulfovibrio* spp.
D. *desulfuricans*	AN	+	+⁻	−	−	+	V	V	D. *desulfuricans*
Wolinella spp.									*Wolinella* spp.
W. *curva*	MA, AN	+	−	+⁻	−	−	S	S	W. *curva*
W. *recta*	AN	−	−	+⁻	−	−	S	S	W. *recta*

Key:
AN   obligate anaerobe
MA   microaerotolerant anaerobe
V   variable
R   resistant
S   sensitive

Source: Based on references 5 and 10.

Table 9-6

Table 9-7. Differentiation of Some Clinically Encountered, Motile, Saccharolytic, Anaerobic, Gram-Negative Bacilli

Species	Relationship to Oxygen	Spreading on BRU/BA	Oxidase	Cellular Morphology	Flagella	Glucose	Lactose	Mannitol	Acids in PYG	Species
Anaerobiospirillum succiniciproducens	AN, MA	−	V	Large, helical rods	Bipolar tufts	+	+⁻	−	A,S	*Anaerobiospirillum succiniciproducens*
Centipeda periodontii	AN	+	−	Large, straight or curved rods	Peritrichous	+	+	+	A,P (L) (S)	*Centipeda periodontii*
Selenomonas sputigena	AN	−	−	Medium-size curved rods	Subpolar	+	+	+	A,P	*Selenomonas sputigena*
Succinivibrio dextrinosolvens	AN	−	−	Medium-size helical rods	Mono-trichous	+	−	V	A,S	*Succinivibrio dextrinosolvens*

Key:
AN obligate anaerobe   A   acetic acid   L   lactic acid
MA microaerotolerant anaerobe P   propionic acid   S   succinic acid
V or ( ) variable

Source: Based on references 5 and 10.

Table 9-8. Characteristics of Some Clinically Encountered, Anaerobic Cocci

Species	Relationship to Oxygen	Gram Stain Reaction	Catalase	Indole	Esculin Hydrolysis	Urease	Nitrate Reduction	Glucose	Lactose	Acids Produced in PYG	Species
Peptostreptococcus spp.											*Peptostreptococcus* spp.
P. anaerobius[a]	AN, MA	+	−	−	−	−	−[+]	+	−	A, IB, IV, IC	*P. anaerobius*[a]
P. asaccharolyticus	AN	+	−[+]	+	−	−	−	−	−	A, B	*P. assacharolyticus*
P. indolicus[b]	AN	+	−	+	−	−	+	−	−	A, P, B	*P. indolicus*[b]
P. magnus	AN	+	−[+]	−	−	−	−	−	−	A	*P. magnus*
P. micros	AN	+	−	−	−	−	−	−	−	A	*P. micros*
P. prevotii	AN	+	−[+]	−	−	−	−	−	−	A, P, B	*P. prevotii*
P. tetradius	AN	+	−	−	−	+	−	+	+	A, B, L	*P. tetradius*
Sarcina spp.											*Sarcina* spp.
S. ventriculi[c]	AN, MA	+	V	−	+	+	+	+	+	A	*S. ventriculi*[c]
Staphylococcus spp.											*Staphylococcus* spp.
S. saccharolyticus	AN, MA, AT	+	+	−	−	−	+	+	−	A	*S. saccharolyticus*
Veillonella spp.											*Veillonella* spp.
V. parvula	AN, MA	−	+	−	−	−	+[−]	−	−	A, P	*V. parvula*

Key:
a	inhibited by SPS disk	AT	aerotolerant anaerobe	IB	isobutyric acid
b	coagulase positive	FA	facultative anaerobe	B	butyric acid
c	form endospores	V	variable	IV	isovaleric acid
AN	obligate anaerobe	A	acetic acid	IC	isocaproic acid
MA	microaerotolerant anaerobe	P	propionic acid	L	lactic acid

Source: Based on references 5 and 10.

Citron et al. have described a modified "gold standard" PRAS biochemical method for identifying bile-resistant *Bacteroides* spp.[9] Their abbreviated protocol involves preparing inoculum directly from pure culture plates, inoculating PRAS biochemical tubes with a tuberculin syringe, two tubes (arabinose and trehalose) plus catalase for identifying indole-negative species, five tubes (rhamnose, salicin, sucrose, trehalose, and xylan) for identifying indole-positive species, overnight incubation, and adding a pH indicator (bromthymol blue) after growth occurs. The investigators reported that 98% of 189 clinical isolates were correctly identified using this protocol.

Table 9-8

Biochemical-Based Minisystems

The first commercially available alternatives to conventional tubed media were the biochemical-based identification systems manufactured by Analytab Products (API 20A®, Plainview, NY) and BBL (Minitek®, Cockeysville, MD). These systems provide many of the same tests as the conventional systems, but in the form of a small, plastic strip or tray (hence the term "minisystem"). The API 20A®, for example, contains the tests listed in Table 9-9.

Table 9-9. Tests Included in the API 20A® Biochemical-Based Minisystem

---

Indole production
Catalase
Urease
Esculin hydrolysis
Gelatinase
Fermentation of:

Glucose	Glycerol
Mannitol	Cellobiose
Lactose	Mannose
Sucrose	Melizitose
Maltose	Raffinose
Salicin	Sorbitol
Xylose	Rhamnose
Arabinose	Trehalose

The biochemical-based minisystems are easier and faster to inoculate than a conventional system. Although they can be inoculated aerobically ("at the bench"), they require **anaerobic** incubation. The larger model BBL anaerobic jars and some of the commercially available bags and pouches can be used to incubate biochemical-based minisystem trays and strips if an anaerobic chamber is not available. After 24 to 48 hours of incubation, test results are read, a code number is generated for each isolate, and the numbers are looked up in a manufacturer-supplied code book. The data bases from which the code books were developed do not contain all the anaerobes that can potentially be isolated from clinical specimens. Because these systems must be incubated **anaerobically** and **require 24–48 hours,** many previous users of these systems have switched to the newer, preexisting enzyme–based systems (described in the next section). Figure 9-2 illustrates one of the biochemical-based minisystems, and Appendix C contains a list of published evaluations of such systems.

Preexisting Enzyme–Based Minisystems

A number of the newer commercial systems are based upon the presence of preexisting enzymes (i.e., enzymes already present in the bacterial cells). Because these minisystems do not depend upon enzyme induction, there is virtually no lag time, and results are available in four hours. The small, plastic panels or cards are easy

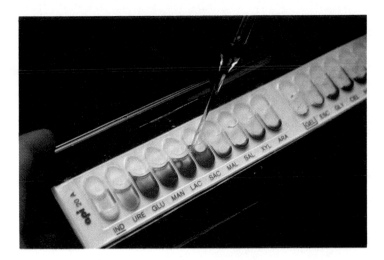

Figure 9-2. Biochemical-based minisystem for anaerobe identification (API 20A®; courtesy of Analytab Products, Plainview, NY). Following preparation of an organism suspension (in pure culture) equal in turbidity to a No. 3 McFarland Standard, the suspension is added to the wells of the API 20A test strip (as is being done in this photograph). The strip is then incubated anaerobically for 24 hours. Following incubation, reagents are added to appropriate wells, and test results are interpreted and recorded. A seven-digit profile number is generated; it identifies the organism when the number is found in the Analytical Profile Index.

to inoculate, can be inoculated at the bench, and do **not** require anaerobic incubation. Most of the systems generate code numbers, which are looked up in a manufacturer-supplied code book. Like the biochemical-based minisystems, these systems are primarily of value for commonly isolated anaerobes. The data bases from which the code books are developed do not contain all the anaerobes that can potentially be isolated from clinical specimens. Examples of preexisting enzyme–based minisystems include RapID ANA II® (Innovative Diagnostic Systems, Atlanta, GA), AN-IDENT® (Analytab Products, Plainview, NY), Rapid Anaerobe Identification Panel® (Baxter Healthcare Corp., MicroScan Division, West Sacramento, CA), ABL Anaerobe Identification System® (Austin Biological Laboratories, Austin, TX), and the ANI Card® (Vitek Systems, Hazelwood, MO). Representative preexisting enzyme–based minisystems are shown in Figure 9-3, and Appendix C contains a list of published evaluations of such systems.

Table 9-10 is a list of specific tests contained in several of the preexisting enzyme–based minisystems. In general, the systems use the same or similar substrates. There are a number of nitrophenyl and naphthylamide compounds, which are colorless substances that produce yellow or red products, respectively, in the presence of appropriate enzymes.

A

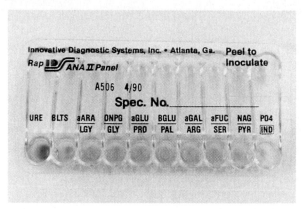

B

Figure 9-3. Preexisting enzyme–based minisystems for anaerobe identification (courtesy of Analytab Products, Plainview, NY, and Innovative Diagnostic Systems, Inc., Atlanta, GA). These two anaerobe identification minisystems are based on the presence of preexiting enzymes (i.e., enzymes already present within the organisms). **(A)** AN-IDENT® strip manufactured by Analytab Products (Plainview, NY). **(B)** RapID ANA II® panel manufactured by Innovative Diagnostic systems (Atlanta, GA). Following preparation of a heavy suspension (No. 3 McFarland or greater for RapID ANA II; No. 5 McFarland for AN-IDENT) of the organism (in pure culture), the suspension is added to the wells of the panel/strip. Inoculation can safely be performed at the bench because further organism growth is not necessary in either system. The strip is then incubated aerobically for four hours at 35–37° C in a non-CO_2 incubator. Following incubation, reagents are added to appropriate wells, and test results are interpreted and recorded. A six-digit (RapID ANA II) or seven-digit (AN-IDENT) number is generated; it identifies the organism when the number is found in the appropriate section of the Code Compendium (RapID ANA II) or Analytical Profile Index (AN-IDENT).

Table 9-10. Tests Incorporated into Three Preexisting Enzyme–Based Mini-systems

Substrate	AN-IDENT® (21 tests)	MicroScan® (24 tests)	RapID ANA II® (18 tests)
Nitrophenol compounds			
bis-p-Nitrophenyl-phosphate	PHS	BPO$_4$	
o-Nitrophenyl-beta-D-galactoside	NPG		ONPG
o-Nitrophenyl-beta-D-glucopyranoside		BGL	
p-Nitrophenyl-alpha-L-arabinofuranoside	ARB		
p-Nitrophenyl-alpha-L-arabinoside			aARA
p-Nitrophenyl-beta-D-disaccharide			BLTS
p-Nitrophenyl-alpha-L-fucopyranoside		AFU	
p-Nitrophenyl-alpha-L-fucoside	FUC		aFUC
p-Nitrophenyl-alpha-D-galactopyranoside		AGAL	
p-Nitrophenyl-beta-D-galactopyranoside		BGAL	
p-Nitrophenyl-alpha-D-galactoside	GAL		aGAL
p-Nitrophenyl-alpha-D-glucopyranoside		AGL	
p-Nitrophenyl-N-acetyl-beta-D-glucosaminide	NGS	NGLU	NAG
p-Nitrophenyl-alpha-D-glucoside	ADG		aGLU
p Nitrophenyl-beta-D-glucoside	BDG		BGLU
p-Nitrophenyl-alpha-D-mannopyranoside		MNP	
p-Nitrophenyl-phosphate		PO$_4$	PO$_4$
Naphthylamide compounds			
L-Alanyl-L-alanyl-beta-naphthylamide	ALA		
L-Arginine-beta-naphthylamide	ARL	ARG	
Arginine-beta-naphthylamide			ARG
Glycine-beta-naphthylamide		GLY	GLY
L-Glycine-beta-naphthylamide	GLY		
Glycylglycine-beta-naphthylamide		GGLY	
L-Histidine-beta-naphthylamide	HIS		
L-Leucine-beta-naphthylamide		LEU	
Leucyl-glycine-beta-naphthylamide			LGY
L-Leucyl-4-methoxy-beta-naphthylamide	LEU		
L-Lysine-beta-naphthylamide (alkaline)		LYB	
L-Lysine-beta-naphthylamide (acid)		LYA	
DL-Methionine-beta-naphthylamide		MET	
L-Phenylalanine-beta-naphthylamide	PHA		

Table 9-10. Tests Incorporated into Three Preexisting Enzyme-Based Minisystems (*continued*)

Substrate	AN-IDENT® (21 tests)	MicroScan® (24 tests)	RapID ANA II® (18 tests)
Phenylalanine-beta-naphthylamide			PAL
L-Proline-beta-naphthylamide	PRO		
Proline-beta-naphthylamide			PRO
L-Prolyl-beta-naphthylamide		PRO	
L-Pyrrolidonyl-beta-naphthylamide	PYR	PYR	
Pyrrolidonyl-beta-naphthylamide			PYR
Serine-beta-naphthyamide			SER
L-Tryptophan-beta-naphthylamide		TRY	
L-Tyrosine-beta-naphthylamide	TYR		
Other tests			
Arginine	ARG		
Catalase	CAT		
Indole	IND	IND	IND
Indoxyl-acetate	INA		
3-Indoxyl phosphate		IDX	
Nitrate		NIT	
Trehalose		TRE	
Urea		URE	URE

Note: In some cases, manufacturers have used different nomenclature for the same compound/
substrate.

As pointed out by Finegold and Edelstein, the preexisting enzyme–based minisystems "are designed for the most frequently isolated or well characterized
anaerobes from clinical sources. Unusual and rarely isolated bacteria, or bacteria
from non-human sources, may be misidentified because the organism is not in the
[manufacturer's] database."[14] These authors stress the importance of using results of
simple tests "to check that the kit-generated identification correlates with the expected
result." In this regard, it was refreshing to find the following statements in the
manufacturer's product insert that accompanies the RapID ANA II panel (Innovative
Diagnostic Systems): "The use of the Rapid ANA II System and the interpretation of
results requires a competent laboratorian who is trained in general microbiological
methods and who should judiciously make use of knowledge, experience, specimen
information, and other pertinent procedures before reporting the identity of isolates
tested using the RapID ANA II System. Specimen source, aerotolerance, Gram-stain
characteristics, and growth on selective agars should be considered when using the
RapID ANA II System."

An automated reading system for the preexisting enzyme–based identification
systems would eliminate the subjectivity associated with interpretations of their
color reactions. AN-IDENT® panels have been read on an automated video image

processing device (the ALADIN®), manufactured by Analytab Products, but this method was not commercially available at press time.[30] Rapid Anaerobe Identification Panels®, manufactured by MicroScan, can be read on a commercially available automated reader (the AutoScan-4®), manufactured by the same company. A recent evaluation of the MicroScan anaerobe identification panels and the AutoScan-4 revealed correct identification of 70% of 237 strains of anaerobes when the panels were read visually, compared to 66% when the panels were instrument-read.[31] Forty-seven strains (20%) were misidentified by visual reading, and 55 strains (23%) were misidentified by automated reading. Thus, with this particular system, automated readings proved to be less accurate than visual readings.

Gas-Liquid Chromatography (GLC) [8, 18, 20, 25]

Laboratories having a routine need to definitively identify anaerobic isolates that cannot be identified using one of the previously mentioned, commercially available minisystems may wish to incorporate gas-liquid chromatography (GLC) into the identification protocol.

Principles of GLC. Gas-liquid chromatography is used in clinical laboratories for a variety of applications (e.g., analysis of nonprotein hormones, drugs, fatty acids, amino acids, and toxic substances). It is a fairly rapid and sensitive technique, requiring only a small quantity of sample. Specific uses of GLC for identifying anaerobic bacteria are discussed later in this section.

Like other types of chromatographic techniques (e.g., gas-solid chromatography, thin-layer chromatography, ion-exchange chromatography, etc.), GLC uses the chemical and physical properties of a particular component to separate it from other components in a mixture. All chromatographic techniques have a **mobile phase** and a **stationary phase.** In GLC, an inert gas serves as the mobile phase, and a liquid as the stationary or separating phase. An unknown mixture is usually prepared for separation, volatilized, and carried as a gas through a packed chromatographic column. Components of the mixture are then separated, and they elute from the column, in order, based on their chemical and physical properties. As the individual components elute from the column, they are detected by one of several types of detecting devices (detectors). A graphic display of the separated components is then printed on a recorder in the form of a chromatogram. Identification of an unknown component is based on its position on the chromatogram (i.e., its relative retention time) as compared to a standard. Quantitation of an unknown component is based on the area under the peak with respect to a known standard.

Components of a Gas-Liquid Chromatograph. The basic components of a gas-liquid chromatograph include the carrier gas, gas control valves, sample inlet or injection port, column, column oven, detector, and recorder (illustrated in Figure 9-4). Figure 9-5 depicts the Capco gas-liquid chromatograph (Dodeca, Freemont, CA), specifically manufactured for and commonly used in anaerobic bacteriology laboratories.

In most instances, the sample to be analyzed on the gas chromatograph must be prepared for injection. **Sample preparation** may be a rather simple or a fairly complicated, lengthy procedure, depending on the compound to be analyzed. Also, some substances may be so similar in structure that a derivative must first be

stationary liquid phase depends primarily on the polarity and volatility of the compounds to be separated. The liquid phase must not react with the solid support and must be stable at high temperatures. Commonly used liquid phases include polyethylene glycol, silicone oil, and polyesters.

The **detector** of a gas chromatograph is an electrical device that measures components by producing electrical currents proportional to their concentrations. Table 9-11 lists the various types of detectors used for GLC.

Table 9-11. Types of Detectors Used in Gas-Liquid Chromatography

Thermal conductivity
Flame ionization
Alkali flame ionization
Electron capture

Thermal conductivity and flame ionization detectors are the most commonly used detectors. The thermal conductivity detector (TCD) is rugged, sensitive, and responds to many different types of compounds. However, it is sensitive to temperature changes and flow rate. In this type of detector, two thin wires are placed at the end of the column and heated by means of an electric current. One of the wires acts as a reference electrode over which only the carrier gas passes. The volatilized sample passes over the second wire. When a compound comes off (elutes from) the column, it passes over the second wire, thereby changing the temperature of the wire and the flow of current in it (its resistance). The change in electrical current is detected, amplified, and printed as a peak by the recorder. The chromatograph pictured in Figure 9-5 contains a thermal conductivity detector.

In the flame ionization detector (FID), two electrodes with a current crossing between them are suspended in a flame at the end of the column. As the sample comes off the column, it passes through the flame, producing ions. The higher number of ions increases the current across the electrodes. This slight increase in current is measured, amplified, and sent to the recorder. The FID is very sensitive, particularly to carbon-containing compounds, and is not affected as much as the TCD by temperature and flow rate. A stream or sample splitter may be used to preserve some of the sample so it is not all destroyed by the flame.

As compounds elute from the column and are detected by one of the different types of detectors, the recorder produces peaks that correspond to the various unknown compounds in the sample. Identification of a particular peak is based on its retention time (the length of time from injection to detection of a peak under specified conditions) as compared to a standard. Quantitative **peak assessment** is based on its height and width; the larger the peak, the greater the quantity of a particular component.

Identifying Anaerobes via Metabolic End Product Analysis by GLC. The use of GLC techniques for analyzing metabolic end products and identifying anaerobes was pioneered by W. E. C. Moore, E. P. Cato, L. V. Holdeman, and others at the Virginia Polytechnic Institute (VPI) in Blacksburg. Their GLC procedure, details of which can be found in the *VPI Anaerobe Laboratory Manual*, is still widely used in laboratories throughout the world.[21] The VPI guidelines for GLC were used for many years at the Centers for Disease Control (CDC), but with time, they were modified by G. L. Lombard, V. R. Dowell, Jr., and their colleagues. The CDC guidelines for GLC can be found in *Gas-Liquid Chromatography Analysis of the Acid Products of Bacteria.*[26] A brief summary of the CDC procedure follows.

The anaerobe to be tested is first grown in a tube of peptone-yeast extract-glucose medium (PYG), which, in addition to those three ingredients, contains cysteine hydrochloride, resazurin, and a salts solution. PYG broth is available commercially from Carr-Scarborough, Scott, Remel, and other media manufacturers. Two milliliters of the PYG culture are used to analyze short-chain, volatile acids. One milliliter is used to analyze nonvolatile, low-molecular-weight, aliphatic and aromatic acids.

Short-chain, volatile acids produced by anaerobes include formic, acetic, propionic, isobutyric, butyric, isovaleric, valeric, isocaproic, caproic, and heptanoic acids (see Table 9-12). After acidification of the 2 ml aliquot of PYG culture with 50% aqueous sulfuric acid, the volatile acids are extracted using ethyl ether. (See reference 34 for a discussion of the use of safer methyl tert-butyl ether in place of ethyl ether.) Fourteen microliters of the ether extract are injected into the appropriate chromatograph column (containing SP-1220 packing material). Volatile acids are identified by comparing the elution times of products in the ether extract with those of known acids in a standardized volatile acid mixture that has been chromatographed in the same manner on the same day. Sample chromatograms of volatile acids are shown in Figure 9-6.

Table 9-12. Acid End Products Produced by Anaerobic Bacteria

Name	No. of Carbons	Formula
Volatile fatty acids		
Formic	1	$HCOOH$
Acetic	2	CH_3COOH
Propionic	3	CH_3CH_2COOH
Isobutyric	4	$CH_3(CH_2)_2COOH$
Butyric	4	same as above
Isovaleric	5	$CH_3(CH_2)_3COOH$
Valeric	5	same as above
Isocaproic	6	$CH_3(CH_2)_4COOH$
Caproic	6	same as above
Heptanoic	7	$CH_3(CH_2)_5COOH$
Caprylic	8	$CH_3(CH_2)_6COOH$
Nonvolatile acids		
Pyruvic	3	$CH_3COCOOH$
Lactic	3	$CH_3CHOHCOOH$
Fumaric	4	$HOOCCH{=}CHCOOH$
Succinic	4	$HOOC(CH_2)_2COOH$

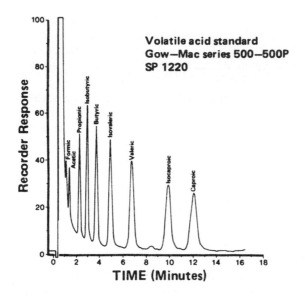

A

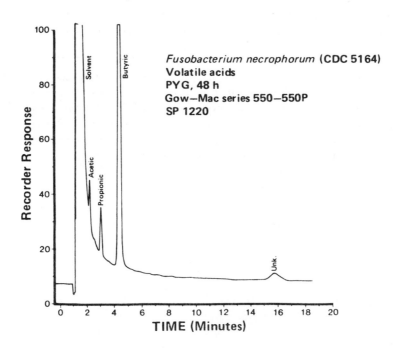

B

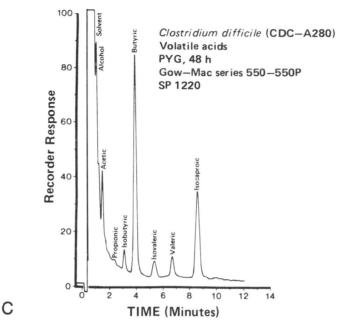

Figure 9-6. Chromatograms of volatile acids. **(A)** Obtained from analysis of a standard volatile acid mixture using a Gow-Mac gas-liquid chromatograph (Gow-Mac Instrument Co., Bound Brook, NJ) and a 6 foot x ¼ inch stainless steel column packed with 15% SP-1220/1% H_3PO_4 on 100/120 Chromosorb W/AW (Supelco Inc., Bellefonte, PA). Helium was used as the carrier gas. The relative locations (retention times) of nine volatile fatty acids are shown in the order they elute from the column. Formic acid (with the shortest carbon chain) is the first to elute, and caproic acid (with the longest carbon chain) is the last. **(B)** Volatile acids extracted from a 48-hour PYG culture of a strain of *Fusobacterium necrophorum*, including a very large butyric acid peak. **(C)** Volatile acids extracted from a 48-hour PYG culture of a strain of *Clostridium difficile*, including a very large butyric acid peak and a large isocaproic acid peak. These chromatograms have been reproduced from the previously referenced CDC lab manual, which should be consulted for details of the extraction procedures and interpretation of actual chromatograms. Chromatograms of specific anaerobes may differ from those presented here, depending upon the particular instrument and column being used, oven temperature, carrier gas flow rate, and other variables.

Nonvolatile, low-molecular-weight, aliphatic and aromatic acids produced by anaerobes include pyruvic, lactic, oxalacetic, oxalic, malonic, fumaric, succinic, benzoic, phenylacetic, and hydrocinnamic acids (see Table 9-12). After acidification of the 1 ml aliquot of PYG culture with aqueous sulfuric acid, methanol is added to methylate the nonvolatile acids, and the aliquot is incubated overnight in a 55°C water bath. After addition of water, the methylated nonvolatile acids are extracted using chloroform. Fourteen microliters of the chloroform extract are injected into the appropriate chromatograph column (containing SP-1000 packing material). Methylated nonvolatile acids (methyl esters) are identified by comparing the elution times of products in the chloroform extract with those of known acids in a standardized, nonvolatile acid mixture that has been chromatographed in the same manner on the same day. Sample chromatograms are shown in Figure 9-7.

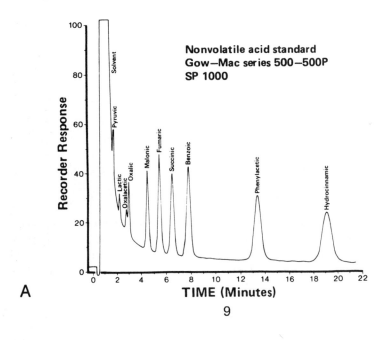

A

9

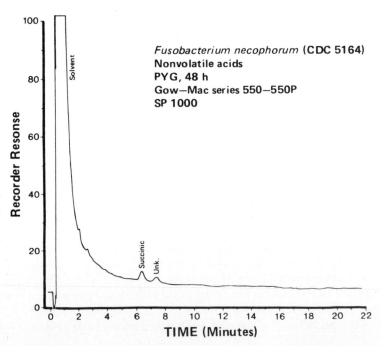

B

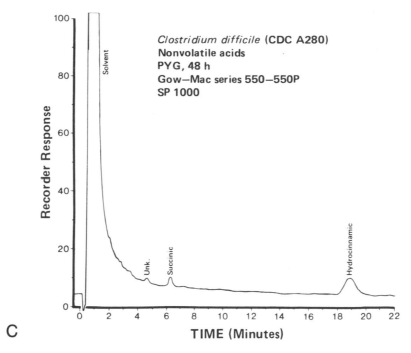

C

Figure 9-7. Chromatograms of nonvolatile acids. **(A)** Obtained from analysis of a standard nonvolatile acid mixture using a Gow-Mac gas-liquid chromatograph (Gow-Mac Instrument Co., Bound Brook, NJ) and a 6 foot x ¼ inch stainless steel column packed with 10% SP-1000/1% H_3PO_4 on 100/120 Chromosorb W/AW (Supelco Inc., Bellefonte, PA). Helium was used as the carrier gas. The relative locations (retention times) of ten methyl esters of nonvolatile acids are shown in the order they elute from the column. Pyruvic acid (with the shortest carbon chain) is the first to elute, and hydrocinnamic acid (with the longest carbon chain) is the last. **(B)** Nonvolatile acids extracted from a 48-hour PYG culture of a strain of *Fusobacterium necrophorum*, showing a small succinic acid peak. **(C)** Nonvolatile acids extracted from a 48-hour PYG culture of a strain of *Clostridium difficile*, including a small, succinic acid peak and a larger, hydrocinnamic acid peak. These chromatograms have been reproduced from the previously referenced CDC lab manual, which should be consulted for details of the extraction procedures and interpretation of actual chromatograms. Chromatograms of specific anaerobes may differ from those presented here, depending upon the particular instrument and column being used, oven temperature, carrier gas flow rate, and other variables.

For valid comparisons and quality assurance, microbiologists must process cultures under uniform conditions and should make every effort to standardize media, reagents, and identification techniques as carefully as possible. Detailed procedures for preparing standard acid mixtures and quality control are contained in the previously referenced CDC publication.

Information regarding acid end products and other available information about the isolate (e.g., colony characteristics, gram reaction, cell morphology, etc.) are used to identify the organism. Tables 9-13 through 9-16 contain information on volatile and nonvolatile acid products of various anaerobes in PYG cultures.

Table 9-13. Volatile and Nonvolatile Acid Products of Various Clostridia in PYG Cultures

	Acid Products	
Species	**Volatile**	**Nonvolatile**
C. baratii	A, B, (P)	(L)
C. bifermentans	A, IC, (P), (IB), (IV)	HCA
C. butyricum	A, B	-
C. clostridioforme	A	-
C. difficile	A, IB, B, IV, V, IC, (P)	(PAA), (HCA)
C. histolyticum	A	(HCA)
C. innocuum	A, B	(L)
C. limosum	A	-
C. novyi type A	A, P, B, (V)	-
C. paraputrificum	A, B	(L)
C. perfringens	A, B, (P)	(L)
C. ramosum	A, L	(PY)
C. septicum	A, B	-
C. sordellii	A, IC, (P), (IB), (IV)	-
C. sporogenes	A, IB, B, IV, (P), (V), (IC)	HCA
C. subterminale	A, IB, B, IV, (P), (V), (IC)	(PAA)
C. tertium	A, B	(L), (HCA)
C. tetani	A, P, B	(HCA)

Key:
A acetic acid	V valeric acid	HCA hydrocinnamic acid
P propionic acid	IC isocaproic acid	PAA phenylacetic acid
IB isobutyric acid	PY pyruvic acid	( ) variable; if produced, there is
B butyric acid	L lactic acid	usually only a trace amount
IV isovaleric acid		

Source: Based on references 5 and 26.

Table 9-14. Volatile and Nonvolatile Acid Products of Various Nonspore-forming, Anaerobic, Gram-Positive Bacilli in PYG Cultures

| Species | Acid Products | |
	Volatile	Nonvolatile
Actinomyces spp.		
A. israelii	A	L, S
A. meyeri	A	L, S
A. naeslundii	A	L, S
A. odontolyticus	A	L, S
A. viscosus	A	L, S
Bifidobacterium spp.		
B. eriksonii	A	L
Eubacterium spp.		
E. alactolyticum	A, B, C	-
E. lentum	(A)	-
E. limosum	A, B	-
Lactobacillus spp.		
L. catenaforme	A	L
Propionibacterium spp.		
P. acnes	A, P	L, S
P. avidum	A, P	(L), S
P. granulosum	A, P	(L), S
P. propionicus	A, P	(L), (S)

Key:
A acetic acid   C caproic acid   ( )   variable; if produced, there is
P propionic acid   L lactic acid   usually only a trace amount
B butyric acid   S succinic acid

Source: Based on reference 26.

Table 9-15. Volatile and Nonvolatile Acid Products of Various Anaerobic, Gram-Negative Bacilli in PYG Cultures

	Acid Products	
Species	**Volatile**	**Nonvolatile**
Bacteroides spp.		
B. distasonis	A, P, (IB), (IV)	(L), S
B. fragilis	A, P, (IB), (IV)	(L), S
B. thetaiotaomicron	A, P, (IB), (IV)	(L), S
B. uniformis	A, P, (IB), (IV)	(L), S
B. ureolyticus	A	S
B. vulgatus	A, P, (IB), (IV)	(L), S
Fusobacterium spp.		
F. mortiferum	A, P, B	(S)
F. necrophorum	A, P, B	(S)
F. nucleatum	A, P, B	(S)
F. varium	A, P, B	(S)
Porphyromonas spp.		
P. asaccharolytica	A, P, IB, B, IV	(S)
P. gingivalis	A, P, IB, B, IV	S, PAA
Prevotella spp.		
P. bivia	A, IV	S
P. disiens	A, IV	S
P. intermedia		

Key:
A	acetic acid	IV	isovaleric acid	PAA	phenylacetic acid
P	propionic acid	L	lactic acid	( )	Variable; if produced, there is
IB	isobutyric acid	S	succinic acid		usually only a trace amount
B	butyric acid				

Source: Based on reference 26.

Table 9-16. Volatile and Nonvolatile Acid Products of Various Anaerobic Cocci in PYG Cultures

Species	Acid Products	
	Volatile	Nonvolatile
Acidaminococcus fermentans	A, P, B, (IV)	-
Gemella morbillorum (*Streptococcus morbillorum*)	(A)	L
Megasphaera elsdenii	A, IB, B, IV, V, C	(L), HCA
Peptostreptococcus spp.		
P. anaerobius	A, (P), (IB), (B), (IV), IC	HCA
P. asaccharolyticus	A, (P), B	-
P. indolicus	A, P, B	-
P. magnus	A	(L)
P. micros	A	(L)
P. prevotii	A, (P), B	(L)
P. productus	A	-
P. tetradius[a]	A, B	L
Sarcina ventriculi[b]	A	-
Staphylococcus saccharolyticus[c]	A	-
Streptococcus parvulus	(A)	L
Veillonella parvula	A, P	-

a Formerly "Gaffkya anaerobia" (urease positive)
b Sporeformer
c Formerly *Peptococcus saccharolyticus*; the only *Staphylococcus* species that does not produce lactic acid in PYG

Key:
A acetic acid   IV isovaleric acid   L   lactic acid
P propionic acid   V   valeric acid   HCA hydrocinnamic acid
IB isobutyric acid   IC isocaproic acid   ( )   variable; if produced, there is
B butyric acid   C   caproic acid   usually only a trace amount

Source: Based on reference 5.

Identifying Anaerobes via Cellular Fatty Acid Analysis by High-Resolution GLC. The previous section described the use of acid end product analysis in identifying anaerobes. The acids referred to there are metabolic by-products produced by anaerobic bacteria in broth culture. Some of the acid end products are volatile fatty acids. A different type of fatty acid analysis available as an aid in the identification of bacteria, including anaerobes, is known as cellular fatty acid analysis.

The term "cellular fatty acids" refers to fatty acids and related compounds (aldehydes, hydrocarbons, dimethyl acetals) present within organisms as cellular components. Over 200 such fatty acids have been identified, and various taxonomic groups have rather predictable fatty acid profiles or patterns. Cellular fatty acids are coded for on bacterial chromosomes, as opposed to plasmids, and are not affected by simple mutations or plasmid loss. Thus, the fatty acid composition of a particular organism is relatively stable when grown using standardized growth conditions (medium, incubation temperature, and time). Although fatty acid profiles can be identified manually, computerized, high-resolution gas chromatography and specialized software programs are now available to automatically analyze cellular fatty acids of unknown bacteria and compare the results to patterns of known species.

The anaerobe to be identified is grown in pure culture in PYG broth or other standardized medium. The cells are removed by centrifugation and then saponified to release the fatty acids from the bacterial lipids. The fatty acids are next methylated under acidic conditions, creating volatile methyl esters. The methyl esters are extracted from the aqueous phase into an organic solvent, base washed to remove components that could harm the chromatographic column, and transferred to a septum-capped vial. They are analyzed by GLC using a fused silica capillary column and flame ionization detector. Hydrogen is the carrier gas, nitrogen is the "makeup" gas, and a mixture of air and hydrogen supports the flame.

A chromatogram depicting the unknown organism's fatty acid composition is generated (Figure 9-8), as is a computer printout or report (Figure 9-9). The report follows computer comparison of the unknown organism's fatty acids to a data base or "library" of fatty acids of known anaerobes that have been grown and analyzed under identical conditions. The report includes computer-generated statistical values or "similarity indices," which are based upon deviations in the unknown organism's fatty acid composition from the known profiles in the library. Because no subjective interpretations are required, the identifications are objective and highly reproducible.

The complete system, called the MIDI Microbial Identification System (MIS) (Figure 9-10), uses an autosampler, gas chromatograph, integrator, computer, and printer from Hewlett-Packard (Avondale, PA, or Palo Alto, CA) and software from Microbial ID, Inc. (Newark, DE). The MIDI-MIS system permits up to 100 samples to be loaded onto an autosampler, from which samples are automatically injected into the instrument. According to the manufacturer, the MIDI-MIS system is capable of running about 48 samples per day (approximately 30 minutes per sample). It takes about six minutes of technologist time and four reagents to prepare each extract. Expendables cost about $1.30 per sample. In addition to the anaerobe library containing more than 375 species, libraries are also available for facultative, gram-positive and gram-negative bacteria; nonfermentative, gram-negative bacilli; mycobacteria; and yeasts.

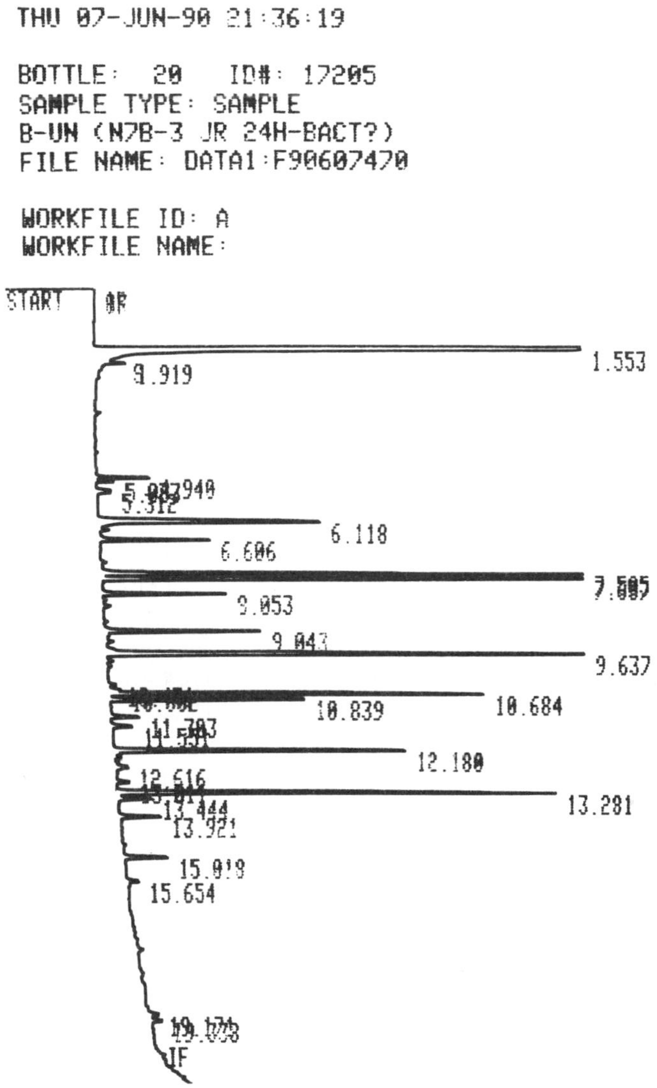
THU 07-JUN-90 21:36:19
BOTTLE: 20 ID#: 17205
SAMPLE TYPE: SAMPLE
B-UN (N7B-3 JR 24H-BACT?)
FILE NAME: DATA1:F90607470
WORKFILE ID: A
WORKFILE NAME:
START
1.553
6.118
6.606
9.053
9.637
10.839
10.684
12.180
12.616
13.281
15.654

Figure 9-8. MIDI-MIS chromatogram (courtesy of Ms. Dianne Wall, Virginia Polytechnic Institute). This is an actual chromatogram produced by a MIDI-MIS system at the Department of Anaerobic Microbiology, Virginia Polytechnic Institute, Blacksburg. Each peak (e.g., 1.553) represents a fatty acid that has eluted from the column and has been detected by the flame ionization detector. An equivalent chain length (ECL) value for each fatty acid is derived by a computer as a function of the elution time for that particular fatty acid in relation to the elution times obtained for straight chain saturated fatty acids in a standard mixture.

```
--------------------------------------------------------------------------------------------
ID:    17205    B-UN (N7B-3 JR 24H-BACT?)                    Date of run: 07-JUN-90 21:40:50
Bottle: 20      SAMPLE [ANAER1]
```

RT	Area	Ar/Ht	Respon	ECL	Name	%	Comment 1	Comment 2
1.553	37995000	0.076	. . .	7.019	SOLVENT PEAK	. . .	< min rt	
1.919	882	0.028	. . .	7.849	. . . .	. . .	< min rt	
4.940	3385	0.036	1.034	12.615	13:0 ISO FAME . . .	0.65	ECL deviates 0.001	Reference -0.003
5.033	1090	0.035	1.031	12.704	13:0 ANTEISO FAME .	0.21	ECL deviates 0.001	Reference -0.003
5.312	2888	0.108	. . .	12.971	. . .	. . .		
6.118	31685	0.081	1.005	13.614	14:0 ISO FAME . . .	5.87	ECL deviates -0.004	Reference -0.007
6.606	7913	0.041	0.996	14.000	14:0 FAME	1.45	ECL deviates 0.000	Reference -0.002
7.505	58694	0.043	0.982	14.623	15:0 ISO FAME . . .	10.63	ECL deviates -0.000	Reference -0.003
7.637	204160	0.043	0.980	14.714	15:0 ANTEISO FAME .	36.87	ECL deviates 0.000	Reference -0.003
8.053	10116	0.046	0.974	15.002	15:0 FAME	1.82	ECL deviates 0.002	Reference -0.001
9.043	13308	0.048	0.963	15.626	16:0 ISO FAME . . .	2.36	ECL deviates -0.001	Reference -0.003
9.637	90366	0.048	0.957	16.001	16:0 FAME	15.94	ECL deviates 0.001	Reference -0.002
10.471	860	0.062	0.949	16.502	15:0 3OH FAME . . .	0.15	ECL deviates -0.004	
10.602	978	0.047	. . .	16.580	. . . .	. . .		
10.684	34088	0.051	0.947	16.630	17:0 ISO FAME . . .	5.95	ECL deviates -0.000	Reference -0.003
10.839	17518	0.050	0.946	16.723	17:0 ANTEISO FAME .	3.06	ECL deviates -0.000	
11.303	2624	0.053	0.942	17.001	17:0 FAME	0.46	ECL deviates 0.001	Reference -0.002
11.551	2330	0.065	0.940	17.146	Sum In Feature 9 . .	0.40	ECL deviates -0.004	16:0 ISO 3OH FAME
12.180	27345	0.052	0.935	17.514	16:0 3OH FAME . . .	4.72	ECL deviates -0.008	
12.616	2092	0.086	0.932	17.768	18:1 CIS 9 FAME . .	0.36	ECL deviates -0.003	
13.011	1800	0.074	0.929	17.999	18:0 FAME	0.31	ECL deviates -0.001	Reference -0.005
13.281	44045	0.056	0.927	18.157	Sum In Feature 11 .	7.53	ECL deviates -0.006	17:0 ISO 3OH FAME
13.444	3238	0.056	0.926	18.252	17:0 ANTE 3OH FAME .	0.55	ECL deviates -0.008	
13.921	4074	0.057	0.923	18.531	17:0 3OH ? FAME . .	0.69	ECL deviates -0.008	
15.018	4423	0.058	. . .	19.173	. . . .	. . .		
15.654	1159	0.059	. . .	19.544	. . . .	. . .		
19.171	1721	0.093	. . .	21.610	. . . .	. . .	> max rt	
19.338	1654	0.077	. . .	21.708	. . . .	. . .	> max rt	
*******	2330	. . .	. . .	. . .	SUMMED FEATURE 9 . .	0.40	16:0 ISO 3OH FAME	UN 17.157 DMA
*******	44045	. . .	. . .	. . .	SUMMED FEATURE 11 .	7.53	17:0 ISO 3OH FAME	18:2 DMA

Solvent Ar	Total Area	Named Area	% Named	Total Amnt	Nbr Ref	ECL Deviation	Ref ECL Shift
37995000	570179	560731	98.34	542315	12	0.004	0.003

```
--------------------------------------------------------------------------------------------
        B0503 [Rev 1.0]   BACTEROIDES . . . . . . . . . . . . . . . . . . 0.670
                          B. ORIS . . . . . . . . . . . . . . . . . . . . 0.670
                          B. D34 . . . . . . . . . . . . . . . . . . . . 0.466
                          B. MELANINOGENICUS . . . . . . . . . . . . . . 0.248
                          B. M. 2381 . . . . . . . . . . . . . . . . . . 0.248
                          B. M. 9343 . . . . . . . . . . . . . . . . . . 0.242
        T0503 [Rev 1.0]   * NO MATCH *
        CALIB [Rev 1.0]   * NO MATCH *
```

Figure 9-9. MIDI-MIS computer printout (courtesy of Ms. Dianne Wall, Virginia Polytechnic Institute). This is an actual computer printout or report generated by a MIDI-MIS system at the Department of Anaerobic Microbiology, Virginia Polytechnic Institute, Blacksburg. This is the printout for the same organism that produced the chromatogram shown in Figure 9-8. Analysis of an unknown strain by the MIDI-MIS system results in an automatic comparison of its fatty acid composition to a stored data base ("library"). The printout indicates the best library match (in this case, *Bacteroides oris*) and, if appropriate, closely related organisms. A statistical value or "similarity index" is assigned to each identification.

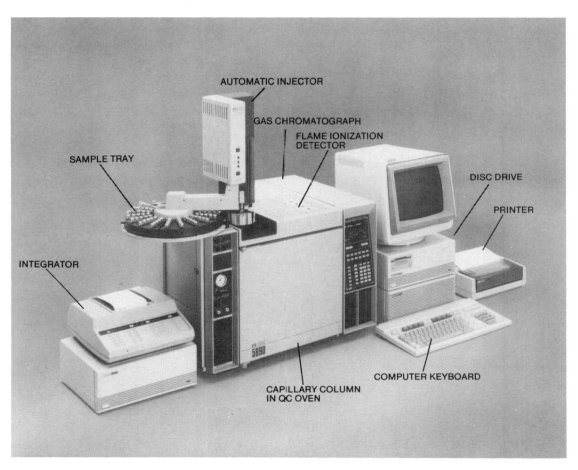

Figure 9-10. Principal components of the MIDI-MIS system (courtesy of Microbial ID, Inc., Newark, DE)

A potentially important third way of using GLC to identify anaerobes in clinical specimens is described in Chapter 14.

Table 9-17 summarizes the various options available to microbiologists for identifying anaerobic bacteria, and Table 9-18 presents a philosophical approach to their use.

Table 9-17. Various Options Available for Identifying Anaerobic Bacteria

Identification Technique	Comments
Presumptive (and sometimes definitive) identifications via colony morphology and Gram stain observations and the results of simple tests (e.g., disks, catalase, spot indole)	Identifications on the **same day** that the PC/SC plate becomes available; not all anaerobes can be identified using PID criteria
Definitive identifications using commercially available, preexisting enzyme–based identification procedures (e.g., ABL, AN-IDENT, MicroScan, RapID ANA II, Vitek ANI card, etc.)	Identifications on the **same day** that the PC/SC plate becomes available; not all anaerobes can be identified using such commercially available systems
Definitive identifications using commercially available, biochemical-based identification systems (e.g., API 20A, Minitek, Sceptor)	Identifications **24–48 hours after** the PC/SC plate becomes available; not all anaerobes can be identified using such commercially available systems
Definitive identifications via cellular fatty acid analysis by high-resolution GLC	Identifications **24–48 hours after** the PC/SC plate becomes available; most clinically encountered anaerobes can be identified using this technique
Definitive identifications via "conventional" tubed biochemical tests and fatty acid analysis by GLC	Identifications **24–72 hours after** the PC/SC plate becomes available; most clinically encountered anaerobes can be identified using this technique

Valuable Tests for Definitive Identification of Motile Anaerobes

Flagella Staining. Although not technically difficult, flagella staining is infrequently performed at the hospital level. It is not necessary for identifying species of *Bacteroides, Fusobacterium, Porphyromonas,* and *Prevotella* (the gram-negative bacilli most frequently encountered in the anaerobe laboratory) because most species are nonmotile. *Mobiluncus* spp. are commonly encountered in specimens from the female genitourinary tract. Although technically gram-positive, these curved, motile bacilli often stain pink with the Gram staining procedure. Flagella staining is of most value in identifying relatively infrequently encountered motile, gram-negative anaerobes such as members of the genera *Anaerobiospirillum, Butyrivibrio, Campylobacter,*

Table 9-18. Recommended Three-Step Approach to Anaerobe Identification

Step	Action
1	Attempt to presumptively identify the isolate using a combination of colony and Gram stain observations and results of simple tests.
2	If the isolate cannot be presumptively identified, inoculate a minisystem that provides same day results.
3	If the minisystem fails to identify the isolate, take one of the following actions*:
	a. Report all information about the isolate that you have obtained, and then concentrate your efforts and resources elsewhere.
	b. Send the isolate to a reference laboratory for ID.
	c. Attempt to identify the isolate in-house using "traditional/conventional" techniques (e.g., tubed PRAS media and GLC).

* The choice will usually be determined by the clinical significance of the isolate and the size and capabilities of the laboratory.

Desulfomonas, Desulfovibrio, Selenomonas, Succinimonas, Succinivibrio, and *Wolinella.* Between Janauary, 1973, and July, 1985, only 13 of these organisms were isolated from a total of 1523 specimens received by the Wadsworth Anaerobic Bacteriology Research Laboratory.[23] Tables 9-19 and 9-20 indicate the number of *Desulfomonas pigra* and *Desulfovibrio desulfuricans* isolates identified at the CDC Anaerobe Reference Laboratory between 1983 and 1990.[5] Various flagellar arrangements of anaerobes are depicted in Figure 9-11, and specific examples can be found in Table 9-21.

Table 9-19. Clinical Source and Geographic Location of *Desulfomonas pigra* Isolates Identified by the CDC Anaerobe Reference Laboratory, 1983–1990

Strain #	Source	Geographic Location
850671	Blood	FL
860035	Peritoneal fluid	TN
860227	Pelvic abscess	NC
860230	Blood	NC
860232	Blood	NC
860649	Liver abscess	MO
880077	Blood	MO
880208	Abscess	TN
890000	Blood	MI

Source: Based on reference 5.

Table 9-20. Clinical Source and Geographic Location of *Desulfovibrio desulfuricans* Isolates Identified by the CDC Anaerobe Reference Laboratory, 1983–1990

Strain #	Source	Geographic Location
850747	Blood	TN
860142	Blood	IA
860177	Blood	MD
860263	Hip abscess	VA
860339	Blood	LA
860343	Blood	CA
860401	Blood	AL
860434	Pleural fluid	Chile
870012	Blood	NM
870337	Blood	CT
890246	Vulval abscess	New Zealand
900223	Blood	SC

Source: Based on reference 5.

Monotrichous organisms possess only a single flagellum.

Lophotrichous organisms have two or more flagella at one end.

Amphitrichous organisms have one flagellum at each end.

Peritrichous organisms have flagella around their entire surface.

Figure 9-11. Flagella terminology

Table 9-21. Examples of Flagellar Arrangements of Anaerobes

Flagellar Arrangement	Examples
No flagella (nonmotile)	Virtually all *Bacteroides* and *Fusobacterium* spp., all *Porphyromonas* and *Prevotella* spp., *Clostridium perfringens, C. ramosum,* some strains of *C. novyi* type A, all anaerobic nonsporeforming, gram-positive bacilli (except *Mobiluncus* spp.), all anaerobic cocci
Monotrichous (single polar flagellum)	*Campylobacter sputorum, C. concisus, Desulfovibrio desulfuricans, Succinivibrio dextrinosolvens, Wolinella curva, W. recta*
Lophotrichous (tuft of polar flagella)	Some *Desulfovibrio* spp., some strains of *Anaerobiospirillum succiniciproducens*
Bipolar tufts of flagella	Some strains of *Anaerobiospirillum succiniciproducens*
Single subpolar flagellum	Some strains of *C. difficile*
Subpolar tuft of flagella	*Selenomonas sputigena*
Tuft of flagella on concave side of curved cells	*Mobiluncus* spp., some *Selenomonas* spp.
Peritrichous flagella	*Centipeda periodontii,* some strains of *C. difficile,* some strains of *C. novyi* type A, *C. septicum, C. sordellii, C. sphenoides, C. tertium, C. tetani*

Whenever a motile anaerobe (either gram-negative or gram-positive) is encountered using PID procedures, the number and position of flagella can be determined by performing a flagella stain. Although several flagella staining procedures have been described, the Centers for Disease Control recommend the modified Ryu procedure, which is no more technically difficult than the Gram stain procedure.[24] Detailed instructions for performing a modified Ryu stain are contained in Appendix D, and anaerobes stained using this procedure are illustrated in Figure 9-12.

A simplified, wet-mount flagella staining procedure has been described by Heimbrook et al.[19] Although permanent stained smears are not produced, the procedure is less technically demanding than the previously mentioned modified Ryu staining technique. Detailed instructions for performing this relatively simple, wet-mount procedure are contained in Appendix D.

Desulfoviridin Test. An anaerobic, gram-negative bacillus suspected of being a *Desulfovibrio, Desulfomonas,* or *Bilophila* species can be tested for the presence of desulfoviridin, an enzyme (sulfite reductase) produced by certain sulfate-reducing organisms. Most, but not all, of the nine different species of *Desulfovibrio* produce desulfoviridin, as does *Desulfomonas pigra.* Like flagella staining, the test for desulfoviridin is not technically difficult (see Appendix D for detailed instructions). It

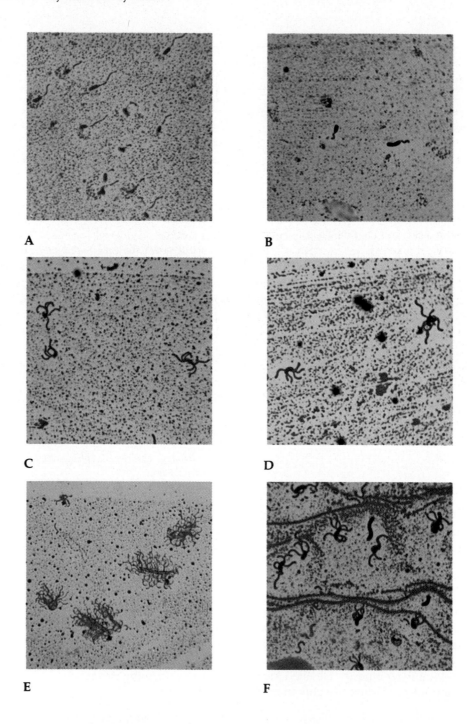

Figure 9-12. Anaerobes stained by the modified Ryu flagellar staining procedure (courtesy of Suzette L. Bartley, James D. Howard, and Ray Simon, Centers for Disease Control, Atlanta). **(A)** *Desulfovibrio desulfuricans.* **(B)** *Wolinella recta.* (*Campylobacter rectus*). **(C)** *Selenomonas sputigena.* **(D)** *Mobiluncus mulieris.* **(E)** *Centipeda periodontii.* **(F)** *Clostridium septicum.*

is, however, not usually performed at the hospital level due to the relative infrequency of isolation of clinically significant, desulfoviridin-positive anaerobes at any given hospital. For example, one group of investigators reported recovering only one *Desulfovibrio* isolate (*D. vulgaris*) over a 12.5-year period at a major medical center; no *Desulfomonas* isolates were mentioned.[23]

Desulfovibrio spp. are found in freshwater and marine sediments and the gastrointestinal tracts of humans and other animals. *Desulfovibrio* cells are typically curved (sometimes s-shaped or helical) but may be straight, and all are motile by means of a single, polar flagellum or a tuft of polar flagella. *Desulfomonas pigra*, which can be found in the human intestinal tract, is quite large, straight, and nonmotile, resembling a member of the *Enterobacteriaceae* family in appearance. Thus, a clinical isolate that is a motile, desulfoviridin-positive, anaerobic, gram-negative bacillus (either curved or straight) can be identified as a *Desulfovibrio* sp., whereas a nonmotile, desulfoviridin-positive, anaerobic, gram-negative, nonpleomorphic bacillus with straight cells can be identified as *Desulfomonas pigra*. *Desulfomonas pigra* is strongly H_2S positive in TSI, iron milk, or H_2S semisolid medium—a key reaction in its identification.

Another nonmotile, anaerobic, gram-negative bacillus that is sometimes desulfoviridin-positive has been described by Baron et al.[3] This fastidious organism, *Bilophila wadsworthia*, is apparently common in fecal specimens and intraabdominal specimens from patients with gangrenous and perforated appendicitis. It is strongly catalase-positive, nitrate-positive, usually urease-positive, and indole-negative, and its growth is stimulated by 20% bile. It grows slowly on BBE agar, producing either irregular, low convex, black, opaque colonies or erose, circular, umbonate, translucent colonies with black centers. On Gram stain, *B. wadsworthia* cells are very pleomorphic (unlike *Desulfomonas*) and unevenly stained, with swollen portions and irregular walls. Table 9-22 summarizes some key characteristics of desulfoviridin-positive, anaerobic, gram-negative bacilli.

Table 9-22. Characteristics of Desulfoviridin-Positive, Anaerobic, Gram-Negative Bacilli

Organism	Desulfoviridin Production	Motility	Flagella Arrangement	Cell Shape
Bilophila wadsworthia	Variable	Nonmotile	N/A	Very pleomorphic
Desulfomonas pigra	Positive	Nonmotile	N/A	Straight
Desulfovibrio spp.	Variable	Motile	Single polar flagellum or polar tuft of flagella	Curved (usually) or straight

Laboratory Confirmation of Clostridial Diseases

Botulism

Diagnosis of foodborne botulism is initially based upon the patient's signs and symptoms, often corroborated by the patient's food history and epidemiological evidence.[17] Both serum (10–15 ml) and stool (25–50 g) specimens should be sent to the laboratory. Laboratory confirmation of foodborne botulism requires isolating *C. botulinum* from the stool specimen or detecting botulinal toxin in the serum, stool, or epidemiologically implicated food. Food items should remain in their original containers whenever possible and should be stored at 4°C.[29]

Laboratory confirmation of wound botulism requires isolating *C. botulinum* from a wound specimen or detecting botulinal toxin in the patient's serum or stool. Serum and stool specimens should be stored at 4°C. Fluid collected from wounds should be submitted in anaerobic transport containers, processed rapidly, and not refrigerated.[29]

Infant botulism diagnosis can be confirmed by detecting *C. botulinum* in the stool of infants demonstrating the characteristic signs and symptoms produced by the paralyzing botulinal toxin. In addition, botulinal toxin is usually present in the infant's stool.[1]

Botulinal toxin is usually identified at city or state health department laboratories or at the CDC by demonstrating toxicity to mice (mouse bioassay); the specific toxin type is determined via neutralization tests using type-specific antitoxins.[29] Details of these procedures can be found in CDC publications and appropriate package inserts.

Tetanus

The diagnosis of tetanus is most often based on the characteristic clinical picture, which often involves trismus, lockjaw, and risus sardonicus. Laboratory results are rarely required to confirm the diagnosis. When *C. tetani* is isolated from the wound(s) of patients with tetanus, toxicity and neutralization tests can be performed by intramuscular injection of culture supernatant or whole culture into untreated mice and mice protected with tetanus antitoxin.[17] Failure to isolate *C. tetani* from wound cultures does not eliminate the possibility of tetanus.[15]

Clostridium perfringens Foodborne Illness

Although foodborne illness (food poisoning) due to *Clostridium perfringens* is extremely common, single cases often go undiagnosed due to the absence of a distinct clinical syndrome.[22] However, group outbreaks of *C. perfringens* food poisoning are frequently suspected from a combination of clinical and epidemiological data.[22] Laboratory confirmation requires one or more of the following findings:[17]

- $>10^5$ colony forming units (CFU) of *C. perfringens* per gram of implicated food

- A median *C. perfringens* spore count of $>10^6$ per gram of stool from affected patients (Such counts have been found in **healthy** individuals.)[32]

- Isolation of the same serotype of *C. perfringens* from stools of affected patients as isolated from the suspected food (Isolates are not always serotypable.)

A fourth alternative is to demonstrate the presence of *C. perfringens* enterotoxin in stool specimens from ill persons and its absence in stool specimens from well persons.[22]

Stool specimens and food samples should be collected and cultured as soon as possible and should be refrigerated (4°C) if a delay is anticipated.[22] Anaerobic transport containers should be used to ship rectal swabs and feces (Chapter 5).

C. perfringens enterotoxin may be detected in a variety of ways, including animal bioassay, tissue culture, counterimmunoelectrophoresis (CIE), double-gel diffusion, enzyme-linked immunosorbent assay (ELISA), and reverse passive latex agglutination (RPLA).[22] The ELISA method is simple, sensitive (able to detect 5 ng of toxin per gram of feces), and rapid (results within 24 hours).[4] The RPLA procedure is as sensitive as ELISA, more specific, and available commercially from Oxoid, USA.[7]

Clostridium difficile Diseases

Clostridium difficile–induced diarrhea should be suspected in any patient who develops diarrhea during or following antimicrobial therapy. Observation of colonic pseudomembranes by lower gastrointestinal tract endoscopy is diagnostic for pseudomembranous colitis, in which case laboratory confirmation is unnecessary.[16] However, proctoscopic examination is unpleasant for the patient and is probably not cost-effective.

A variety of selective media is available for culture of *C. difficile* from stool specimens, including CCFA and others described in Chapter 6. Because many people harbor *C. difficile* as part of their intestinal flora, recovery of the organism from patients with antimicrobial-associated diarrhea (AAD) is supportive information but cannot be considered diagnostic.[16]

The best means for laboratory diagnosis/confirmation of *C. difficile*–induced diarrhea is demonstrating a cytotoxic effect by tissue culture.[16, 28] A commercial tissue culture kit is available for this purpose (Baxter Healthcare Corporation, Bartels Diagnostics Division, Bellevue, WA). In this assay, *C. difficile* toxins are easily and rapidly detected by inoculating stool filtrates onto monolayers of sensitive human fibroblast cells. Cells affected by the toxin demonstrate a characteristic cytopathic effect (CPE): They round up and become asteroidlike in appearance (see Figure 9-13). Neutralization of the toxin by *C. difficile* antitoxin provides evidence that the cell changes were due to *C. difficile* toxins.

Many other methods have been developed for detecting *C. difficile* toxins A and B and *C. difficile* antigens, including CIE, ELISA, latex agglutination, and a dot immunobinding assay. The Culturette Brand Rapid Latex Test (Becton Dickinson Microbiology Systems, Cockeysville, MD) may be suitable as an initial screening test, but positive results should be confirmed by tissue culture assay.[6, 27, 28] Positive latex test results may be considered presumptive evidence of the presence of *C. difficile*, but because the test does not differentiate between toxin-positive and toxin-negative strains, a cytotoxin assay should then be performed to confirm the presence of cytotoxins. The latex test could be performed at the hospital level and the tissue culture assay at reference laboratories. A similar latex agglutination test (MERITEC-*C. difficile*®) is available from Meridian Diagnostics, Inc. (Cincinnati, OH).

A recent comparative study noted a 92% agreement between the results of a dot immunobinding assay (*C. diff*-CUBE®, Difco Laboratories, Ann Arbor, MI) and those of a cytotoxin assay (CTA), compared to a 90% agreement between latex agglutination

(Becton Dickinson Microbiology Systems, Cockeysville, MD) and CTA results.[35] The authors concluded that the *C. diff*-CUBE test could be effectively used as a screening procedure for presumptive diagnosis of *C. difficile*–associated disease but that stools yielding negative results should be further tested by cytotoxin assay.

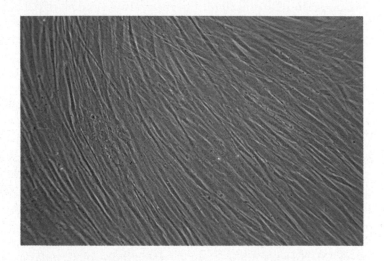

A

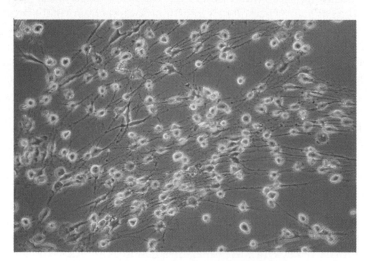

B

Figure 9-13. *Clostridium difficile* cytotoxin assay (courtesy of Bartels Diagnostics Division, Baxter Healthcare Corporation, Bellevue, WA). **(A)** Normal, elongated appearance of the human foreskin fibroblast cells used in the Bartels cytotoxicity assay for *Clostridium difficile* toxins. A blue filter has been used to enhance contrast. **(B)** Effect of *C. difficile* cytotoxin on the human foreskin fibroblast cells. Affected cells round up; have stringy, vacuolated cytoplasm; and become somewhat refractile.

Chapter in Review

- A wide variety of techniques for definitively identifying anaerobic isolates is available. Examples include prereduced, anaerobically sterilized (PRAS) and non-PRAS tubed biochemical test media, biochemical-based minisystems, preexisting enzyme–based minisystems, gas-liquid chromatographic analysis of metabolic end products, and cellular fatty acid analysis by GLC.

- In general, conventional or traditional systems for definitive identification of anaerobic isolates employ several to many large test tubes containing PRAS or non-PRAS biochemical test media. Tests include a variety of carbohydrate fermentations, 20% bile, catalase, esculin hydrolysis, gas production, gelatin hydrolysis, H_2S production, indole production, iron milk (proteolysis), motility, nitrate reduction, and urease.

- The first commercially available alternatives to conventional tubed media were biochemical-based identification systems (e.g., API 20 A® and Minitek®). These provide many of the same tests as the conventional systems, but in the form of a small plastic strip or tray (hence the term "minisystem"). The biochemical-based minisystems are easier and faster to inoculate than a conventional system. Although they can be inoculated aerobically (at the bench), they require **anaerobic** incubation for 24–48 hours. Following incubation, test results are read, a code number is generated for each isolate, and the numbers are looked up in a manufacturer-supplied code book.

- Incubation requirements of the biochemical-based systems have caused many previous users to switch to the newer, preexisting enzyme–based systems (e.g., RapID ANA II®, AN-IDENT®, MicroScan Rapid Anaerobe Identification Panel®, ABL Anaerobe Identification System®, and the ANI Card®). These detect enzymes already present in the bacterial cells. Because they do not depend upon enzyme induction, there is virtually no lag time, and results are available in four hours. The small plastic panels or cards are easy to inoculate, can be inoculated at the bench, and do **not** require anaerobic incubation. Most of the systems generate code numbers, which are looked up in a manufacturer-supplied code book.

- **None** of the commercially available biochemical-based or enzyme-based minisystems will identify **all** the anaerobes that can potentially be isolated from clinical specimens. These systems are primarily of value for commonly isolated and well-characterized anaerobes (e.g., members of the *B. fragilis* group, pigmented species of *Porphyromonas* and *Prevotella*, *C. perfringens*, *F. nucleatum*, and the anaerobic cocci). The data bases from which the code books are developed do not contain anaerobes only rarely isolated from human clinical specimens, nor do they contain unique anaerobes from veterinary sources. Such organisms may be misidentified using minisystems.

- Regardless of which rapid minisystem you use, you are likely to encounter anaerobes that can be identified only by using more sophisticated techniques (e.g., tubes of PRAS or non-PRAS media and/or GLC). The decision to employ time-consuming, labor-intensive procedures to identify anaerobes of questionable clinical significance is one that must be weighed very carefully. In view of the current emphasis on cost containment in clinical laboratories, a detailed description of the organisms might suffice— especially when susceptibility test results are also furnished. Rarely isolated organisms that are difficult and/or time-consuming to identify might best be sent to reference laboratories (See Appendix A).

- Using tubed biochemical systems and GLC procedures is often necessary at the reference laboratory level because these facilities frequently receive isolates deemed clinically significant that cannot be identified at the hospital level. Veterinary laboratories may also require tubed media and GLC to identify unique veterinary isolates not contained in the data bases of the rapid minisystems. A number of reference and veterinary laboratories use the Hungate technique or VPI system.

- Gas-liquid chromatography (GLC) is frequently used in reference and veterinary laboratories as an adjunct to other identification methods. GLC utilizes the chemical and physical properties of a particular component to separate it from other components in a mixture. In GLC, an inert gas serves as the mobile phase and a liquid as the stationary or separating phase. An unknown mixture is prepared for separation, volatilized, and carried as a gas through a packed chromatographic column. Mixture components are then separated, and they elute from the column in order, based on their chemical and physical properties. As the individual components elute from the column, they are detected by one of several types of devices (detectors). A graphic display of the separated components is then printed on a recorder in the form of a chromatogram. Identification of an unknown component is based on its position on the chromatogram (i.e., its relative retention time) as compared to a standard. Quantitation of an unknown component is based on the area under the peak with respect to a known standard.

- Two ways to use GLC to identify anaerobes are (1) analysis of acid endproducts of metabolism and (2) analysis of cellular fatty acids. In the first technique, metabolic end products or their methyl esters are first extracted from broth cultures, then injected into the gas-liquid chromatograph. In the second technique, bacterial cells are saponified to release the fatty acids from lipids within the cells. The fatty acids are next methylated, creating volatile methyl esters. The methyl esters are then extracted from the aqueous phase into an organic solvent, base washed, and injected into a gas-liquid chromatograph.

- Two additional valuable methods of identifying motile bacilli are flagella staining and the desulfoviridin test. Flagella staining provides information as to the number and location of flagella. A motile, desulfoviridin-positive, gram-negative bacillus can be identified as a *Desulfovibrio* sp. A nonmotile, desulfoviridin-positive, gram-negative bacillus is either *Desulfomonas pigra* (straight rods) or *Bilophila wadsworthia* (very pleomorphic).

- Laboratory confirmation of foodborne botulism requires isolating *C. botulinum* from the stool specimen or detecting botulinal toxin in the serum, stool, or epidemiologically implicated food. Laboratory confirmation of wound botulism requires isolating *C. botulinum* from a wound specimen or detecting botulinal toxin in the patient's serum or stool. Diagnosis of infant botulism can be confirmed by detecting *C. botulinum* and/or botulinal toxin in the stool of infants who are demonstrating the characteristic signs and symptoms produced by the paralyzing botulinal toxin. Botulinal toxin is usually identified at city or state health department laboratories or the CDC.

- The diagnosis of tetanus is most often based on the characteristic clinical picture, and laboratory results are rarely required to confirm it. When *C. tetani* is isolated from the wound(s) of patients with tetanus, toxicity and neutralization tests can be performed at city or state health department laboratories or the CDC. Failure to isolate *C. tetani* from wound cultures does not eliminate the possibility of tetanus.

- Laboratory confirmation of *Clostridium perfringens* food poisoning requires two or more of the following: (1) high colony counts of *C. perfringens* in the implicated food, (2) high *C. perfringens* spore counts in stool specimens from affected patients, (3) isolation of the same serotype of *C. perfringens* from stools of affected patients as isolated from the suspected food, (4) demonstrating the presence of *C. perfringens* enterotoxin in stool specimens from ill persons and its absence in stool specimens from well persons. *C. perfringens* enterotoxin is usually detected at city or state health department laboratories or at the CDC.

- Several different types of selective media are available for culture of *Clostridium difficile* from stool specimens, including CCFA. Because many people harbor *C. difficile* as part of their intestinal flora, mere recovery of the organism from patients with antimicrobial-associated diarrhea cannot be considered diagnostic. The "gold standard" for laboratory diagnosis/confirmation of *C. difficile*–induced diarrhea is demonstration of a cytotoxic effect by tissue culture assay. Neutralization of the toxin by *C. difficile* antitoxin provides evidence that the cell changes were due to *C. difficile* toxins. Other methods for detecting *C. difficile* toxins A and B and *C. difficile* antigens include CIE, ELISA, latex agglutination, and a dot immunobinding assay.

Self-Assessment Exercises

1. Explain what is meant by a conventional or traditional system for definitive identification of anaerobes.

   _____

   _____

   _____

2. Identify two advantages and one disadvantage of biochemical-based minisystems for anaerobe identification, when compared to a conventional identification system.

   Advantages: _____

   _____

   Disadvantage: _____

3. Describe four advantages and one disadvantage of preexisting enzyme–based minisystems for anaerobe identification, when compared to a conventional identification system.

   Advantages: a. _____

   b. _____

   c. _____

   d. _____

   Disadvantage: _____

4. Identify two advantages of preexisting enzyme–based minisystems for anaerobe identification, when compared to biochemical-based minisystems.

   Advantages: a. _____

   b. _____

5. State three components of the Hungate technique or VPI system for isolation and identification of anaerobes.

   a. _____

   b. _____

   c. _____

6. Explain the principles of gas-liquid chromatography; include the meaning of "motile phase" and "stationary phase."

   _____

   _____

   _____

   _____

   _____

7. Describe two ways in which gas-liquid chromatography can be used to identify anaerobes.

   a. _____

   b. _____

8. List five genera of anaerobes that contain flagellated species.

   a. _____

   b. _____

   c. _____

   d. _____

   e. _____

9. Name two desulfoviridin-producing genera of anaerobes.

   a. _____

   b. _____

10. Briefly discuss laboratory confirmation of the following clostridial diseases: botulism, tetanus, *C. perfringens* food poisoning, *C. difficile*–induced diarrhea.

References

1. Arnon, S. S. 1989. Infant botulism, p. 601–609. In S. M. Finegold and W. L. George (eds.), Anaerobic infections in humans. Academic Press, San Diego.
2. Baron, E. J., and S. M. Finegold. 1990. Bailey & Scott's diagnostic microbiology, 8th ed. C. V. Mosby, St. Louis.
3. Baron, E. J., P. Summanen, J. Downes, M. C. Roberts, H. Wexler, and S. M. Finegold. 1989. *Bilophila wadsworthia*, gen. nov. and sp. nov., a unique gram-negative anaerobic rod recovered from appendicitis specimens and human faeces. J. Gen. Microbiol. 135:3405–3411.
4. Bartholomew, B. A., M. F. Stringer, G. N. Watson, and R. J. Gilbert. 1985. Development and application of an enzyme linked immunosorbent assay for *Clostridium perfringens* type A enterotoxin. J. Clin. Pathol. 38:222–228.
5. Bartley, S. L. (Centers for Disease Control). 1990. Personal communication.
6. Bennett, R. G., B. E. Laughon, L. M. Mundy, L. D. Bobo, C. A. Gaydos, W. B. Greenough III, and J. G. Bartlett. 1989. Evaluation of a latex agglutination test for *Clostridium difficile* in two nursing home outbreaks. J. Clin. Microbiol. 27:889–893.
7. Birkhead, G., R. L. Vogt, E. M. Heun, J. T. Snyder, and B. A. McClane. 1988. Characterization of an outbreak of *Clostridium perfringens* food poisoning by quantitative fecal culture and fecal enterotoxin measurement. J. Clin. Microbiol. 26:471–474.
8. Bishop, M. L., J. L. Duben-Engelkirk, and E. P. Fody (eds.). 1991. Clinical chemistry: principles, procedures, correlations, 2nd ed. J. B. Lippincott, Philadelphia.
9. Citron, D. M., E. J. Baron, S. M. Finegold, and E. J. C. Goldstein. 1990. Short prereduced anaerobically sterilized (PRAS) biochemical scheme for identification of clinical isolates of bile-resistant *Bacteroides* species. J. Clin. Microbiol. 28:2220–2223.
10. Dowell, V. R., Jr. 1988. Characteristics of some clinically encountered anaerobic bacteria. Unpublished document that accompanied a lecture presented at the IV National Symposium on Clinical Microbiology of the Sociedade Brasileira de Microbiologia, Rio de Janeiro, Brazil. Centers for Disease Control, Atlanta.
11. Dowell, V. R., Jr., and T. M. Hawkins. 1974. Laboratory methods in anaerobic bacteriology: CDC laboratory manual. Centers for Disease Control, Atlanta.

12. Edelstein, M. A. C. 1990. Processing clinical specimens for anaerobic bacteria: isolation and identification procedures, p. 477–507. In E. J. Baron and S. M. Finegold (eds.), Bailey & Scott's diagnostic microbiology, 8th ed. C. V. Mosby, St. Louis.

13. Finegold, S. M. 1987. Anaerobic bacteria: their role in infection and their management. Postgrad. Med. 81:141–147.

14. Finegold, S. M., and M. A. C. Edelstein. 1988. Coping with anaerobes in the 80s, p. 1–10. In J. M. Hardie and S. P. Borriello (eds.), Anaerobes today. John Wiley & Sons, New York.

15. Furste, W., A. Aguirre, and D. J. Knoepfler. 1989. Tetanus, p. 611–627. In S. M. Finegold and W. L. George (eds.), Anaerobic infections in humans. Academic Press, San Diego.

16. George, W. L. 1989. Antimicrobial agent-associated diarrhea and colitis, p. 661–678. In S. M. Finegold and W. L. George (eds.), Anaerobic infections in humans. Academic Press, San Diego.

17. Hatheway, C. L. 1990. Toxigenic clostridia. Clin. Microbiol. Rev. 3:66–98.

18. Heftmann, E. 1975. Chromatography: a laboratory handbook of chromatographic and electrophoretic methods, 3rd ed. Van Nostrand Reinhold, New York.

19. Heimbrook, M. E., W. L. L. Wang, and G. Campbell. 1989. Staining bacterial flagella easily. J. Clin. Microbiol. 27:2612–2615.

20. Hicks, M. R., M. C. Haven, J. R. Schenken, and C. A. McWhorter. 1987. Laboratory instrumentation. J. B. Lippincott, Philadelphia.

21. Holdeman, L. V., E. P. Cato, and W. E. C. Moore (eds.). 1977 (with 1987 update). VPI Anaerobe Laboratory Manual, 4th ed. Virginia Polytechnic Institute and State University, Blacksburg.

22. Johnson, C. C. 1989. *Clostridium perfringens* food poisoning, p. 629–638. *In* S. M. Finegold and W. L. George (eds.), Anaerobic infections in humans. Academic Press, San Diego.

23. Johnson, C. C., and S. M. Finegold. 1987. Uncommonly encountered, motile, anaerobic, gram-negative bacilli associated with infection. Rev. Infect. Dis. 9:1150–1162.

24. Kodaka, H., A. Y. Armfield, G. L. Lombard, and V. R. Dowell, Jr. 1982. Practical procedure for demonstrating bacterial flagella. J. Clin. Microbiol. 16: 948–952.

25. Lee, L. W., and L. M. Schmidt. 1983. Elementary principles of laboratory instruments. C. V. Mosby, St. Louis.

26. Lombard, G. L., and V. R. Dowell, Jr. 1982. Gas liquid chromatography and analysis of the acid products of bacteria. Centers for Disease Control, Atlanta.

27. Lyerly, D. M., D. W. Ball, J. Toth, and T. D. Wilkins. 1988. Characterization of cross-reactive proteins detected by Culturette brand rapid latex test for *Clostridium difficile*. J. Clin. Microbiol. 26:397–400.

28. Lyerly, D. M., H. C. Krivan, and T. D. Williams. 1988. *Clostridium difficile*: its disease and toxins. Clin. Microbiol. Rev. 1:1–18.

29. MacDonald, K. L. 1989. Botulism in adults, p. 591–600. In S. M. Finegold and W. L. George (eds.), Anaerobic infections in humans. Academic Press, San Diego.

30. Navarro, M. C., M. A. Shulman, J. A. Bahrenburg, L. Klints, and D. Schreier. 1987. Comparison of ALADIN™ video image processing to manual interpretation of An-IDENT®. Abstract C-59, Annual Meeting of the American Society for Microbiology, Atlanta.

31. Stoakes, L., T. Kelly, K. Manarin, B. Schieven, R. Lannigan, D. Groves, and Z. Hussain. 1990. Accuracy and reproducibility of the MicroScan rapid anaerobe identification system with an automated reader. J. Clin. Microbiol. 28:1135–1138.

32. Stringer, M. F., G. N. Watson, R. J. Gilbert, J. G. Wallace, J. E. Hassall, E. I. Tanner, and P. P. Webber. 1985. Fecal carriage of *Clostridium perfringens*. J. Hyg. 95:277–288.

33. Sutter, V. L., D. M. Citron, M. A. C. Edelstein, and S. M. Finegold. 1985. Wadsworth anaerobic bacteriology manual, 4th ed. Star Publishing, Belmont, CA.

34. Thomann, W. R., and G. B. Hill. 1986. Modified extraction procedure for gas-liquid chromatography applied to the identification of anaerobic bacteria. J. Clin. Microbiol. 23:392–394.

35. Woods, G. L., and P. C. Iwen. 1990. Comparison of a dot immunobinding assay, latex agglutination, and cytotoxin assay for laboratory diagnosis of *Clostridium difficile*-associated diarrhea. J. Clin. Microbiol. 28:855–857.

CHAPTER 10

SUSCEPTIBILITY TESTING

Anaerobe Resistance to Antimicrobial Agents
Susceptibility Testing of Anaerobes
    When To Test
    What To Test
    Generally Effective Antimicrobial Agents
    Susceptibility Testing Options
        Agar Methods
        Broth methods
Beta-Lactamase Testing
Treating Anaerobe-Associated Infectious
    Processes/Diseases
        Antimicrobial Agents Active Against
            Anaerobes

At the conclusion of this chapter, you will be able to:

1. Give two valid reasons why antimicrobial susceptibility testing should **not** be performed routinely on all anaerobic isolates.
2. Cite three clinical situations in which antimicrobial susceptibility testing of anaerobic isolates **should** be performed.
3. List three species of anaerobes for which susceptibility testing should routinely be performed whenever the isolates are deemed clinically significant.
4. Name three antimicrobial agents that should be included in susceptibility testing of clinically significant anaerobic isolates.
5. Cite three antimicrobial agents essentially always effective against anaerobes.
6. List three methods of antimicrobial susceptibility testing of anaerobes that employ agar plates as part of the procedure and three methods that do not.
7. Briefly describe the principles of the Epsilometer, the spiral gradient endpoint, and broth-disk elution methods of antimicrobial susceptibility testing.
8. Name the method of antimicrobial susceptibility testing of anaerobes to which other methods should be compared whenever other methods are being evaluated.

9. Briefly describe three mechanisms by which anaerobes may become resistant to penicillins.
10. Briefly describe the three major approaches to treating anaerobe-associated infectious processes.

Definitions

Breakpoint (or breakpoint concentration): An achievable serum concentration of a particular antimicrobial agent at or below which organisms are considered susceptible. Organisms not susceptible at a particular breakpoint level are considered resistant.

MIC: Minimal inhibitory concentration—the lowest concentration of antimicrobial agent showing no visible growth of the organism in broth dilution tests. In an agar dilution method, the MIC may be a haze, a single colony, or multiple tiny colonies. Although the MIC is not an absolute value, it does give the physician an indication of the concentration of antimicrobial agent required to inhibit growth of the infecting organism.

NCCLS: National Committee for Clinical Laboratory Standards—interdisciplinary, nonprofit organization that promotes the development and voluntary use of national and international standards and guidelines for improved operation of clinical laboratories (See Appendix B for its address.)

Susceptibility testing of anaerobes is currently a very controversial subject. Complete agreement has not yet been reached as to **when** anaerobic isolates should be tested or **which method** should be used. Matters are in a state of flux, and no "bottom line" is presently available. This chapter, which contains the information, philosophies, and guidelines available at press time, should not be considered the definitive word on susceptibility testing of anaerobes. Readers are advised to contact the National Committee for Clinical Laboratory Standards (NCCLS) to obtain up-to-date guidance on the subject.

Traditionally, the isolation, identification, and susceptibility testing of obligate anaerobes have been quite slow, when compared to other groups of bacteria. As a result, antimicrobial therapy of anaerobe-associated infectious processes has frequently been initiated on **an empirical basis,** that is, when physicians suspected the presence of anaerobes, they selected an antimicrobial agent they **thought** would be effective. Although not a very scientific approach, empirical therapy has undoubtedly saved many lives.

Dr. Stephen Allen of Indiana University Medical Center has pointed out, however,

> Treatments that fail, based upon incorrect empirical clinical decisions, are likely to be associated with prolonged, **costly** hospitalizations and unacceptable morbidity and mortality. . . . Such instances justify the cost of good anaerobic bacteriology. Isolation, identification, and susceptibility tests are required for anaerobes for the same reasons they are required for nonanaerobic microorganisms that are involved in infectious diseases. . . . susceptibility patterns of anaerobes have shifted so that the old rules of "predictability" are no longer true. Antimicrobial susceptibility testing is no longer considered "superfluous," but is now mandatory on anaerobes isolated from serious infections and from patients who fail to respond to empirical treatment.[2]

Thus, for many years, authorities in anaerobic bacteriology have been recommending that susceptibility testing of anaerobes be performed. The following are additional opinions on the subject expressed by other recognized authorities:

> Antimicrobial susceptibility testing of anaerobic bacteria has assumed greater importance in recent years. . . . studies reveal the changing antimicrobial patterns of anaerobic bacteria. . . . patterns vary from one medical center to another. These observations should alert physicians and microbiologists to the need for frequent susceptibility testing of clinically important anaerobic isolates.[32]

> Routine susceptibility testing of all anaerobic isolates is not recommended, but there are a number of circumstances when determination of the susceptibility of individual isolates is of great importance.[33]

Despite these recommendations, susceptibility testing of anaerobes remains a controversial topic. Some physicians and microbiologists are not convinced of the need for it. Arguments frequently heard **against** susceptibility testing of anaerobes include: (1) the questionable significance of the anaerobic isolate in the first place, (2) the "predictability" of anaerobe susceptibilities, (3) the cost, and (4) the difficulty of performing susceptibility testing on anaerobes.

Questions frequently asked by microbiologists include: Should susceptibility testing of anaerobes be performed at all in clinical microbiology laboratories? If so, which isolates should be tested? Which antimicrobial agents should be tested? Which susceptibility test method should be used? These are all valid questions, and each is addressed in this chapter.

Anaerobe Resistance to Antimicrobial Agents

Although anaerobe susceptibility patterns were once thought to be quite predictable, the results of numerous studies have shown this not to be the case. For example, Dr. Jon Rosenblatt from the Mayo Clinic reported the following in 1984:[32]

- Within the *Bacteroides fragilis* group, 60% of the strains tested were considered resistant to tetracycline, 51% to cefoperazone, 45% to cefotaxime, and 19% to moxalactam. Smaller percentages (<10%) were resistant to piperacillin, clindamycin, and cefoxitin.

- With respect to *Bacteroides* spp. other than *B. fragilis* group organisms, 65% of the isolates tested were considered resistant to penicillin, 56% to cephalothin, 56% to tetracycline, and <10% to chloramphenicol, cefoxitin, clindamycin, and cefoxitin.

- Many strains of *Clostridium clostridioforme* and *C. innocuum* were considered resistant to tetracycline (55%), cefoxitin (48%), chloramphenicol (33%), clindamycin (30%), penicillin (26%), and metronidazole (12%).

- Many isolates of *C. ramosum* and other clostridia were considered resistant to tetracycline (39%), clindamycin (23%), cefoxitin (22%), penicillin (16%), metronidazole (11%), and chloramphenicol (10%).

Thus, as early as 1984, many anaerobic isolates were demonstrating resistance to even some of the antimicrobial agents most commonly used today to treat anaerobe-associated infectious processes (e.g., clindamycin, cefoxitin, and metronidazole).

Because *B. fragilis* group organisms (1) are commonly isolated, (2) are frequently involved in serious, life-threatening infectious processes, and (3) of all anaerobes, are among the most resistant to antimicrobial agents, we present more recent information about them. Results of testing 1229 strains of *B. fragilis* group organisms revealed that susceptibility patterns varied among different species and different hospitals.[11] The findings illustrated the need for individual hospitals to determine local susceptibility patterns for this important group of pathogens (to provide guidance for empirical therapy) and to speciate within the group.

In general, the most active of the beta-lactam antibiotics tested were imipenem and ticarcillin-clavulanic acid, each of which had a <1% resistance rate. Cefoxitin, piperacillin, and moxalactam also possessed good activity. The most active non-beta-lactam drugs were chloramphenicol and metronidazole; no isolates were found to be resistant to these. Approximately 5% of the isolates were resistant to clindamycin. Finegold reported clindamycin resistance in *B. vulgatus* (57% of 7 strains tested), *B. thetaiotaomicron* (20% of 15 strains tested), and *B. distasonis* (20% of 5 strains tested).[13] As a rule, *B. fragilis* is more susceptible to certain agents than *B. thetaiotaomicron*, *B. vulgatus*, and *B. ovatus*.[28]

Another study revealed that 33–45% of the *B. fragilis* group isolates were species other than *B. fragilis*.[17] Because the non–*B. fragilis* isolates were frequently resistant to cefotetan and cefmetazole, the authors felt "the empiric use of these agents in seriously ill patients in whom these species are potential pathogens is cause for alarm." Imipenem was the only agent tested that showed consistently good activity against virtually all *B. fragilis* isolates. The authors recommended that clinical laboratories identify **all** *B. fragilis* group isolates to the species level and perform *in vitro* susceptibility testing on them.

Cuchural et al. reported comparative activities of ten of the newer beta-lactam agents against members of the *B. fragilis* group.[12] A total of 557 isolates was tested in this nationwide survey, conducted in 1986. The most active beta-lactam agents were imipenem and ticarcillin-clavulanic acid, which had 0.2 and 1.7% resistance rates, respectively. The rank order of the remaining beta-lactam agents was cefoxitin, piperacillin, moxalactam, ceftizoxime, cefotetan, cefotaxime, cefoperazone, and ceftazadime. No isolates were found to be resistant to metronidazole or chloramphenicol, but a 5% resistance rate was found for clindamycin. Because resistance rates vary significantly among medical centers, the investigators stressed the importance of establishing **local** susceptibility patterns for use in guiding empiric therapy of infectious processes involving members of the *B. fragilis* group. Such infectious processes are frequently serious and life threatening.

Although much of the original publicity regarding resistance of anaerobes to antimicrobial agents involved the resistance of members of the *B. fragilis* group to penicillin and much of the more recent publicity has involved clindamycin and cefoxitin resistance in this group, it is important to note that **resistance occurs among many different anaerobes and involves many different antimicrobial agents.** In addition to members of the *B. fragilis* group, especially resistant anaerobes include *B. gracilis*, *Clostridium ramosum*, and *Fusobacterium varium*.[13]

Additional information regarding resistance of anaerobes to antimicrobial agents can be found in Table 10-1. The results contained in this table reflect susceptibility patterns obtained at the Wadsworth Veterans Administration Medical Center in Los Angeles. Locally determined patterns should be used to guide empirical therapy at other institutions.

Susceptibility Testing of Anaerobes

When To Test

The NCCLS Working Group on Anaerobic Susceptibility Testing (hereafter referred to as the NCCLS Working Group) reported four major reasons for susceptibility testing of anaerobic isolates: [28]

- To determine patterns of susceptibility of anaerobes to **new** antimicrobial agents

- To monitor susceptibility patterns periodically in various medical centers in the United States and other countries

- To monitor susceptibility patterns periodically in local communities and hospitals

- To help manage infections in individual patients

Until all anaerobic isolates are considered clinically significant, no authority on the subject is likely to advocate antimicrobial susceptibility testing of **all** anaerobic isolates. As long as improperly selected or collected specimens continue to be processed in microbiology laboratories, "contaminant" anaerobes (i.e., components of the indigenous microflora that are not participating in the infectious process) will continue to be isolated. Performing susceptibility testing on such contaminants would represent an enormous waste of time, effort, and money, which obviously could not be condoned in the cost-conscious environment of today's clinical micro- biology laboratories. Thus, antimicrobial susceptibility testing should be limited to those isolates considered clinically significant. The following are recommendations made by recognized authorities in anaerobic bacteriology:

> The following should receive high priority: i) anaerobes isolated in pure culture, ii) blood culture isolates (not suspected of being contaminants), and iii) isolates (whether or not they are in pure culture) from properly selected specimens from the central nervous system, lower respiratory tract, abdomen, pelvis, muscle, bone and joints. [2]

> Susceptibility tests are usually needed for patients with serious infections such as endocarditis or brain abscess or infections requiring prolonged therapy such as osteomyelitis, or in infections that persist or recur despite appropriate empirical antimicrobial therapy. [33]

> Routine testing should be limited to those anaerobes isolated from blood cultures, bone and joint infections, and central nervous system infections, as well as to anaerobic bacteria isolated in pure culture. [34]

Table 10-1. Percentage of Anaerobes Susceptible at Breakpoint Concentrations Using a Brucella Blood Agar Dilution Technique

Antimicrobial Agent	Breakpoint (μg/ml)	Anaerobe						
		Bacteroides fragilis	*Bacteroides fragilis* group	*Bacteroides gracilis*	*Fusobacterium* spp.	Anaerobic Cocci	*Clostridium* spp.	Nonspore-forming Gram-Positive Bacilli
Ampicillin/ sulbactam*	16	100	100	Not tested	97	100	100	100
Cefoperazone	32	57	51	67	88	100	100	84
Cefotaxime	32	50	50	78	100	100	100	90
Cefotetan	32	85	56	Not tested	81	100	100	100
Cefoxitin	32	92	75	78	99	100	65	95
Ceftizoxime	32	43**	45**	89	100	100	100	87
Chloramphenicol	16	100	100	100	100	100	100	97
Clindamycin	4	93	81	61	92	97	90	86
Imipenem	8	100	100	100	95	100	100	100
Metronidazole	16	100	100	93	100	98.5	99	63
Moxalactam	16	78	55	78	77	100	100	73
Penicillin G	8	5	6	67	100	100	100	97
Piperacillin†	128	84	85	78	99	100	100	100

* Amoxicillin/clavulanic acid and ticarcillin/clavulanic acid have comparable activities against anaerobes.

** When a broth microdilution technique is used, 90–95% of strains are susceptible at breakpoint.

† Carbenicillin, ticarcillin, and mezlocillin have similar activities against anaerobes.

Source: Based on reference 14.

The NCCLS Working Group believes that susceptibility testing is **not** required for most anaerobes isolated from patients. They cite the following three situations warranting susceptibility testing of anaerobes: [28]

- The usual therapeutic regimens have failed, and the infectious process persists.

- The role of antimicrobial agents is pivotal in determining the outcome.

- It is difficult to make an empiric decision based on precedent (i.e., no precedent may exist).

The NCCLS Working Group also suggests that the following six specific infectious processes warrant susceptibility testing of anaerobic isolates: [28]

- Brain abscess

- Endocarditis

- Infection of a prosthetic device or vascular graft

- Infectious process involving a joint

- Osteomyelitis

- Refractory or recurrent bacteremia

Because many anaerobe-associated infectious processes are polymicrobial, deciding which anaerobic isolates warrant susceptibility testing is frequently difficult. The NCCLS Working Group suggests that anaerobes that are recognized as virulent and/or commonly resistant to antimicrobial agents should be considered for testing. The following are examples of these: [28]

- Members of the *Bacteroides fragilis* group

- Other *Bacteroides* spp. (including *B. gracilis*)

- Pigmented species of *Porphyromonas* and *Prevotella*

- *Clostridium perfringens*

- *Clostridium ramosum*

- Certain *Fusobacterium* spp. (including *F. varium*)

- *Bilophila wadsworthia*

What To Test

The NCCLS Working Group suggests that the following antimicrobial agents be considered for use when testing clinically significant anaerobic isolates: [28]

- Penicillin G

- One or more of the broad-spectrum, antipseudomonal penicillins (e.g., ticarcillin or piperacillin)

- Clindamycin

- Cefoxitin (and, although not specifically mentioned, probably also cefotetan)

- Certain other cephalosporins with known activity against anaerobes

Generally Effective Antimicrobial Agents

The following antimicrobial agents need **not** be tested at present because they are essentially always effective against anaerobic isolates: [28]

- Chloramphenicol

- Combinations of a beta-lactam drug and a beta-lactamase inhibitor (e.g., amoxicillin plus clavulanic acid, ampicillin plus sulbactam, and ticarcillin plus clavulanic acid), except for many strains of *Bilophila wadsworthia*

- Imipenem

- Metronidazole (except for most nonsporeforming, gram-positive bacilli and some strains of *Peptostreptococcus* spp.)

The selection of agents to be tested will, of course, be influenced by their availability (i.e, by the hospital pharmacy's formulary). Dr. Sydney Finegold states that many clinicians will use the agents in the previous list rather than submit requests for susceptibility testing, especially when "a patient—particularly one who is seriously ill—fails to respond well to an empiric regimen."[14] According to the NCCLS, reports of susceptibility testing results should indicate that the agents in the previous list "are not tested against anaerobes because virtually all strains are susceptible at present (except for most nonsporeforming, anaerobic, gram-positive rods and some strains of *Peptostreptococcus* in the case of metronidazole and many strains of *Bilophila wadsworthia* in the case of imipenem and the beta-lactam/beta-lactamase inhibitors)."[28]

Susceptibility Testing Options

Individuals who remain unconvinced that susceptibility testing of selected anaerobes is necessary often cite the cost and/or difficulty of performing such testing as reasons for not doing it. But susceptibility testing of anaerobes need not be expensive or difficult. Of far greater significance are considerations of the **accuracy** of the methods and their **correlation** with the *in vivo* or clinical situation.

Microbiologists will find a wide variety of methods for susceptibility testing of anaerobes. We divide them into two major categories: agar methods (requiring plates of agar medium) and broth methods (not requiring plates of agar medium). In the first category are agar dilution techniques, the spiral gradient endpoint method, the Epsilometer or E test, and disk-diffusion techniques. To date, no "Kirby-Bauer-type" of disk-diffusion method has been standardized to the point that it can be recommended for susceptibility testing of anaerobes. In the second category (broth methods not requiring plated agar medium) are broth macrodilution, broth micro-dilution, and broth-disk elution techniques.

There is, unfortunately, no general agreement as to a "gold standard" technique because problems are encountered with all available methods.[16] The problems include a lack of reproducibility, failure of some anaerobes to grow on or in particular media, difficulty in reading endpoints with certain methods, and a lack of compara-bility between methods.[16]

> Ideally, [an antimicrobial susceptibility test for anaerobes] should be simple, quick, inexpensive, reproducible, and commercially available, and the media used should support good growth of essentially all anaerobes of clinical importance. [In addition,] it should have a good predictive value on the basis of *in vivo* correlations.[16]

Numerous studies have demonstrated that organism susceptibility to antibacterial agents differs depending upon whether the organisms are grown in liquid or on solid media. According to Lorian, "the identical ultrastructures of bacteria found *in vivo* and organisms grown *in vitro* on a surface support the conclusion that *in vitro* experiments aimed at duplicating *in vivo* conditions must be done on solid media."[25]

Regardless of whether agar or broth methods are employed, the NCCLS recommends that at least two of the following quality control (QC) organisms be used each time an anaerobic susceptibility test is set up:[28]

- *Bacteroides fragilis* (ATCC 25285)

- *Bacteroides thetaiotaomicron* (ATCC 29741)

- *Clostridium perfringens* (ATCC 13124)

- *Eubacterium lentum* (ATCC 43055) (Barry and Zabransky recently described the rationale for including this organism in agar dilution QC protocols.[6] This organism may be used as a substitute for *C. perfringens* when higher ranges are desired or when *C. perfringens* is not recommended.)

This list is provided solely to illustrate QC standards in effect at press time. Use only the most recent editions of any NCCLS standards.

Agar Methods. In general, **agar dilution** methods of susceptibility testing involve a large number of agar plates, each containing a specific concentration of a certain antimicrobial agent. To obtain minimal inhibitory concentration (MIC) values, eight or more plates (i.e., eight or more different concentrations) are required for each antimicrobial agent being tested. A large number of isolates is inoculated simultaneously to each plate, using some type of a replicator, such as a Steers replicator. Usually reserved for institutions testing large numbers of isolates simultaneously, the agar dilution method is not very practical for use in smaller clinical laboratories.

In 1985, the NCCLS published an "approved standard" reference agar dilution procedure, using Wilkins-Chalgren (W-C) agar, for antimicrobial susceptibility testing of anaerobic bacteria.[27] Revised versions of the procedure were later published in a "tentative standard"[29] and an "approved standard."[28] The reference agar dilution procedure has been described as "laborious, complex, time-consuming, and relatively expensive."[18]

Although this procedure cannot be recommended for use in small clinical laboratories having relatively few significant anaerobic isolates, the NCCLS standardized agar dilution method is useful for evaluating alternative procedures—procedures more practical for use in small laboratories. However, as has been pointed out by Finegold and Wexler, the NCCLS agar dilution method does **not** allow growth of all anaerobes, and further standardization is needed.[15] Anaerobes reported to be difficult to grow on W-C agar include strains of pigmented *Porphyromonas* and *Prevotella, Fusobacterium* spp. (including *F. nucleatum*), and anaerobic cocci. Consult the most recent NCCLS approved standared for specific details as to how to perform the reference agar dilution procedure and alternative techniques.

The major difference between the **Wadsworth agar dilution procedure** and the reference agar dilution procedure is substitution of supplemented Brucella blood agar (BRU/BA) plates for W-C agar plates. The BRU/BA medium supports growth of essentially all anaerobes, although MICs tend to be higher on BRU/BA than on W-C agar.

The **limited agar dilution procedure** of Hauser et al. is more economical than the reference agar dilution procedure and more practical for use in clinical laboratories.[19] Because the procedure employs divided plates (quadrant plates or biplates), up to 16 different organisms can be tested against 12 antimicrobial agents in one anaerobe jar. Either BRU/BA or W-C agar plates can be used.

Hill and Schalkowsky described the development and evaluation of the **spiral gradient endpoint (SGE) method** for susceptibility testing of *Bacteroides* and *Fusobacterium* spp.[20] In this method, an instrument called a spiral plater (Spiral System Instruments, Bethesda, MD) is used to produce a drug concentration gradient of up to eight twofold dilutions on a single agar plate. The instrument accomplishes continual dilution by precisely depositing a small amount of liquid (the solution of antimicrobial agent) at a variable rate along a spiral track produced on the surface of a rotating agar plate. The concentration gradient decreases in the agar as deposition progresses from near the center of the plate outward.

In the SGE procedure described by Molitoris et al., 150 mm Brucella blood agar plates containing 5% laked blood, hemin, and vitamin K_1 were used, and antimicrobial agent solutions were allowed to diffuse into the medium for three to four hours prior to inoculation (the "diffusion interval").[26] Inocula were adjusted to a turbidity equivalent to a 0.5 McFarland standard. Suspensions of test organisms were applied to the plates in radial lines across the concentration gradient, using an automated inoculator.

After 48 hours of incubation at 37°C, distances were measured from the plate center to the endpoints of growth. Visible growth along the radial streak ceases where the concentration of antimicrobial agent has reached the bactericidal and/or bacteriostatic concentration (Figure 10-1). These measurements were then entered into a computer, and gradient endpoint concentrations were determined using software provided by Spiral System Instruments. Endpoint concentrations are based upon the position of the endpoint and diffusion characteristics of the antimicrobial agents.

Molitoris et al.[26] found that SGE concentrations correlated well with conventional agar dilution MIC values and concluded that spiral gradient endpoints offer a reliable alternative to the conventional agar dilution method for determining antimicrobial susceptibility of anaerobic bacteria.[26] As of press time, however, the SGE method had not yet received an endorsement from the NCCLS for such testing.

The SGE method requires purchasing a spiral plater instrument (~$13,000) (Figure 10-2), a replicator (~$4,000) (Figure 10-3), computer software (~$2,000), and additional accessories. It requires the user to prepare spiral gradient plates, which can be stored for only several hours prior to use. Until plates can be stored for extended periods of time and/or are available commercially, this procedure seems unlikely to be popular in clinical microbiology laboratories.

A standardized **disk diffusion** method of antimicrobial susceptibility testing for *Enterobacteriaceae*, enterococci, staphylococci, and nonfermentative, gram-negative bacilli has been available since the mid 1960s. It is frequently referred to as the "Bauer-Kirby" or "Kirby-Bauer" technique, named for two of the individuals who initially described it. For results to be valid, many aspects of the procedure must be rigidly controlled, with strict adherence to NCCLS guidelines.

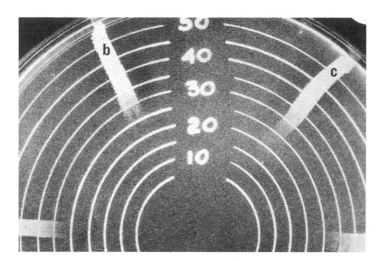

Figure 10-1. Spiral gradient endpoint (SGE) method (courtesy of Spiral System Instruments, Bethesda, MD). This photograph illustrates organisms growing on a spiral gradient plate and the points where visible growth ceases. Distances from the center of the plate to the endpoints of growth are measured and then entered into a computer, which calculates gradient endpoint determinations.

Figure 10-2. Spiral plater for use in the SGE method (courtesy of Spiral System Instruments, Bethesda, MD). This photograph illustrates the instrument (spiral plater) used to prepare spiral gradient plates for use in the spiral gradient endpoint (SGE) procedure. While the plate rotates, a small amount of antimicrobial agent is accurately and repeatedly deposited at a variable rate along a spiral track on the agar surface, creating a wide range of antimicrobial concentrations in the agar. Test organisms are then streaked radially on the agar surface.

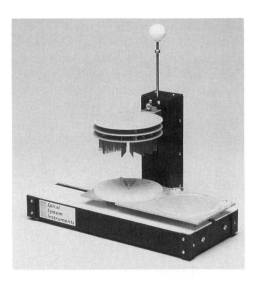

Figure 10-3. Replicator for use in the SGE method (courtesy of Spiral System Instruments, Bethesda, MD). This photograph illustrates the instrument (replicator) that can be used to deposit up to 15 test strains across the antimicrobial agent gradient via its 15 rows of hollow pins.

In this technique, antimicrobial-impregnated disks are placed on the surface of an inoculated Mueller-Hinton agar plate. During incubation, the antimicrobial agents diffuse from the disks in a circular pattern. If the organism is to any extent susceptible to an agent, a zone of no growth (or inhibition of growth) will be evident following the 18-hour incubation period. The diameter of each zone is then measured, and the value obtained is compared to a published table of values. Depending upon the zone diameter, the organism is reported as "susceptible," "moderately susceptible," "intermediate," or "resistant"—terms that relate to obtainable serum levels for these antimicrobial agents.

Investigators have experimented for many years with disk diffusion methods for susceptibility testing of anaerobes, but to date no NCCLS standardized technique exists. The results of a 14-laboratory evaluation of a disk diffusion technique using W-C agar was recently published.[7] Nine different antimicrobial agent disks were evaluated: ampicillin, ampicillin/sulbactam, cefoperazone, cefoperazone/sulbactam, cefoxitin, chloramphenicol, clindamycin, metronidazole, and tinidazole. Anaerobes that fail to grow adequately after overnight incubation on unsupplemented W-C agar could not be tested by this procedure. The investigators concluded that the disk diffusion method described by Horn et al.[21] can be performed as precisely as the standard reference agar dilution method for rapidly growing members of the *B. fragilis* group. Additional studies, involving other antimicrobial agents and other anaerobes, will have to be performed before this method can be recommended.

Although no NCCLS standardized "Kirby-Bauer type" disk diffusion method is available for susceptibility testing of anaerobic isolates, a diffusion method of susceptibility testing called the **PDM Epsilometer method or "E test"** has been used successfully for anaerobes.[8, 10]

The term "Epsilometer" refers to a thin, 5 x 50 mm, inert, nonporous strip (or "carrier") with a continuous exponential gradient of antimicrobial agent immobilized on one side and a reading and interpretive scale on the other. The antimicrobial agent gradient covers a broad concentration range, corresponding to approximately 20 twofold dilutions. The slopes and concentration ranges are optimally designed to correspond to clinically relevant MIC ranges and breakpoints selected for categorization of susceptibility groups (see Figure 10-4).

An agar plate containing a suitable test medium is inoculated with a test organism according to the manufacturer's instructions. Test strips are then applied in an optimal pattern (see Figure 10-5) so that the maximum concentration on each strip is nearest to the outer edge of the petri dish. The plate is immediately incubated anaerobically for the prescribed time period.

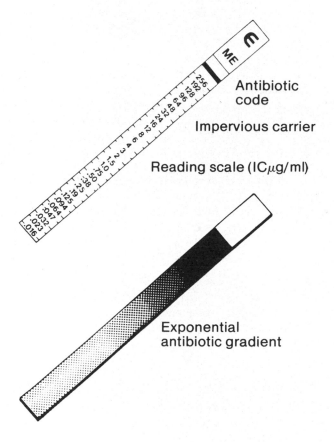

Figure 10-4. Appearance of an E test carrier (the Epsilometer) (courtesy of AB Biodisk, Solna, Sweden). These diagrams illustrate one of the nonporous carrier strips that give the Epsilometer or E test its name. One side of the strip contains a continuous exponential gradient of antimicrobial agent, and the other side contains a reading and interpretive scale. One carrier strip is used for each antimicrobial agent to be tested.

1. Inoculation of agar
   plate according to
   standard practice

2. Application of test
   carriers and incubation
   for 24–48 hours

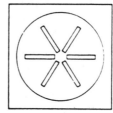

3. Reading of inhibitory
   concentrations (IC)
   and clinical interpretation

Figure 10-5.   The E test procedure (courtesy of AB Biodisk, Solna, Sweden). To perform the E test, the agar surface is first inoculated with the test organism. Carrier strips are then applied to the agar surface. Following incubation, inhibitory concentrations are determined and recorded.

When applied to an inoculated agar plate, the antimicrobial gradient is immediately released from the test strip into the agar, creating a continuous and exponential gradient of antimicrobial agent concentrations beneath the linear axis of the carrier. After incubation, an inhibition ellipse is seen (see Figure 10-6). The zone edge intersects the carrier at the antimicrobial concentration no longer inhibitory to growth. The point of intersection gives the "inhibitory concentration" (IC) in μg/ml—a direct measure of the susceptibility of the microorganism to the tested antimicrobial agent. ICs are read directly from the scale on the carrier. Figure 10-7 depicts results obtained when a strain of *Clostridium perfringens* was tested by the Epsilometer method. Figure 10-8 provides a close-up view of the point where the zone edge intersects the test carrier.

Citron et al. tested a total of 106 strains of anaerobes (including *Bacteroides, Clostridium, Fusobacterium,* and *Peptostreptococcus* spp.) by E test and agar dilution.[10] As the inoculum, a suspension of each organism equivalent to a 0.5 McFarland standard was prepared by adding growth from a 72-hour plate to a tube of Brucella broth. The antimicrobial agents tested were cefotaxime (CT), cefoxitin (FX), clindamycin (CC), imipenem (IM), metronidazole (MZ), and penicillin (PN). Both Brucella blood agar plates and W-C agar plates were tested, and fewer major discrepancies were obtained using W-C agar. A major discrepancy was defined as greater than or equal to a two dilution difference that resulted in a change of interpretation from susceptible to resistant or vice versa. Only three major discrepancies between E test and agar dilution results were noted when 66 rapid growers were read at 24 hours, using W-C agar—one for CT, one for CC, and one for PN. Agreement within plus or minus 2 dilutions ranged from a low of 84% (for FX) to 100% (for MZ).

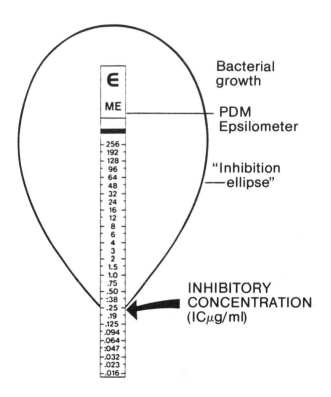

Figure 10-6. The E test principle (courtesy of AB Biodisk, Solna, Sweden). If organism growth was to any extent inhibited by the antimicrobial agent, an inhibition ellipse (such as the one shown) is seen. Using the interpretive scale on the carrier, an inhibitory concentration (IC) is determined and recorded.

The authors concluded that the E test "is simple to perform and read, and offers an easy way for clinical laboratories to determine MICs on anaerobes." Indeed, this technique shows great promise. It is commercially available, is relatively easy to perform, and bears some similarity to a procedure already familiar to most microbiologists (the "Kirby-Bauer" test for facultative bacteria). Additional studies will be needed, however, especially with respect to standardization of all steps in the test procedure, before the E test can be recommended or receive NCCLS endorsement.

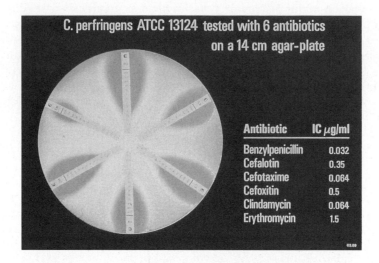

Figure 10-7. Susceptibility testing of *Clostridium perfringens* using the E test procedure (courtesy of AB Biodisk, Solna, Sweden). This photograph depicts results obtained when a strain of *Clostridium perfringens* (ATCC 13124) was tested against six different antimicrobial agents using the Epsilometer or E test method.

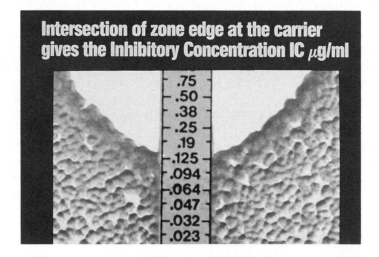

Figure 10-8. E test: intersection of the inhibition ellipse and interpretive scale (courtesy of AB Biodisk, Solna, Sweden). This photograph provides a close-up view of the point where the edge of the zone of no growth (inhibition ellipse) intersects the interpretive scale on the carrier strip. The inhibitory concentration (IC) is read directly from the interpretive scale.

Broth Methods.   The NCCLS has published guidelines for alternative methods for susceptibility testing of anaerobic bacteria, including a broth dilution (or broth macrodilution) technique and a broth microdilution technique.[28] A previously endorsed third type of broth method, called the broth-disk (or broth-disk elution) technique, is not currently recommended by the NCCLS.

There are essentially two types of broth dilution techniques: a macro method and a micro method. The macro method, referred to here as the **broth dilution** method, consists of a large number of test tubes, each containing a different concentration of antimicrobial agent in a broth that will support growth of the anaerobe. A separate row of tubes is required for each antimicrobial agent being tested. Thus, testing one isolate would require eight or so test tubes for each agent being tested.

The NCCLS recommends the use of Brucella broth containing hemin, $NaHCO_3$, and vitamin K_1 or some other acceptable broth (e.g., Schaedler's, West-Wilkins, brain-heart infusion, or W-C broth). Endpoints (minimum inhibitory concentrations or MICs) are read after 48 hours of anaerobic incubation at 35–37°C. The MIC is the lowest concentration of drug showing no visible growth. An inoculated broth containing no antimicrobial agent serves as a growth control. The broth dilution technique is more dependable than the microdilution technique but is less practical due to the commercial availability of frozen or lyophilized microtiter MIC trays for susceptibility testing of anaerobes.[28]

A **broth microdilution** procedure uses a plastic microtiter tray in place of the large numbers of test tubes used in the macro method. Such methods are popular in clinical laboratories because they are commercially available (e.g., from MicroScan, Becton Dickinson [Sceptor®], and MSI/Micro-Media), can be stored frozen or lyophilized for extended periods of time, and results can sometimes be read by instruments. Most of the commercial systems require 48 hours of incubation before results are available. The antimicrobial agents used in the MicroScan, MSI/Micro-Media, and Sceptor® MIC panels are shown in Table 10-2. MicroScan panels contain West-Wilkins broth; MSI/Micro-Media panels contain W-C broth.

A relatively new microtiter MIC system, developed by MicroTech Medical Systems (Aurora, CO) and described by Ratkovich and Kilgore, is currently marketed by Innovative Diagnostic Systems, Inc. (Atlanta, GA).[31] The system (shown in Figure 10-9) uses W-C broth and permits simultaneous testing of 14 or 15 different antimicrobial agents, depending upon which panel is used. The ANA MIC-20 panel, which requires storage at -20°C, contains ampicillin/sulbactam (1–32 μg/ml), cefotaxime (2–64 μg/ml), cefotetan (2–64 μg/ml), cefoxitin (2–64 μg/ml), ceftizoxime (2–64 μg/ml), ceftriaxone (2–64 μg/ml), chloramphenicol (1–32 μg/ml), clindamycin (0.5–16 μg/ml), metronidazole (0.5–16 μg/ml), mezlocillin (4–128 μg/ml), penicillin (0.06–32 μg/ml), piperacillin (4–128 μg/ml), tetracycline (0.5–16 μg/ml), and ticarcillin (4–128 μg/ml). The ANA MIC-70 panel, which requires storage at -70°C, contains the same antimicrobial agents, except that imipenem (0.5–16 μg/ml) is included, and ticarcillin/clavulanic acid (4–128 μg/ml) takes the place of ticarcillin alone.

According to Zabransky, many experienced microbiologists favor the broth microdilution method of anaerobe susceptibility testing.[37] He points out, however, that standardization of the procedure is required, including choice of medium, method of inoculum preparation, and inoculum size. Broth microdilution methods require a 48-hour incubation period in an anaerobic environment and well volumes of 0.1 milliliters.

Table 10-2. Antimicrobial Agents Used in Three Commercial Anaerobe MIC Panels

Agent	Concentration (μg/ml)		
	MicroScan	MSI/Micro Media	Sceptor®*
Carbenicillin	8–128	8–512	8–512
Cefotaxime	N/A	N/A	0.5–32
Cefoxitin	1–16	1–64	0.5–32
Cephalothin	N/A	N/A	0.5–32
Chloramphenicol	0.5–8	0.5–32	0.5–32
Clindamycin	0.25–4	0.25–16	0.25–16
Erythromycin	N/A	N/A	0.5–32
Penicillin	0.06–4	0.06–4	0.12–16
Tetracycline	N/A	0.25–16	0.5–32
Metronidazole	0.5–8	0.25–16	1–64

* In addition to these antimicrobial agents, the Sceptor® MIC panel also contains Nitrocefin for beta-lactamase testing.

The **broth-disk elution** (BDE) or broth-disk method of anaerobe susceptibility testing has been described as the simplest, least expensive, most rapid, and most flexible procedure for daily use in clinical laboratories of all sizes.[9, 35] Unfortunately, it is unreliable with certain "bug-drug" combinations and is **not** currently recommended by the NCCLS.

In the BDE procedure, a designated number of antimicrobial disks of a prescribed potency is placed into a broth that will support the growth of the anaerobic isolate being tested. The agent elutes from the disk and diffuses uniformly throughout the broth, resulting in an appropriate concentration of the agent. One concentration of each agent (the breakpoint concentration) is tested. One problem with this test is that there is no real consensus on breakpoint concentrations for anaerobes.[28]

Following inoculation and incubation of the tubes, the amount of growth in each tube is compared to the amount of growth in an antimicrobial agent–free control tube. An organism is considered **susceptible** to a particular agent if there is growth in the control tube but none in the antimicrobial agent–containing tube. Growth in both tubes indicates that the organism is **resistant** to that particular agent.

The original BDE test, described by Wilkins and Theil, required prereduced media, roll-tube inoculations, and anaerobic conditions.[36] Kurzynski et al. modified the procedure so that supplemented thioglycollate broth and aerobic incubation could be used.[23] The modified procedure requires neither an anaerobic chamber nor prereduced media, so it can be used in even the smallest laboratories. Because the procedure is far less cumbersome than broth macrodilution or agar dilution procedures, it rapidly gained popularity. When properly performed, it can provide clinically useful information for many antimicrobial agents. Depending upon the growth rate of the isolate being tested, results are often available within 24 hours.

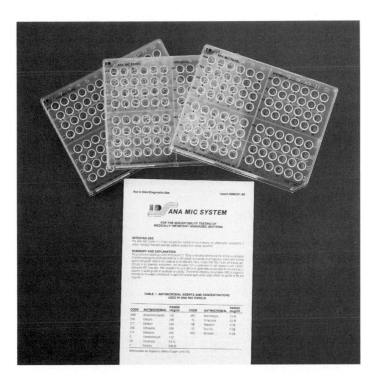

Figure 10-9. Commercial broth microdilution system (courtesy of Innovative Diagnostic Systems, Inc., Atlanta, GA). This photograph depicts the ANA MIC SYSTEM, a commercial frozen broth microdilution method. The panels are manufactured by MicroTech Medical Systems (Aurora, CO) and distributed by Innovative Diagnostic Systems, Inc. (Atlanta, GA). Each panel contains serial dilutions of 13 antimicrobial agents. Minimal inhibitory concentrations (MICs) are determined by examining wells for presence or absence of visible growth. The MIC of each agent is the lowest concentration yielding no visible growth.

However, like all susceptibility testing systems available for anaerobes, the BDE test has its limitations. For example, discordant results between BDE and broth microdilution test results for some of the cephalosporin-cephamycin agents prompted one group of investigators to recommend discontinuation of the BDE procedure for ceftizoxime, ceftriaxone, and cefotaxime and to suggest a thorough reexamination and standardization of the BDE method.[1] An earlier report had advised that neither cefoxitin nor ceftizoxime be tested by the BDE method.[22] Overinoculation of BDE tubes may result in false-resistance results.[1] Because the actual antimicrobial agent concentration within a disk can vary widely above or below the stated concentration, the final concentration in a BDE tube may be well above or below the theoretical concentration.[1] When several disks are used to attain the desired concentration, particles that separate from the disks may interfere with the reading.[14] Although thioglycollate broth had been recommended by the NCCLS in the past, it has been shown that some anaerobes grow more poorly in it than in other broths (e.g., brain heart infusion, Schaedler, and Anaerobe Broth MIC [Difco]) that can be used in the BDE test.[1] In addition, thioglycollate is said to inactivate imipenem.[5] Thus, although the BDE procedure has been recommended in the past, it is presently not endorsed by the NCCLS. Following a thorough reevaluation and precise standardization of the method, it may return.

To date, **none** of these susceptibility testing methods has been established as fully reliable for predicting the clinical or bacteriological outcome of a given anaerobe-associated infectious process. Nonetheless, the NCCLS-approved methods are generally useful for determining the comparative effectiveness of various antimicrobial agents and for surveillance of resistance patterns.[28]

Beta-Lactamase Testing

Beta-lactamases are enzymes that destroy the beta-lactam ring of penicillins (penicillinases) and cephalosporins (cephalosporinases), thus rendering these antibiotics ineffective. The first anaerobes shown to produce beta-lactamases were members of the *Bacteroides fragilis* group. Those of *B. fragilis* are primarily cephalosporinases.[30] It is now known that strains of all of the following anaerobes produce beta-lactamases:[13, 30]

Bacteroides spp.
  B. capillosus
  B. coagulans
  B. fragilis group
    B. distasonis
    B. fragilis
    B. ovatus
    B. thetaiotaomicron
    B. uniformis
    B. vulgatus
  B. splanchnicus
Clostridium spp.
  C. butyricum
  C. clostridiiforme
  C. ramosum
Fusobacterium spp.
  F. mortiferum
  F. nucleatum
  F. varium

Prevotella spp.
  P. bivia
  P. buccae
  P. disiens
  P. melaninogenica
  P. oralis
  P. oris
  P. ruminicola
Others
  Megamonas hypermegas
  Mitsuokella multiacidus
  Porphyromonas asaccharolytica

The ability of an anaerobe to produce beta-lactamase enzymes can be determined using simple, commercially available methods. Investigators who evaluated three of these methods recommended that the nitrocefin disk test (Cefinase®, manufactured by Becton Dickinson Microbiology Systems) be used for testing anaerobes, based upon accuracy and ease of use.[24] If the Cefinase® test is negative after 15 minutes, Appelbaum et al. recommend that the disk be incubated for one hour at 37°C.[4] A positive Cefinase® test can predict resistance of the organism being tested to penicillin G and ampicillin.[28] A negative test result is of little predictive value.

The absence of beta-lactamase does not necessarily mean that an organism will be susceptible to beta-lactam antibiotics. Anaerobes can be resistant to such agents by mechanisms other than beta-lactamases (e.g., by alteration of the number or type of penicillin-binding proteins or blocked penetration of the drug into the active site via alteration of the bacterial outer membrane pores or "porins").[30] Anaerobes known to be resistant to beta-lactam drugs by mechanisms other than production of beta-lactamases include *Bacteroides gracilis, Bilophila wadsworthia,* and some strains of *B. distasonis* and *B. fragilis.*[28] Thus, beta-lactamase testing may be used as an adjunct to susceptibility testing, but never as a replacement for it.[33] It is not necessary to test all clinically significant anaerobes for beta-lactamase production. Members of the *B. fragilis* group, for example, need not be tested for beta-lactamase because one may assume that these organisms are beta-lactamase producers. Additional guidelines for beta-lactamase testing of anaerobes are contained in Chapter 12.

Appelbaum et al. reported the results of testing 320 strains of non–*B. fragilis* group *Bacteroides* spp. and 129 strains of *Fusobacterium* spp. for beta-lactamase production by the nitrocefin disk method and susceptibility of these strains to amoxicillin, amoxicillin-clavulanate, ticarcillin, ticarcillin-clavulanate, cefoxitin, imipenem, and metronidazole by the NCCLS agar dilution method.[3] The strains had been isolated at 28 U.S. medical centers. Overall, 64.7% of the *Bacteroides* spp. and 41.1% of the *Fusobacterium* spp. were beta-lactamase positive. With the exception of *B. denticola* and *B. gracilis,* all species of *Bacteroides* that were tested contained beta-lactamase positive strains. Most species of *Fusobacterium* tested contained beta-lactamase positive strains. Organisms with the highest beta-lactamase positivity rates were *B. bivius* (*Prevotella bivia*) (85%), *B. splanchnicus* (83.3%), *B. eggerthii* (77.8%), *P. oralis* (77.1%), and *F. mortiferum* (76.9%).

The authors concluded that the significant rates of beta-lactamase production in these two groups of anaerobes illustrate the need for routine beta-lactamase testing of such isolates in clinical laboratories. Based upon the anaerobes' susceptibility testing results, the authors felt that amoxicillin-clavulanate, ticarcillin, cefoxitin, imipenem, and metronidazole should be suitable agents for treating infectious processes involving beta-lactamase-producing strains of these organisms and that most beta-lactamase negative strains should be susceptible to any of the beta-lactam agents used in the study.

Treating Anaerobe-Associated Infectious Processes/Diseases

A detailed discussion of the treatment of anaerobe-associated infectious processes and diseases is beyond the scope of this book. For additional information, see a recent chapter by Finegold, which is the basis of this section.[13]

Treatment of such infectious processes involves a three-pronged attack:[13]

- Create an environment in which anaerobes cannot proliferate. Useful measures include removing dead tissue (debridement), draining pus, eliminating obstructions, decompressing tissues, releasing trapped gas, and improving circulation and oxygenation of tissues. In lesser infections, surgical therapy may be all that is required.

- Arrest the spread of anaerobes into healthy tissues. Antimicrobial agents play an important role here

- Neutralize toxins produced by the anaerobes when such toxins are present. Specific antitoxins can be used.

Initiation of antimicrobial therapy is usually empirical and is based upon the following knowledge:[13]

- The nature and location of the infectious process. Special characteristics of specific diseases significantly affect the approach to therapy. In bacterial endocarditis, for example, a bactericidal drug must be used. The blood-brain barrier makes certain drugs unsuitable and others less effective for treating brain abscess or meningitis (clindamycin, for example, does not cross the blood-brain barrier).

- The pathogens anticipated in infectious processes of the type being treated. Aspiration pneumonia, for example, would be expected to involve mixed flora of the oral cavity. Intraabdominal infectious processes would be expected to involve mixed gastrointestinal flora. Female genitourinary infectious processes would be expected to involve mixed vaginal flora.

- Factors that might have modified the anticipated flora.

- Typical susceptibility patterns of the expected flora (see Table 10-1).

- Gram stain results.

- The severity of the infection.

The clinical outcome in any given case will be influenced by a multitude of factors, including the speed with which antimicrobial therapy is initiated; whether the agent crosses the blood-brain barrier; the concentration of agent achieved in serum, tissues, and body cavities or spaces; the degree of protein binding that occurs; and the effect of microbial and other enzymes on the agent.[28]

Antimicrobial Agents Active Against Anaerobes

As mentioned earlier in this chapter, certain antimicrobial agents are nearly always active against anaerobes. These include metronidazole, chloramphenicol, imipenem, ticarcillin plus clavulanic acid, and other combinations of a beta-lactam drug and a beta-lactamase inhibitor. Brief descriptions of some of the antianaerobe drugs follow.[13]

Penicillin has a poor aerobic spectrum of activity, low toxicity (some hypersensitivity reactions and neurotoxicity), good CSF penetration, and very good bactericidal activity. Although penicillin can be administered both orally and parenterally, the parenteral route is usually preferred for anaerobic infections because higher dosage and blood levels are facilitated.

Clindamycin has a poor aerobic spectrum of activity, low to moderate toxicity (some pseudomembranous colitis), poor CSF penetration, and moderate bactericidal activity. Clindamycin can be administered both orally and parenterally.

Metronidazole has no aerobic spectrum of activity, low toxicity (some neurotoxicity), excellent CSF penetration, and excellent bactericidal activity. Metronidazole can be administered both orally and parenterally.

Chloramphenicol has a good aerobic spectrum of activity, high toxicity (aplastic anemia), excellent CSF penetration, and no bactericidal activity. Chloramphenicol can be administered both orally and parenterally.

Cefoxitin has a good spectrum of aerobic activity, low toxicity (some hypersensitivity reactions, pseudomembranous colitis), moderate CSF penetration, and very good bactericidal activity. Cefoxitin may be administered parenterally, but not orally.

The carboxypenicillins (e.g., carbenicillin and ticarcillin) have a good aerobic spectrum of activity, relatively low toxicity (some hypersensitivity reactions, neurotoxicity, bleeding, and fluid and electrolyte problems), good CSF penetration, and very good bactericidal activity. The carboxypenicillins are poorly absorbed by the oral route and therefore not suitable for therapy of systemic infectious processes. They can be administered parenterally.

Piperacillin and the ureidopenicillins (e.g., mezlocillin and azlocillin) have a very good aerobic spectrum of activity, relatively low toxicity (some hypersensitivity reactions, neurotoxicity, and bleeding), good CSF penetration, and very good bactericidal activity. Piperacillin and the ureidopenicillins are poorly absorbed by the oral route, so are not suitable for therapy of systemic infectious processes. They can be administered parenterally.

Chapter in Review

- Susceptibility testing of anaerobic isolates remains a controversial subject, with a lack of consensus on **which** anaerobic isolates should be tested and **how** they should be tested.

- Antimicrobial resistance occurs among many different anaerobes and involves many different antimicrobial agents. Especially resistant anaerobes include members of the *Bacteroides fragilis* group, *B. gracilis*, *Clostridium ramosum*, and *Fusobacterium varium*.

- The major indications for performing susceptibility testing of anaerobes are (1) to determine patterns of susceptibility of anaerobes to **new** antimicrobial agents, (2) to monitor susceptibility patterns periodically in medical centers, (3) to monitor susceptibility patterns periodically in local communities and hospitals, and (4) when appropriate, to help manage infectious processes in individual patients.

- Clinical situations warranting susceptibility testing of anaerobes include those in which (1) the usual therapeutic regimens have failed, and the infectious process persists, (2) the role of antimicrobial agents is pivotal in determining the outcome, and (3) it is difficult to make an empiric decision due to a lack of precedent.

- Specific infectious processes warranting susceptibility testing of anaerobic isolates include brain abscesses, endocarditis, infection of prosthetic devices or vascular grafts, infectious processes involving joints, osteomyelitis, and refractory or recurrent bacteremia.

- Especially virulent or drug-resistant anaerobes that warrant susceptibility testing include members of the *Bacteroides fragilis* group, other *Bacteroides* spp. (including *B. gracilis*), pigmented species of *Porphyromonas* and *Prevotella*, *Clostridium perfringens*, *C. ramosum*, certain *Fusobacterium* spp., and probably *Bilophila wadsworthia*.

- Antimicrobial agents for use in susceptibility testing of clinically significant anaerobes include penicillin G, one or more of the broad-spectrum, antipseudomonal penicillins (e.g., ticarcillin or piperacillin), clindamycin, cefoxitin, cefotetan, and certain other cephalosporins with known activity against anaerobes.

- Antimicrobial agents that are essentially always effective against anaerobic isolates include chloramphenicol; combinations of a beta-lactam drug and a beta-lactamase inhibitor (e.g., amoxicillin plus clavulanic acid, ampicillin plus sulbactam, and ticarcillin plus clavulanic acid), except for many strains of *Bilophila wadsworthia*; imipenem; and metronidazole, except against nonsporeforming, gram-positive bacilli.

- A variety of methods is available for susceptibility testing of anaerobic isolates, although problems are encountered with all of them. Agar plate methods include agar dilution, Wadsworth agar dilution, limited agar dilution, spiral gradient endpoint (SGE), Epsilometer (E test), and disk diffusion. Broth methods include broth dilution, broth microdilution, and broth-disk elution. Not all these procedures are currently endorsed by the NCCLS.

- Many different anaerobes produce beta-lactamases, including *Bacteroides* spp. (*B. capillosus*, *B. coagulans*, *B. fragilis* group, *B. splanchnicus*), *Clostridium* spp. (*C. butyricum*, *C. clostridiiforme*, *C. ramosum*), *Fusobacterium* spp. (*F. mortiferum*, *F. nucleatum*, *F. varium*), *Megamonas hypermegas*, *Mitsuokella multiacidus*, *Porphyromonas asaccharolytica*, and *Prevotella* spp. (*P. bivia*, *P. buccae*, *P. disiens*, *P. melaninogenica*, *P. oralis*, *P. oris*, *P. ruminicola*). Some anaerobes (e.g., *Bacteroides gracilis*, *Bilophila wadsworthia*, and some strains of *B. distasonis* and *B. fragilis*) are resistant to beta-lactam drugs by mechanisms other than production of beta-lactamases.

- Treatment of anaerobe-associated infectious processes involves a three-pronged attack. First, an environment must be created in which anaerobes cannot proliferate (e.g., by debridement, drainage, elimination of obstructions, decompression of tissues, release of trapped gas, improvement of circulation, and improvement of oxygenation of tissues). Next, the spread of anaerobes into healthy tissues must be arrested (e.g., by use of antimicrobial agents). Third, when applicable, specific antitoxins must be used to neutralize toxins produced by the anaerobes.

Self-Assessment Exercises

1. Give two valid reasons why antimicrobial susceptibility testing should **not** be performed routinely on all anaerobic isolates.

    a. _____

    b. _____

2. Cite three clinical situations in which antimicrobial susceptibility testing of anaerobic isolates **should** be performed.

    a. _____

    b. _____

    c. _____

3. Name three species of anaerobes for which susceptibility testing **should** routinely be performed whenever the isolates are deemed clinically significant.

    a. _____

    b. _____

    c. _____

4. List three antimicrobial agents that should be included in susceptibility testing of clinically significant anaerobic isolates.

    a. _____

    b. _____

    c. _____

5. Name three antimicrobial agents that are essentially always effective against anaerobes.

    a. _____

    b. _____

    c. _____

6. List three methods of antimicrobial susceptibility testing of anaerobes that employ agar plates as part of the procedure and three methods that do not.

    a. _____

    b. _____

    c. _____

    d. _____

    e. _____

    f. _____

7. To what method of antimicrobial susceptibility testing of anaerobes should all other methods be compared whenever other methods are being evaluated?

    _____

8. Briefly describe three mechanisms by which anaerobes may become resistant to penicillins.

    a. _____

    b. _____

    c. _____

9. What are the three major approaches to treating anaerobe-associated infectious processes?

    a. _____

    b. _____

    c. _____

10. TRUE or FALSE: There is no NCCLS standardized "Kirby-Bauer type" of disk diffusion susceptibility testing for anaerobes.

11. TRUE or FALSE: The broth-disk elution method of susceptibility testing provides a minimal inhibitory concentration (MIC) for each antimicrobial agent being tested.

12. TRUE or FALSE: The most practical method for susceptibility testing of anaerobes in small clinical microbiology laboratories is the NCCLS standardized agar dilution technique.

13. TRUE or FALSE: All clinically significant anaerobic isolates should be tested for beta-lactamase production.

14. TRUE or FALSE: Broth microdilution methods for susceptibility testing of anaerobes are popular in clinical laboratories because they are commercially available, can be stored frozen or lyophilized for extended periods of time, and, in some cases, results can be read by instruments.

References

1. Aldridge, K. E., A. Henderberg, D. D. Schiro, and C. V. Sanders. 1990. Discordant results between the broth disk elution and broth microdilution susceptibility tests with *Bacteroides fragilis* group isolates. J. Clin. Microbiol. 28:375–378.
2. Allen, S. D. 1984. Current relevance of anaerobic bacteriology. Clin. Microbiol. Newsl. 6:147–149.
3. Appelbaum, P. C., S. K. Spangler, and M. R. Jacobs. 1990. Beta-lactamase production and susceptibilities to amoxicillin, amoxicillin-clavulanate, ticarcillin, ticarcillin-clavulanate, cefoxitin, imipenem, and metronidazole of 320 non-*Bacteroides fragilis Bacteroides isolates* and 129 fusobacteria from 28 U. S. centers. Antimicrob. Agents Chemother. 34:1546–1550.
4. Appelbaum, P. C., S. K. Spangler, and M. R. Jacobs. 1990. Evaluation of two methods for rapid testing for beta-lactamase production in *Bacteroides* and *Fusobacterium*. Eur. J. Clin. Microbiol. Infect. Dis. 9:47–50.
5. Baron, E. J., D. M. Citron, and H. M. Wexler. 1990. Son of anaerobic susceptibility—revisted. Clin. Microbiol. Newl. 12:69–72.
6. Barry, A. L., and R. J. Zabransky. 1990. *Eubacterium lentum* ATCC 43055, a new reference strain for quality control of anaerobic susceptibility tests. J. Clin. Microbiol. 28:2375–2376.
7. Barry, A. L., P. C. Fuchs, E. H. Gerlach, S. D. Allen, J. F. Acar, K. E. Aldridge, A. -M. Bourgault, H. Grimm, G. S. Hall, W. Heizmann, R. N. Jones, J. M. Swenson, C. Thornsberry, H. Wexler, J. D. Williams, and J. Wust. 1990. Multilaboratory evaluation of an agar diffusion disk susceptibility test for rapidly growing anaerobic bacteria. Rev. Infect. Dis. 12 (Suppl 2):S210–S217.
8. Bolmstrom, A., S. Arvidson, M. Ericsson, and A. Karlsson. 1989. Antimicrobial susceptibility testing of anaerobes with a novel technique—the PDM Epsilometer. Abstract C6, Third European Congress on Anaerobic Bacteria and Infections, Munich, FRG.
9. Citron, D. 1983. Susceptibility testing of anaerobic bacteria. Am. J. Med. Technol. 49:769–772.

10. Citron, D. M., M. I. Ostovari, A. Karlsson, and E. J. C. Goldstein. 1989. Evaluation of the Epsilometer (E test) for susceptibility testing of anaerobic bacteria. Abstract 876, Twenty–Ninth Interscience Conference on Antimicrobial Agents and Chemotherapy, Houston.

11. Cuchural, G. J., F. P. Tally, N. V. Jacobus, K. Aldridge, T. Cleary, S. M. Finegold, G. Hill, P. Iannini, J. P. O'Keefe, C. Pierson, D. Crook, T. Russo, and D. Hecht. 1988. Susceptibility of the *Bacteroides fragilis* group in the United States: analysis by site of location. Antimicrob. Agents Chemother. 32:717–722.

12. Cuchural, G. J., Jr., F. P. Tally, N. V. Jacobus, T. Cleary, S. M. Finegold, G. Hill, P. Iannini, J. P. O'Keefe, and C. Pierson. 1990. Comparative activities of newer beta–lactam agents against members of the *Bacteroides fragilis* group. Antimicrob. Agents Chemother. 34:479–480.

13. Finegold, S. M. 1989. Therapy of anaerobic infections, p. 793–818. In S. M. Finegold and W. L. George (eds.), Anaerobic infections in humans. Academic Press, San Diego.

14. Finegold, S. M. 1990. Anaerobes: problems and controversies in bacteriology, infections, and susceptibility testing. Rev. Infect. Dis. 12 (Suppl 2):S223–S230.

15. Finegold, S. M., and H. M. Wexler. 1988. Therapeutic implications of bacteriologic findings in mixed aerobic-anaerobic infections. Antimicrob. Agents Chemother. 32:611–616.

16. Finegold, S. M., and the National Committee for Clinical Laboratory Standards Working Group on Anaerobic Susceptibility Testing. 1988. Minireview: susceptibility testing of anaerobes. J. Clin. Microbiol. 26:1253–1256.

17. Goldstein, E. J. C., and D. M. Citron. 1988. Annual incidence, epidemiology, and comparative in vitro susceptibilities to cefoxitin, cefotetan, cefmetazole, and ceftizoxime of recent community-acquired isolates of the *Bacteroides fragilis* group. J. Clin. Microbiol. 26:2361–2366.

18. Goldstein, E. J. C., and D. M. Citron. 1989. Susceptibility testing of anaerobes: fact, fancy, and wishful thinking. Clin. Ther. 11:710–723.

19. Hauser, K. J., J. A. Johnston, and R. J. Zabransky. 1975. Economical agar dilution procedure for susceptibility testing of anaerobes. Antimicrob. Agents Chemother. 7:712–714.

20. Hill, G. B., and S. Schalkowsky. 1990. Development and evaluation of the spiral gradient endpoint method for susceptibility testing of anaerobic gram–negative bacilli. Rev. Inf. Dis. 12 (Suppl 2):S200–S209.

21. Horn, R., A. –M. Bourgault, and F. Lamothe. 1987. Disk diffusion susceptibility testing of the *Bacteroides fragilis* group. Antimicrob. Agents Chemother. 31:1596–1599.

22. Jones, R. N., A. L. Barry, P. C. Fuchs, and S. D. Allen. 1987. Ceftizoxime and cefoxitin susceptibility testing against anaerobic bacteria: comparison of results from three NCCLS methods and quality control recommendations for the reference agar dilution procedure. Diagn. Microbiol. Infect. Dis. 8:87–94.

23. Kurzynski, T. A., J. W. Yrios, A. G. Helstad, and C. R. Field. 1976. Aerobically incubated thioglycollate broth disk method for antibiotic susceptibility testing of anaerobes. Antimicrob. Agents Chemother. 10:727–732.

24. LaRocco, M., and A. Robinson. 1986. Evaluation of three commercial tests for rapid detection of beta–lactamase in anaerobic bacteria. Eur. J. Clin. Microbiol. 4:593–594.

25. Lorian, V. 1989. In vitro simulation of in vivo conditions: physical state of the culture medium. J. Clin. Microbiol. 27:2403–2406.

26. Molitoris, E., F. Jashnian, H. M. Wexler, and S. M. Finegold. 1990. Comparison of spiral gradient to conventional agar dilution for susceptibility testing of anaerobic bacteria. Abstract C260, Annual Meeting of the American Society for Microbiology, Annaheim.

27. NCCLS M11–A. 1985. Reference agar dilution procedure for antimicrobial susceptibility testing of anaerobic bacteria: approved standard. Vol. 5, No. 2. National Committee for Clinical Laboratory Standards, Villanova, PA.

28. NCCLS M11–A2. 1990. Methods for antimicrobial susceptibility testing of anaerobic bacteria, 2nd ed. : approved standard. National Committee for Clinical Laboratory Standards, Villanova, PA.

29. NCCLS M11–T2. 1989. Methods for antimicrobial susceptibility testing of anaerobic bacteria: tentative standard. National Committee for Clinical Laboratory Standards, Villanova, PA.

30. Nord, C. E., and M. Hedberg. 1990. Resistance to beta–lactam antibiotics in anaerobic bacteria. Rev. Infect. Dis. 12 (Suppl 2):S231–S234.

31. Ratkovich, M., and J. G. Kilgore. 1989. Evaluation of an anaerobe MIC panel for antimicrobial susceptibility testing of anaerobic bacteria. Abstract C290, Annual Meeting of the American Society for Microbiology, New Orleans.

32. Rosenblatt, J. E. 1984. Antimicrobial susceptibility testing of anaerobic bacteria. Rev. Infect. Dis. 6 (Suppl 1):S242–S248.

33. Sutter, V. L., D. M. Citron, M. A. C. Edelstein, and S. M. Finegold. 1985. Wadsworth anaerobic bacteriology manual, 4th ed. Star Publishing, Belmont, CA.

34. Swenson, R. M. 1986. Rationale for identification and susceptibility testing of anaerobic bacteria. Rev. Infect. Dis. 8:809–813.

35. Wexler, H. M. 1989. Susceptibility testing procedures, p. 715–729. In S. M. Finegold and W. L. George (eds.), Anaerobic infections in humans. Academic Press, San Diego.

36. Wilkins, T. D., and T. Thiel. 1973. Modified broth–disk method for testing the antibiotic susceptibility of anaerobic bacteria. Antimicrob. Agents Chemother. 3:350–356.

37. Zabransky, R. J. 1989. Revisiting anaerobe susceptibility testing. Clin. Microbiol. Newsl. 11:185–188.

CHAPTER **11**

QUALITY ASSURANCE (QA)

At the conclusion of this chapter, you will be able to:

1. Explain the difference between "quality assurance" (QA) and "quality control" (QC) as they relate to anaerobic bacteriology.
2. Differentiate between the terms "accuracy" and "precision" as they relate to laboratory tests.
3. List three important considerations pertaining to the quality of anaerobic bacteriology specimens submitted to the laboratory.
4. Describe three specific actions that laboratory personnel may take to improve the overall quality of specimens submitted to the anaerobic bacteriology laboratory.
5. List ten QA/QC considerations pertaining to events that occur **within** the anaerobic bacteriology section of the laboratory.
6. Relate two actions that laboratory personnel may take to ensure the quality of events that occur **after** results leave the anaerobic bacteriology laboratory.
7. Identify ten common sources of error in the anaerobic bacteriology laboratory.

Definitions

Accuracy: A measure of the "truth" or validity of a laboratory test result

Precision: The expression of the variability of analysis or an indication of the amount of random error that exists in an analytical process or laboratory test.[8] An imprecise test would give completely different results each time it is repeated.

Quality assurance (QA): Any and all actions taken by health care professionals to assure the overall accuracy and reliability of clinical anaerobic bacteriology services within a given health care setting, including actions that improve the quality of (1) specimens sent to the laboratory, (2) procedures performed within the laboratory, and (3) events that occur after results leave the laboratory

Quality control (QC): Any and all actions taken by laboratory personnel to assure the quality of procedures performed within the anaerobic bacteriology section of the microbiology laboratory; a very important, but not the sole, component of QA

Many of the quality assurance (QA) and quality control (QC) measures applicable to anaerobic bacteriology are identical to those applied in the microbiology laboratory in general. Because a discussion of **all** microbiology laboratory QA/QC measures is beyond the scope of this chapter, emphasis will be placed on measures that specifically impact on anaerobic bacteriology. Such measures affect events that occur **before** specimens arrive in the laboratory, events that occur **within** the laboratory, and events that occur **after** results leave the laboratory.

In her discussion of the "new age" laboratory, Barr emphasizes the fact that clinical laboratory scientists must interact with clinicians in an effort to improve the overall quality and clinical relevance of information generated by the laboratory.[2] Microbiologists must be intimately involved in all three phases of the production of laboratory information. Barr refers to these phases as "input," "process," and "output."

The input phase includes the clinical relevance of requested tests and the quality of specimens submitted to the laboratory. For example, if the results of a particular laboratory test will not alter the physician's course of action, the test should **not** be ordered. Tests likely to affect the course of treatment **should** be ordered. Only specimens deemed "appropriate" should be sent to the laboratory, and they must be collected and transported in a proper manner.

The process phase relates to the clinical relevance of procedures performed in the laboratory and the **manner** in which they are performed. **Quality control** measures are enforced within the laboratory to ensure the accuracy and precision of tests being performed there. Accuracy is the measure of "truth" of a result; precision is the expression of the variability of a particular analysis.[8] Accuracy and precision are independent of each other. For example, a procedure can be precise (i.e., the same result may be achieved each time a particular test is repeated), but the result can be inaccurate.

The output phase includes the written/instrument-printed results generated by the laboratory, the timeliness with which results reach the patient's chart, their proper interpretation, and how they affect the course of treatment.

The president of the American Society of Clinical Pathologists has stated,

> The 1990s, more than ever before, will require effectiveness and efficiency in the performance of medical laboratory tests, especially at the beginning of the test—the ordering—and at the end—the interpretation and action.... The 90s is the decade for clinical laboratory people of all kinds to guide doctors and patients so the right tests are ordered and performed, to make sure that the interpretation is reasonable, and to assist in assuring that an appropriate action is taken based on the result of the laboratory test.[4]

This chapter offers guidance regarding actions that clinical laboratory personnel may take to assure the quality of anaerobic bacteriology procedures performed within their institutions. Detailed information concerning QA measures applicable to clinical microbiology in general may be found elsewhere.[1,5]

QA Before Specimens Arrive in the Laboratory

Specimen Selection

Persons responsible for selecting specimens (usually physicians and nurses) must be familiar with the types of specimens that **are** and **are not** suitable for anaerobic bacteriology. This information, together with details on proper specimen collection and transport, should be disseminated by the laboratory in the form of written guidelines and presented by clinical laboratory professionals at appropriate rounds or in-services. Guidelines for specimen selection are contained in Chapter 5.

Request Slips

Information to be provided on the request slip includes patient demographics, specimen type and source, date and time of collection, the physician to whom results are to be sent, and information regarding any antimicrobial agents the patient is receiving. The time of collection is especially important because it will later be used to determine the rapidity with which the specimen was transported to the laboratory. The type of specimen frequently alerts the microbiologist to the most likely types of anaerobes to expect, which often serves as a guide to media selection.

Specimen Collection

Specimens for anaerobic bacteriology must be collected so as to minimize contamination of the specimen with organisms of the indigenous microflora (including anaerobes) and minimize exposure of the specimen to oxygen. Chapter 5 contains guidelines for proper specimen collection.

Specimen Labeling

As a minimum, specimen containers must be labeled with the patient's name and identification number, source of specimen, and date/time of collection. In the laboratory, the name on each specimen container must be cross-checked against the name on the request slip to ensure that specimens have not been inadvertently switched.

Specimen Transport

Specimens for anaerobic bacteriology must be transported rapidly (to minimize the time anaerobes remain at room temperature) and transported so as to minimize further exposure of the specimen to oxygen. They should always be transported in appropriate transport devices whenever a delay is anticipated between collection and processing. Clinical laboratory professionals are responsibile for ensuring that such containers are available, that individuals are instructed in their proper use, and that they are actually used. Chapter 5 contains guidelines for proper specimen transport.

Written Laboratory Directives/Guidelines

The laboratory must furnish written guidelines to clinicians regarding the following:

- **Appropriate and inappropriate specimens**
- **How specimens are to be collected**
- **How specimens are to be transported to the laboratory**
- **Policy regarding duplication of specimens and excessive specimens received on any given day**
- **Criteria for rejection of specimens;** to ensure that results from the anaerobic bacteriology laboratory are clinically relevant and of value to the patient, criteria must be established for rejecting poor quality specimens and should include the following:
  - **Unlabeled specimen container**
  - **Inappropriate specimen for patient's condition**
  - **Incorrectly collected**
  - **Incorrectly transported**
  - **Too great a delay between collection and arrival in the lab**
  - **Dried out specimen**

Education of Clinicians

Whenever possible, clinical laboratory professionals should promote proper anaerobic bacteriology techniques by participating in the education of medical and nursing students and in various hospital rounds and in-services. In the "new age" laboratory, depending upon the setting, patient assessment and test ordering may be performed by a wide variety of practitioners, including nurses, nurse practitioners, physician assistants, dentists, nurse midwives, clinical pharmacists, and physicians.[2] Therefore, QA stategies cannot be addressed solely to physicians.

QA in the Laboratory

Education of Personnel

Individuals working in the anaerobic bacteriology laboratory should have received specialized training in anaerobic bacteriology procedures. Excellent anaerobic bacteriology workshops are frequently presented in conjunction with national and regional microbiology/medical technology meetings. In addition, self-study materials are available from educational organizations (e.g., the Colorado Association for Continuing Medical Laboratory Education [CACMLE], Denver, CO).

Safety Considerations

Examine specimen containers for leakage when they arrive in the laboratory. If leakage has occurred, immediately place the container in a second container to prevent further contamination of the table/bench surface. Handle all clinical specimens as if they were obtained from AIDS patients (a policy known as "universal precautions"). Inoculate primary isolation media in a laminar flow cabinet, and immediately place the inoculated media in an anaerobic system. Some anaerobic chambers, known as "glove boxes," are equipped with permanently attached rubber gloves that offer protection when handling infectious materials. Wear disposable latex gloves while processing specimens in a "gloveless" type of anaerobic chamber or when processing a specimen within a laminar flow cabinet. Should you have to process specimens on an open bench, wear a mask and protective eyewear in addition to gloves. Always take care to avoid splashing or aerosolizing specimens during processing, but especially when working on an open bench. Written safety protocols must be available and must be read by all laboratory employees.

Assessing Specimen Quality

The following items should be assessed and appropriate entries recorded on the accompanying request slip. In certain cases, specimens will be rejected, and new specimens will be requested.

- **Is the specimen container properly labeled?**
- **Is it an appropriate specimen?**
- **Was it collected properly?**
- **Was it transported properly?** (in a transport container)?

- **Was it transported rapidly?** (Check the collection time and record when it arrived at the laboratory.)
- **Has the specimen dried out?**

Speed of Specimen Processing

Make an entry on the worksheet regarding the time the specimen is processed so you can later evaluate the time elapsing between arrival and processing. This is as important a consideration as the time elapsing between collection and arrival.

Objective Criteria for PID of Organisms in Gram-Stained Smears

The laboratory protocol should contain objective criteria regarding the Gram stain appearance and presumptive identification of any organisms observed. The characteristic cell morphology of representative anaerobes is described in Chapter 6.

Preliminary Reports

Record the date/time that a preliminary report is issued on the worksheet and/or request slip. Later, evaluations can be made regarding the timeliness of preliminary reports.

Media Choice and Quality Control (QC)

Failure to use appropriate media for anaerobic bacteriology can prevent or delay isolation of anaerobes contributing to a patient's infectious process. Failure to use appropriate media may result in a need for repeat cultures, which would obviously increase the cost of anaerobic bacteriology (see Chapter 12). Are appropriate media being used for anaerobic bacteriology? Are they known to support growth of the more fastidious anaerobes, such as fusobacteria, gram-positive cocci, and *Porphyromonas* spp.? Are tubed and plated media evaluated in-house by using QC organisms? Is there written evidence of this? Appropriate organisms for use in QC of media are listed in Tables 11-1 and 11-2; they may be obtained directly from the American Type Culture Collection (ATCC) or from companies listed in Appendix B.

Table 11-1. Control Organisms for Use by Manufacturers in QA of Commercially Prepared Anaerobic Bacteriology Culture Media

Medium	Atmosphere and Length of Incubation	Control Organisms (ATCC* No.)	Expected Results
Anaerobic blood agars	Anaerobic, 24–48 hours	*Bacteroides fragilis* (25285)	Growth
		Bacteroides levii (29147)	Growth
		Clostridium perfringens (13124)	Growth, beta hemolysis
		Fusobacterium nucleatum (25586)	Growth
		Peptostreptococcus anaerobius (27337)	Growth

Table 11-1. Control Organisms for Use by Manufacturers in QA of Commercially Prepared Anaerobic Bacteriology Culture Media (*continued*)

Medium	Atmosphere and Length of Incubation	Control Organisms (ATCC* No.)	Expected Results
Anaerobic broths	Anaerobic, 48 hours (tightened cap)		
THIO with or without indicator		*Clostridium novyi* type A (7659)	Growth
		Staphylococcus aureus (25923)	Growth
THIO medium, enriched		*Bacteroides levii* (29147)	Growth
		Bacteroides vulgatus (8482)	Growth
		Clostridium perfringens (13124)	Growth

Note: This table is provided solely to illustrate standards in effect at press time; readers are urged to use only the most recent editions of any NCCLS standards.

* The American Type Culture Collection (see Appendix B for address).

Source: Based on reference 7.

Table 11-2. Control Organisms for In-House QA of Anaerobic Bacteriology Culture Media

Medium	QC Organism	Expected Result(s)	Reference
Bacteroides Bile Esculin (BBE) Agar	*Bacteroides fragilis* (ATCC 25285*)	Growth, dark colonies, browning of the medium due to esculin hydrolysis	9
	Prevotella melaninogenica (ATCC 25845)	No growth (inhibited by the bile in the medium)	9
	Escherichia coli (ATCC 25922)	No growth (inhibited by the gentamicin in the medium)	9
Brucella Blood Agar (BRU)	*Clostridium perfringens* (ATCC 13124)	Growth, double zone of hemolysis	9
	P. melaninogenica (ATCC 25845)	Growth, brown-black pigmented colonies (may require >48 hours incubation)	9

Table 11-2. Control Organisms for In-House QA of Anaerobic Bacteriology Culture Media (*continued*)

Medium	QC Organism	Expected Result(s)	Reference
Cycloserine-Cefoxitin-Fructose Agar (CCFA)	*Clostridium difficile*	Growth	9
	B. fragilis (ATCC 25285)	Inhibition of growth by the cefoxitin in the medium	9
	E. coli (ATCC 25922)	Inhibition of growth by the cycloserine in the medium	9
Egg Yolk Agar (EYA)	*C. perfringens* (ATCC 13124)	Growth, lecithinase reaction	9
	Fusobacterium necrophorum	Growth, lipase reaction	9
	Clostridium novyi type A	Both lecithinase and lipase reactions	3
	B. fragilis	No lecithinase or lipase reactions	3
Kanamycin-Vancomycin-Laked Blood (KVLB) Agar	*P. melaninogenica* (ATCC 25845)	Growth, brown-black pigmented colonies (may require >48 hours incubation)	9
	C. perfringens (ATCC 13124)	No growth (inhibited by the vancomycin in the medium)	9
	E. coli (ATCC 25922)	No growth (inhibited by the kanamycin in the medium)	9
Phenylethyl Alcohol (PEA) Agar	*B. fragilis* (ATCC 25285)	Growth	9
	Proteus mirabilis	Inhibition of swarming	9
Thioglycollate (THIO)	*Bacteroides ovatus* (ATCC 8483)	Growth in 24 hours	9
	C. novyi type A	Growth	3
	Enterococcus sp.	Growth	3
Tryptic Soy Broth Blood Bottles	*C. novyi* type A	Growth	3
	B. fragilis	Growth	3
	Haemophilus influenzae	Growth	3

* The American Type Culture Collection number is shown whenever it was included in the reference.

Holding Systems

If an anaerobic chamber is not available, is a holding system in use at the processing station? If not, are properly submitted specimens batch processed? Is a holding system available at the anaerobe bench? Are the containers (usually jars) being used correctly, by minimizing convection currents within them and minimizing the time the inoculated plates remain at room temperature? Chapter 6 contains information on holding systems.

Anaerobic Incubation Systems

If an anaerobic chamber is being used, is it monitored for anaerobic atmosphere, temperature, humidity, and leaks? Is the catalyst changed/rejuvenated at proper intervals? Catalyst in jars should be rejuvenated after every use (see Chapter 6). Are anaerobic indicators used each time an anaerobic jar, bag, or pouch is set up? Chapter 6 contains information on the various methods available for anaerobic incubation.

Stains, Reagents, and Tests

Are written procedures available for QC of Gram stain reagents, tubed and plated media, and all reagents and kits used in the anaerobic bacteriology laboratory? Are specific individuals designated to perform QC duties? Is there written evidence that QC is being performed? Is there written evidence of corrective actions taken and indication of who took such actions and when? Tables 11-3 and 11-4 contain information regarding appropriate QC organisms for some of the reagents, tests, and kits used in the anaerobic bacteriology laboratory. Information regarding appropriate QC organisms should be contained in product inserts accompanying **all** reagents, tests, and kits for use in the anaerobic bacteriology laboratory.

Table 11-3. Control Organisms for QA of Selected Minisystems for Identifying Anaerobic Isolates

Product	Manufacturer	Manufacturer's Recommendations
API 20A®	Analytab Products	*Bacteroides ovatus* (ATCC 8483*), *Clostridium histolyticum* (ATCC 19401), *C. perfringens* (ATCC 13124), *C. sordellii* (9714), *Propionibacterium acnes* (ATCC 11827)
AN-IDENT®	Analytab Products	*Actinomyces odontolyticus* (ATCC 17929), *Bacteroides fragilis* (ATCC 23745), *B. ovatus* (ATCC 8483), *Clostridium histolyticum* (ATCC 19401), *C. perfringens* (ATCC 13124)
MicroScan Rapid Anaerobe Identification Panel®	Baxter Healthcare Corp., Microscan Division	*Bacteroides fragilis* (ATCC 25285), *Clostridium perfringens* (ATCC 13124), *C. sordellii* (ATCC 9714), *Peptostreptococcus magnus* (ATCC 29328)

Table 11-3. Control Organisms for QA of Selected Minisystems for Identifying Anaerobic Isolates (*continued*)

Product	Manufacturer	Manufacturer's Recommendations
Minitek® Anaerobe II	Becton Dickinson Microbiology Systems	*Bacteroides ovatus* (ATCC 8483), *Clostridium perfringens* (ATCC 13124), *C. sordellii* (ATCC 9714), *Veillonella parvula* (ATCC 10790)
RapID ANA II®	Innovative Diagnostic Systems, Inc.	*Bacteroides distasonis* (ATCC 8503), *Bacteroides uniformis* (ATCC 8492), *Clostridium sordellii* (ATCC 9714)

* The American Type Culture Collection number.

Table 11-4. Control Organisms for QA of Selected Disks and Tests for Identifying Anaerobic Isolates

Disk/Test	QC Organism	Expected Result(s)	Reference
Disks			
Colistin (10 μm)	*Fusobacterium necrophorum*	Susceptible	9
	Bacteroides fragilis (ATCC 25285*)	Resistant	9
Kanamycin (1 mg)	*Clostridium perfringens* (ATCC 13124)	Susceptible	9
	B. fragilis (ATCC 25285)	Resistant	9
Vancomycin (5 μm)	*C. perfringens* (ATCC 13124)	Susceptible	9
	B. fragilis (ATCC 25285)	Resistant	9
Nitrate	*Bacteroides ureolyticus*	Positive (red color due to reduction of nitrate to nitrite)	9
	B. fragilis (ATCC 25285)	Negative (no red color because nitrate was not reduced)	9
Sodium polyanethol sulfonate (SPS)	*Peptostreptococcus anaerobius*	Susceptible (at least a 12 mm zone of inhibition)	9
	P. micros	Resistant (zone of inhibition smaller than 12 mm)	9

Table 11-4. Control Organisms for QA of Selected Disks and Tests for Identifying Anaerobic Isolates (*continued*)

Disk/Test	QC Organism	Expected Result(s)	Reference
Tests			
Indole-nitrate-motility medium	*Bacteroides thetaiotaomicron*	Indole produced	3
	Clostridium novyi type A	Positive motility	3
	B. fragilis	Indole-negative and nonmotile	3
Indole test reagents	*F. necrophorum*	Indole-positive	9
	B. fragilis (ATCC 25285)	Indole-negative	9
Urea broth	*B. ureolyticus*	Urease-positive	9
	B. fragilis (ATCC 25285)	Urease-negative	9

* The American Type Culture Collection number is shown whenever it was included in the reference.

Choice and QC of Susceptibility Testing Methods

Is an appropriate susceptibility testing method being used for anaerobes? Are QC organisms being used to monitor the accuracy of susceptibility testing? The QC organisms recommended by the National Committee for Clinical Laboratory Standards are listed in Chapter 10.[6] These organisms may be obtained directly from the American Type Culture Collection or companies listed in Appendix B.

Proficiency Surveys

The laboratory should participate in some type of ongoing microbiology proficiency survey, such as the one available from the College of American Pathologists (CAP) (see Appendix B for address). The survey should periodically include specimens for anaerobic bacteriology. The following are some of the anaerobes included in recent CAP proficiency surveys:

Actinomyces naeslundii	*Clostridium sporogenes*
Bacteroides fragilis	*Clostridium tertium*
Prevotella melaninogenica	*Fusobacterium mortiferum*
Prevotella oralis	*Fusobacterium nucleatum*
Bacteroides thetaiotaomicron	*Peptostreptococcus anaerobius*
Clostridium perfringens	*Peptostreptococcus magnus*

QA After Results Leave the Laboratory

Laboratory responsibility for QA does not end when written, instrument-printed, or electronically transmitted results leave the laboratory. This is when it can be determined if results generated in the laboratory are being received in sufficient time to be of value and are actually used in the diagnosis or treatment of patients. Through consultations with nursing service personnel and physicians, clinical laboratory professionals will be able to determine the timeliness and value of laboratory testing and results. Were the specific tests requested appropriate for the patient's condition? Did the isolation and identification of anaerobes influence the clinician's management of the patient? Is there evidence that susceptibility test results guided initial therapy or led to a change in therapy?

One individual from the laboratory (or, in large hospitals, perhaps one individual from each of the major sections of the laboratory) should be designated to discuss QA matters with physicians and nursing service personnel. Together, they should review matters such as (1) the time it takes for specimens to reach the laboratory, (2) the condition of specimens at the time of receipt, (3) the time it takes for results to reach patients' charts, (4) the appropriateness of the requests, and (5) whether test results are actually utilized.

By interacting with physicians and nursing service personnel, clinical laboratory professionals are afforded an opportunity to observe and appreciate the varied ways laboratory services contribute to patient care. As August et al. have stated, an effective QA program helps clinical laboratory professionals "realize that they are not producing data in a vacuum; patient lives are ultimately affected by what is done in the laboratory."[1]

Common Sources of Error in the Anaerobic Bacteriology Laboratory

The following list of errors that commonly occur in the anaerobic bacteriology laboratory has been modified from a list of "pitfalls" provided by Sutter et al.:[9]

- Failure to prepare and examine a Gram-stained smear of the clinical specimen (The Gram stain is discussed in detail in Chapter 6.)

- Contamination of the clinical specimen with endogenous anaerobes (Collection of specimens for anaerobic bacteriology is described in Chapter 5.)

- Failure to process specimens rapidly and/or to maintain them in an oxygen-free environment prior to processing (Transport and processing of specimens are described in Chapters 5 and 6, respectively.)

- Reliance upon a liquid medium, such as enriched thioglycollate broth (THIO), as the sole system for recovering anaerobes from clinical specimens (The use and limitations of THIO are described in Chapter 6.)

- Use of inappropriate/inadequate plated media; failure to use PRAS media and media containing supplements such as Vitamin K_1 and hemin; failure to include selective media in the primary setup (Appropriate media for use in anaerobic bacteriology are described in Chapter 6.)

- Use of a toxic gas mixture in conjunction with anaerobic holding jars or anaerobic jar displacement procedures or failure to include CO_2 in the gas mixture (Anaerobic incubation is discussed in Chapter 7.)

- Failure to obtain an anaerobic atmosphere due to cracks in jars, lids, or O-rings or use of "poisoned" catalyst pellets (See Chapter 7.)

- Failure to incorporate a suitable oxidation-reduction indicator and/or QC organism in the anaerobic system to ensure that anaerobic conditions were achieved (See Chapter 7.)

- Failure to hold cultures long enough (five to seven days) for growth of particularly slow-growing anaerobes (See Chapter 7.)

- Failure to recognize a pink-staining bacillus as a gram-positive organism (The use of special potency antimicrobial disks for this purpose is described in Chapters 7 and 8.)

- Failure to use the aerotolerance test to determine whether an isolate is truly an anaerobe or failure to hold the CO_2-incubated chocolate agar plate long enough (48 hours) for slow-growing, fastidious organisms to appear (Aerotolerance testing is described in Chapter 7.)

- Attempting to determine the antimicrobial susceptibility of an anaerobic isolate using an unstandardized disk-diffusion type of test (There are **no** NCCLS-approved disk-diffusion methods available that are suitable for testing anaerobes. Susceptibility testing of anaerobes is described in Chapter 10.)

- Use of nonviable organisms or mixed cultures to inoculate identification systems requiring viable organisms

- Misidentification of coccobacilli as cocci or elongated cocci as coccobacilli

- Failure to use appropriate QC organisms to check isolation and differential media, media supplements, reagents, identification systems, and susceptibility tests

Chapter in Review

- Quality assurance in anaerobic bacteriology begins **before** specimens arrive in the laboratory. In addition to proper selection, collection, and transport of specimens being submitted, request slips must be filled out legibly and completely, and specimen containers must be properly labeled. The laboratory is responsible for furnishing written directives/guidelines regarding submission of specimens for anaerobic bacteriology, and clinical laboratory professionals have a responsiblity for participating in rounds and in-services as often as possible or deemed necessary.

- QA/QC procedures that occur **within** the laboratory include proper training of laboratory personnel, strict adherence to safety policies, assessment of specimen quality, rapid processing of specimens, timely submission of preliminary and final reports, use of appropriate media and methods of achieving anaerobic incubation, and the use of QC organisms to determine the validity of staining procedures, tests, kits, and susceptibility testing.

- QA does not end when the final report is sent from the laboratory. Clinical laboratory professionals must participate in determinations regarding the quality of specimens submitted to the laboratory, the timeliness of specimen submission and processing, the speed with which preliminary and final results are entered into patients' charts, the appropriateness of test requests, and how test results are used in treating patients.

Self-Assessment Exercises

1. Explain the term "quality assurance" (QA) as it relates to anaerobic bacteriology.

   _____

   _____

   _____

2. Explain the term "quality control" (QC) as it relates to anaerobic bacteriology.

   _____

   _____

   _____

3. Differentiate between the terms "accuracy" and "precision" as they relate to laboratory tests.

   _____

   _____

   _____

   _____

4. List three important considerations pertaining to the **quality** of anaerobic bacteriology specimens submitted to the laboratory.

   a. _____

   b. _____

   c. _____

5. Describe three specific actions that laboratory personnel may take to improve the overall quality of anaerobic bacteriology specimens submitted to the laboratory.

   a. _____

   b. _____

   c. _____

6. List ten QA/QC considerations pertaining to events that occur **within** the anaerobic bacteriology section of the laboratory.

   a. _____

   b. _____

   c. _____

   d. _____

   e. _____

   f. _____

   g. _____

   h. _____

   i. _____

   j. _____

7. Describe two actions that laboratory personnel may take to ensure the quality of events that occur **after** anaerobic bacteriology results leave the laboratory.

   a. _____

   b. _____

8. List five common sources of error in the anaerobic bacteriology laboratory.

   a. _____

   b. _____

   c. _____

   d. _____

   e. _____

9. TRUE or FALSE: Laboratories isolating large numbers of *Bacteroides* and *Clostridium* spp. and relatively small numbers of fusobacteria, *Porphyromonas* spp., and anaerobic cocci can pride themselves on their anaerobic bacteriology capabilities.

References

1. August, M. J., J. A. Hindler, T. W. Huber, and D. L. Sewell. 1990. Cumitech 3A, Quality control and quality assurance practices in clinical microbiology. Co-ordinating ed., A. S. Weissfeld. American Society for Microbiology, Washington, D. C.

2. Barr, J. T. 1989. Clinical laboratory utilization: the role of the clinical laboratory scientist, p. XXXI–XLVI. *In* B. G. Davis, M. L. Bishop, and D. Mass (eds.), Clinical laboratory science: strategies for practice. J. B. Lippincott, Philadelphia.

3. Gavan, T. L. 1991. Quality control in microbiology. *In* J. B. Henry (ed.), Clinical diagnosis and mangement by laboratory methods. W. B. Saunders, Philadelphia.

4. Lundberg, G. D. 1990. Editorial: completing the laboratory test loop. Lab. Med. 21:215.

5. Miller, J. M., and B. B. Wentworth. 1985. Methods for quality control in diagnostic microbiology. American Public Health Association, Washington, D. C.

6. NCCLS M11-A2. 1990. Methods for antimicrobial susceptibility testing of anaerobic bacteria, 2nd ed.: approved standard. National Committee for Clinical Laboratory Standards, Villanova, PA.

7. NCCLS M22-A. 1990. Quality assurance for commercially prepared microbiological culture media: approved standard. National Committee for Clinical Laboratory Standards, Villanova, PA.

8. Stewart, C. E., and J. A. Koepke. 1987. Basic quality assurance practices for clinical laboratories. J. B. Lippincott, Philadelphia.

9. Sutter, V. L., D. M. Citron, M. A. C. Edelstein, and S. M. Finegold. 1985. Wadsworth anaerobic bacteriology manual, 4th ed. Star Publishing, Belmont, CA.

CHAPTER **12**

COST CONTAINMENT

Guidelines Pertaining to Specimens
    Specimen Quality
    Inappropriate Specimens
    Request Slips
    The Clinician's Needs
    Repeat Cultures
Guidelines Pertaining to Media
    High-Quality Media
    Appropriate Selective and Differential Media
Guidelines Pertaining to Identification
    The Appropriate Level of Identificaiton
    The Least Expensive, Most Accurate Means of
        Identification
    Clostridium difficile Cultures
Guidelines Pertaining to Beta-Lactamase and
    Susceptibility Testing
    Beta-Lactamase Testing
    Susceptibility Testing
Guidelines Pertaining to Laboratory Reports

At the conclusion of this chapter, you will be able to:

1. State two of the most exasperating problems facing microbiologists in the anaerobic bacteriology laboratory.
2. Identify three ways microbiologists could interact with physicians and/or nurses to reduce the costs associated with anaerobic bacteriology.
3. Name four parameters that should be considered when determining the extent of the anaerobic workup for a given specimen.
4. State at least two important cost containment guidelines regarding each of the following: specimens submitted for anaerobic bacteriology, request slips accompanying specimens for anaerobic bacteriology, media used in the anaerobic bacteriology laboratory.

5. State at least one important cost containment guideline regarding each of the following: *Clostridium difficile* cultures, beta-lactamase testing, reports sent from the anaerobic bacteriology laboratory.
6. Cite at least three clinical situations in which susceptibility testing of an anaerobic isolate would be warranted.

Definitions

Carbuncle: A necrotizing infectious process of skin and subcutaneous tissue composed of a cluster of boils (furuncles) with multiple formed or incipient drainage sinuses; usually due to *Staphylococcus aureus*

Cholecystitis: Inflammation of the gall bladder

Decubitus ulcer: An ulceration caused by prolonged pressure in a patient allowed to lie too still in bed for a long period of time; also called a decubitus, bed sore, or pressure sore

Fistula: An abnormal passage or communication, usually between two internal organs or leading from an internal organ to the body surface

Furuncle: A painful nodule formed in the skin by circumscribed inflammation of the central corium and subcutaneous tissue, enclosing a central slough or "core," usually caused by staphylococci; also called a boil or furunculus

Perirectal abscess: An abscess in the areolar tissue around the rectum

Pilonidal cyst: A hair-containing dermoid cyst or sinus in the region of the sacrum and coccyx that often opens at a postanal dimple

Subphrenic abscess: An abscess beneath the diaphragm

Two of the most exasperating problems in anaerobic bacteriology are (1) how to improve the quality of specimens submitted to and processed by the laboratory and (2) how far to go in the workup of anaerobic cultures. Contamination of specimens with indigenous microflora organisms increases the amount of time and money spent in the workup of cultures and generates clinically irrelevant and potentially misleading information. Most infectious processes involving anaerobes contain a mixture of organisms (i.e., they are polymicrobic), and the mixture is often quite complex. Investigators at the V.A. Wadsworth Medical Center in Los Angeles have indicated that the workup of a single culture could cost as much as $500 to $1000 or more.[2] They cite examples of subphrenic abscesses yielding 20 or more different organisms and a root canal–related infectious process that produced over 50 different isolates.

Unless attention is paid to cost containment, anaerobic bacteriology could become prohibitively expensive. This would result in a catch-22 situation. Looking for ways to reduce a laboratory's budget, administrators might be tempted to eliminate or drastically reduce anaerobic bacteriology services. In many cases, this would prolong the time it takes to establish a valid diagnosis and initiate appropriate therapy. This, in turn, would increase the length of the patient's hospital stay, which would drive up the costs of caring for that particular patient—costs that in all likelihood would not be reimbursed by third party providers. This chapter offers guidelines for reducing or minimizing costs in the clinical anaerobic bacteriology laboratory.

In 1984, the College of American Pathologists conducted a three-day conference in Aspen, Colorado, to discuss cost-effective microbiology and provide guidance to clinical microbiologists on this increasingly important topic. The guidelines were published in 1985 in a book entitled *The Role of Clinical Microbiology in Cost-Effective Health Care.*[9] This chapter contains many of the recommendations contained in sections of that book dealing with wound and anaerobic cultures. Additional information has been extracted from a more recent publication by Finegold and Edelstein.[2]

Guidelines Pertaining to Specimens

Specimen Quality

As previously mentioned, anaerobic cultures can be costly and time-consuming. An anaerobic workup of a carelessly collected specimen, contaminated with indigenous microflora, may provide potentially erroneous and misleading information. For these reasons, some authorities feel that anaerobic cultures should be considered specialized procedures, that they require **specific** requests, and that culture request slips should have a **defined area** for such requests.[7] Smith suggests that physicians be required to **sign** each request slip to indicate they have carefully considered the appropriateness of the request and that specimens for anaerobic bacteriology always be collected by the most qualified person available.[9]

Chapter 5 described specimens considered inappropriate for routine anaerobic culture. Examples include throat and nasopharyngeal swabs, gingival swabs, expectorated sputum, feces and rectal swabs, voided and catheterized urine, and vaginal and cervical swabs. **Aspirates** are the preferred type of specimens for anaerobic bacteriology, but if pus cannot be obtained by needle and syringe, biopsy material from the edges of ulcers or walls of abscess cavities should be submitted.[7] Detailed information concerning specimens considered appropriate and inappropriate for anaerobic bacteriology should be contained in laboratory directives that describe specimen submission and are furnished to physicians and nurses. Microbiologists are responsible for providing acceptable materials for use in collection and transport of specimens for anaerobic bacteriology and for educating those responsible for selecting, collecting, and transporting specimens.[10]

Although an extensive anaerobic workup is justified for certain types of specimens, it is unnecessary for others. Expensive workups need not be performed for uncomplicated infectious processes that will respond well to surgical management (e.g., appendicitis without abscess, resection of gangrenous small bowel without complications, or acute cholecystitis).[2] Unless culture results will be of epidemiological value, clinicians should not routinely culture clinical materials from infectious processes that can be cured by simple surgical incision and drainage or removal of foreign material (e.g., furuncles or carbuncles). Culturing fistulae, sinus tracts, perirectal abscesses, and pilonidal cysts is generally unrewarding because they often contain mixtures of indigenous microflora that do not reflect the true nature of the underlying infectious process. Anaerobic cultures should not be performed on specimens contaminated with oral, GI, or GU flora, such as expectorated sputum, fecal material, or voided urine.[9] Routine anaerobic cultures of tooth sockets or surface cultures from decubitus ulcers and burn wounds should not be accepted by the laboratory.[5]

Inappropriate Specimens

Only appropriate specimens should be accepted for anaerobic bacteriology (see Chapter 5). Inappropriate specimens should be rejected, but the microbiologist should always **consult** with the clinician whenever an inappropriate specimen would be difficult or impossible to replace. The physician and nursing station should be notified by written memo and telephone as rapidly as possible, and information should be provided regarding (1) reason(s) for rejection and (2) guidelines for proper specimen selection, collection, and transport.[9] If the physician is unavailable, leave a message that provides names and telephone numbers of laboratory personnel the physician may contact for information. Policies regarding rejection of inappropriate specimens should always be (1) developed with physician input and (2) clearly stated in the laboratory directives describing specimen submission that are furnished to physicians and nurses.

Request Slips

To guide them in processing specimens, microbiologists should be provided with information regarding the patient's clinical condition and the physician's diagnostic considerations.[5] Clinicians and microbiologists alike should insist that culture requests provide an accurate definition of the type of lesion being cultured.[9] Rather than simply writing "wound," physicians should specify the **type** of wound on the request slip (e.g., abscess, skin lesion; surgical, trauma, or burn wound) and should state the **anatomical site** of the infectious process.[5]

The Clinician's Needs

A Gram stain should be performed on all specimens submitted to the anaerobic bacteriology laboratory, and results should be conveyed to clinicians as rapidly as possible.[9] At that time, a determination should be made as to exactly what information the physician is seeking. He or she may simply need to know whether or not anaerobes are present. Macroscopic and microscopic observations of the specimen and the results of simple tests (e.g., odor, fluorescence under UV light, purulence, presence of necrotic tissue and gas, Gram stain observations, and spores) can often provide such information (described in Chapter 6). In certain infections, the specific pathogens are fairly predictable. In many situations, only a limited anaerobic workup is necessary. Speciation is not required for certain anaerobes, although it is recommended for others (e.g., members of the *Bacteroides fragilis* group). The clinician may not require susceptibility test results. Therefore, microbiologists should **communicate** with the physician to determine what information he or she really needs.

Repeat Cultures

Microbiologists should consult with physicians when they receive requests for repeat cultures. The clinician may merely be interested in knowing if a particular organism is still present, in which case a complete workup is not required. Repeat cultures are usually not needed if the patient is doing well.[2] Repeated, routine culturing of healing wounds should be discouraged, and unless there is evidence of recrudescence of the infectious process, repeat cultures should be limited to one per week.[9] Cultures on surgical drainage sites should not be performed more than once per week, and routine anaerobic cultures on these specimens should not be performed at all.[5] The policy regarding repeat cultures should be contained in the laboratory directives describing specimen submission that are furnished to physicians and nurses.

Guidelines Pertaining to Media

High-Quality Media

The routine use of high-quality media in the primary isolation setup often eliminates the need to reculture (see Chapter 6). Studies have demonstrated considerable variation in the ability of commercial media to support the growth of fastidious anaerobes.[6, 8] Use of less expensive, lower quality media represents one of the **worst ways** to reduce costs. If anaerobes are participating in a given infectious process, that information must be obtained as rapidly as possible. To do otherwise would (1) extend the length of the patient's hospital stay, (2) increase total costs associated with that patient, and (3) most important, jeopardize the patient's health.

Appropriate Selective and Differential Media

When incorporated into the primary isolation setup, high-quality selective and differential media are very helpful in anaerobic bacteriology (Chapter 6). BBE plates, for example, often provide evidence within 24 hours of inoculation that a specimen contained a member or members of the *Bacteroides fragilis* group. Because the quality of selective and differential media varies from one manufacturer to another, take care to purchase media only from reliable sources. Because the antimicrobial agents within selective media deteriorate with age, use only in-date, properly stored selective media.

Guidelines Pertaining to Identification

The Appropriate Level of Identification

Anaerobic isolates should be identified only to a level appropriate for (1) the specimen source, (2) the types and numbers of different anaerobes present, (3) the patient's status, and (4) the needs of the clinician.[2] Infectious processes frequently contain many different organisms, and it is seldom practical to speciate them all. In such cases, it would be difficult, if not impossible, to know the relative importance of the various isolates. Microbiologists should contact the attending physician whenever it is unclear as to what information is needed. Very ill patients may require an extensive workup, whereas such a workup for recovering patients could represent a monumental waste of time, effort, and money.[2]

As reported by Rosenblatt, only Gram stain results were used at the Mayo Clinic to identify isolates from specimens yielding more than three anaerobes or specimens yielding more than four organisms of any type (unless the culture was from a critical area, such as blood or a brain abscess).[7] When large numbers of mixed flora were present, the likelihood of contamination increased, and the value of specific identification of individual isolates diminished.

Based on a limited survey of physician needs, Westerman proposed the following simplified reporting scheme for anaerobes isolated as predominant organisms from wound specimens:[10]

- For anaerobic, gram-negative bacilli, report either *Bacteroides* spp. (which might be penicillin-resistant) or gram-negative bacilli, not *Bacteroides*.

- For anaerobic, gram-positive bacilli, report *Clostridium* spp., *Actinomyces* spp., or anaerobic, gram-positive bacilli, not *Clostridium* or *Actinomyces*.

- For anaerobic cocci, report either anaerobic, gram-positive cocci or anaerobic, gram-negative cocci.

Although Westerman's recommendations would indeed reduce the costs associated with a more extensive workup of anaerobic isolates, we cannot endorse such an abbreviated scheme for the following reasons: (1) It would be inappropriate in many clinical settings. (2) Many *Bacteroides* spp. have been reclassified as *Porphyromonas* or *Prevotella* spp. since the scheme was proposed. (3) Antimicrobial resistance occurs in a wide variety of anaerobes and varies from species to species and strain to strain. (4) It would not generate data regarding the frequency with which various species are involved in infectious processes. (5) Speciation of anaerobes frequently provides information regarding the original nidus of infection (e.g., above or below the waist).

When diplomates of the American Board of Medical Microbiology were surveyed in 1984, they offered a variety of suggestions for limiting the workup of and costs associated with anaerobic cultures.[1] Of 97 completed questionnaires, the most frequent response (37) was to base the extent of workup on the type of specimen received. Twelve respondents suggested discussions with physicians as a way of limiting such workups. Eleven of the diplomates felt that the workup should be based upon the number of anaerobic isolates, most citing one to three isolates per culture as the limit. Less frequent responses included limiting the workup to isolates in the *Bacteroides fragilis* group (seven), only to genus level (six), only to common pathogens (three), or only to presumptive identifications (three).

The Least Expensive, Most Accurate Means of Identification

Whenever possible, use simple, inexpensive tests (e.g., colony and cellular morphology, spot indole, catalase, urease, nitrate disk, bile disk, special potency antimicrobial disks, lipase and lecithinase reactions, and growth stimulation tests) for presumptive and in some cases definitive identification of anaerobes (see Chapter 8). Finegold and Edelstein recommend limiting complete identifications to the following:[2]

- Organisms suspected of being members of the *B. fragilis* group, *C. perfringens*, *C. septicum*, and *Actinomyces* spp., when isolated from an appropriate setting

- Anaerobes isolated from seriously ill patients (e.g., patients with bacteremia)

- Whenever fewer than three anaerobes are isolated from the specimen

Easy-to-use, relatively inexpensive, commercially available biochemical- and enzyme-based ID systems are often useful in making definitive identifications (Chapter 9). In other situations, you may merely need to **describe** the anaerobic isolates or provide presumptive identifications based upon the results of inexpensive observations and tests. Table 9-18 contains a 3-step approach to the identification of anaerobic isolates. A rough quantitation of the various organisms isolated from a given specimen might provide an insight to the relative importance of the different isolates.[2]

Clostridium difficile Cultures

"The most definitive diagnosis of pseudomembranous colitis (PMC) is by the endoscopic detection of pseudomembranes or microabscesses in antibiotic-treated patients with diarrhea, who have *C. difficile* toxin in their stools."[4] Isolation of high numbers (10^7 or greater) of toxigenic *C. difficile* from patients with watery or bloody diarrhea has been used for presumptive diagnosis of antibiotic-associated diarrhea (AAD) and/or PMC due to this organism. However, keep in mind that *C. difficile* can be isolated from the feces of healthy individuals, and even the isolation of toxigenic strains of *C. difficile* from patients with AAD is **not** a definitive diagnosis.[4]

Limit *C. difficile* cultures to watery or bloody diarrheal specimens. Do not culture formed stool specimens for this organism.[2] Because the finding of colonic pseudomembranes is virtually diagnostic for *C. difficile*–induced disease, stool cultures for this organism are unnecessary if such pseudomembranes have already been observed.[3]

Guidelines Pertaining to Beta-Lactamase and Susceptibility Testing

Beta-Lactamase Testing

Beta-lactamase testing should not be routinely performed on all anaerobic isolates. Although the Cefinase® disk test should be performed on all unidentified, anaerobic, gram-negative bacilli and clostridia isolated in pure culture from sterile body sites and ill patients, it need not be performed on members of the *B. fragilis* group or *C. perfringens*; you may assume the former to be **resistant** to beta-lactam agents and (for now) the latter to be **susceptible** to them.[2] A negative beta-lactamase test is no guarantee that the organism being tested will be susceptible to beta-lactam agents; some anaerobes (e.g., *B. gracilis* and some *B. fragilis* group strains) are resistant despite the absence of beta-lactamases.[2]

Susceptibility Testing

Although problems exist with **all** the methods of susceptibility testing of anaerobes (Chapter 10), practical tests like the commercially available microtiter MIC tests **can** often provide clinically useful information. However, susceptibility testing should **not** be performed on **all** anaerobic isolates. According to Finegold and Edelstein, situations where susceptibility testing is warranted include the following:[2]

- Isolates with known variability in susceptibility patterns (e.g., *B. fragilis* group, other anaerobic, gram-negative bacilli, *Clostridium* species other than *C. perfringens*)

- Organisms isolated in pure culture

- Organisms isolated from seriously ill patients (see Chapter 10)

- Organisms isolated from patients undergoing long-term therapy

- Organisms isolated from patients who fail to respond to empiric therapy or are relapsing

Larger hospitals and medical centers also have a responsibility to monitor their anaerobic isolates for changes in susceptibility patterns and to test newly available antimicrobial agents.

Guidelines Pertaining to Laboratory Reports

Finegold and Edelstein state that microbiologists have a responsibility, in the course of providing reports, to provide sufficient information to enable clinicians to effectively **interpret** and **utilize** the information contained in the reports to manage patients' infectious diseases.[2] Microbiologists must possess sufficient judgment skills and knowledge to enable them to (1) interact with physicians, (2) furnish clinically relevant information, (3) provide information in a timely manner, and (4) provide information in a form that clinicians can use.

Chapter in Review

- Two of the most exasperating problems in anaerobic bacteriology are (1) how to improve the quality of specimens submitted to and processed by the laboratory and (2) how far to go in the workup of anaerobic cultures.

- Only appropriate, good quality specimens should be submitted for anaerobic bacteriology. Microbiologists must provide guidelines regarding appropriate and inappropriate specimens by participating in educational activities (e.g., rounds and in-services) and publishing laboratory directives. The processing of inappropriate specimens wastes valuable resources and generates inaccurate, misleading, and potentially dangerous information. Inappropriate specimens should be rejected.

- Physicians must provide sufficient information about specimens to guide microbiologists in their appropriate processing. The exact anatomical site from which the specimen was collected, the nature of the lesion, and the clinician's diagnosis of the problem are extremely valuable pieces of information. The microbiologist needs to know the extent of anaerobic workup that the physician requires.

- The use of less expensive, lower quality media represents one of the most foolish ways to reduce costs in the anaerobic bacteriology laboratory. High-quality media permit isolation of fastidious obligate anaerobes and often eliminate the need to reculture. The routine use of high-quality, in-date, properly stored selective and differential media in the primary isolation setup frequently permits rapid presumptive identification of clinically significant anaerobes.

- Anaerobic isolates should be identified only to a level appropriate for (1) the specimen source, (2) the types and numbers of different anaerobes present, (3) the patient's status, and (4) the clinician's needs. If the clinician hasn't provided clear instructions, the microbiologist should contact him or her to determine the extent to which to identify the anaerobic isolate(s). In certain situations, a clinician may merely need to know if anaerobes are present or absent or whether a specific anaerobe or "group" of anaerobes is still present. A clinician may require only presumptive identifications based upon simple observations and the results of inexpensive tests and may have no need for antimicrobial susceptibility information.

- *Clostridium difficile* cultures should be performed only on appropriate specimens and only when a diagnosis of pseudomembranous colitis has not already been established using proctoscopic or histologic observations.

- Beta-lactamase and antimicrobial susceptibility testing should be performed only on appropriate isolates. One should assume that members of the *Bacteroides fragilis* group are beta-lactamase positive and routinely perform susceptibility testing on such isolates whenever they are deemed clinically significant. As long as *Clostridium perfringens* remains susceptible to penicillin, it does not require routine beta-lactamase testing.

- Susceptibility testing should be limited to isolates with known variability in susceptibility patterns and to organisms isolated in pure culture, from seriously ill patients, from patients undergoing long-term therapy, or from patients who fail to respond to empirical therapy or are relapsing. Large hospitals and medical centers have a responsibility to monitor their anaerobic isolates for changes in susceptibility patterns and to test newly available antimicrobial agents.

- Microbiologists have a responsibility, in the course of providing reports, to supply sufficient information to enable clinicians to effectively interpret and utilize the information contained in the reports to manage patients' infectious diseases.

Self-Assessment Exercises

1. State two of the most exasperating problems facing microbiologists in the anaerobic bacteriology laboratory.

   a. _____

   b. _____

2. Identify three ways in which microbiologists could interact with physicians and/or nurses to reduce costs associated with anaerobic bacteriology.

   a. _____

   b. _____

   c. _____

3. Name four parameters that should be considered when determining the extent of anaerobic workup for a given specimen.

    a. _____

    b. _____

    c. _____

    d. _____

4. State at least two important cost containment guidelines regarding each of the following: specimens submitted for anaerobic bacteriology, request slips accompanying specimens for anaerobic bacteriology, media used in the anaerobic bacteriology laboratory.

    a. _____

    b. _____

    c. _____

5. Give an important cost containment guideline regarding each of the following: *Clostridium difficile* cultures, beta-lactamase testing, reports sent from the anaerobic bacteriology laboratory.

    a. _____

    b. _____

    c. _____

6. Cite at least three clinical situations in which susceptibility testing of an anaerobic isolate would be warranted.

    a. _____

    b. _____

    c. _____

References

1. Coyle, M. B. 1984. DRGs and the practice of clinical microbiology. Clin. Microbiol. Newsl. 6:175–179.

2. Finegold, S. M., and M. A. C. Edelstein. 1988. Coping with anaerobes in the 80s, p. 1–10. *In* J. M. Hardie and S. P. Borriello (eds.), Anaerobes today. John Wiley & Sons, New York.

3. George, W. L. 1989. Antimicrobial agent–associated diarrhea and colitis, p. 661–678. *In* S. M. Finegold and W. L. George (eds.), Anaerobic infections in humans. Academic Press, San Diego.

4. Lyerly, D. M., H. C. Krivan, and T. D. Wilkins. 1988. *Clostridium difficile*: its disease and toxins. Clin. Microbiol. Rev. 1:1–18.

5. MacLowry, J. D. 1985. General laboratory aspects of wound infections, p. 341–348. *In* J. W. Smith (ed.), The role of clinical microbiology in cost-effective health care. College of American Pathologists, Skokie, IL.

6. Mangels, J. I., and B. P. Douglas. 1989. Comparison of four commercial Brucella agar media for growth of anaerobic organisms. J. Clin. Microbiol. 27:2268–2271.

7. Rosenblatt, J. E. 1985. Anaerobic cultures, p. 349–357. *In* J. W. Smith (ed.), The role of clinical microbiology in cost–effective health care. College of American Pathologists, Skokie, IL.

8. Sheppard, A., C. Cammarata, and D. H. Martin. 1990. Comparison of different medium bases for the semiquantitative isolation of anaerobes from vaginal secretions. J. Clin. Microbiol. 28:455–457.

9. Smith, J. W. 1985. Wound and anaerobic infections: overview, p. 321–324. In J. W. Smith (ed.), The role of clinical microbiology in cost–effective health care. College of American Pathologists, Skokie, IL.

10. Westerman, E. L. 1985. Clinical aspects of wound and anaerobic infections: medical perspective, p. 325–332. *In* J. W. Smith (ed.), The role of clinical microbiology in cost-effective health care. College of American Pathologists, Skokie, IL.

CHAPTER 13

VETERINARY ANAEROBIC BACTERIOLOGY

Endogenous Anaerobes of Animals
The Role of Anaerobes in Veterinary Diseases
Frequency of Anaerobe Isolation from Veterinary
    Clinical Specimens
Selection, Collection, and Transport of Specimens
Specimen Processing
Processing and Identifying Anaerobic Isolates
Treating Anaerobe-Associated Diseases of Animals
       Surgical Debridement and Drainage
       Antimicrobial Therapy
Susceptibility Testing

At the conclusion of this chapter, you will be able to:

1. State the names of five anaerobes of significance in veterinary medicine that are of little or no significance in human medicine.
2. Name an anaerobe associated with each of the following diseases and an/the animal species in which each occurs: big head, black disease, black leg, lockjaw, lumpy jaw, malignant edema, red water disease, swine dysentery.
3. State the three types of *Clostridium botulinum* of greatest importance in animals.
4. State the three types/subtypes of *Clostridium perfringens* of greatest importance in animals.
5. Identify the essential elements in selection, collection, and transport of veterinary clinical specimens for anaerobic bacteriology.
6. List the three major components in the treatment of anaerobe-associated infectious processes in animals.

Definitions

Big head: An acute, infectious disease of rams caused by *Clostridium novyi* or *C. sordellii* and, rarely, *C. chauvoei*; follows the frequent head-butting activities of young rams

Black disease (infectious necrotic hepatitis): An acute, infectious disease of sheep, sometimes cattle, and, rarely, pigs and horses; caused by *Clostridium novyi* type B

Black leg: An acute, febrile disease of cattle and sheep caused by *Clostridium chauvoei*; characterized by emphysematous swelling, usually in the heavy muscles

Lockjaw: A manifestation of tetanus caused by *Clostridium tetani* neurotoxin (tetanospasmin) in which head muscle spasms make it difficult for the animal to grasp and chew food

Lumpy jaw (bovine actinomycosis): A chronic disease of the jaw bones of cattle caused by *Actinomyces bovis*; characterized by swelling, abscesses, fistulous tracts, extensive fibrosis, osteitis, granuloma, and viscous, yellow pus containing "sulfur granules"

Malignant edema: A highly fatal infection in cattle caused by *Clostridium septicum*; characterized by toxemia

Overeating disease: Enterotoxemia in sheep caused by *Clostridium perfringens* type D

Red water disease (bacillary hemoglobinuria): An acute, infectious, toxemic disease primarily of cattle caused by *Clostridium haemolyticum* (*C. novyi* type D)

Sulfur granules: Round, oval, or irregularly shaped, 2–5 mm masses of bacteria (*Actinomyces bovis* in bovine actinomycosis) cemented together by a polysaccharide-protein complex; approximately 50% of each granule is host-derived calcium phosphate. Sulfur granules have an amorphous center and a radiating fringe of clubs thought to represent bacteria and bacterial products that have become mineralized by calcium phosphate.[13]

Swine dysentery: A common, mucohemorrhagic, diarrheal disease of swine that affects the colon and is caused by *Treponema (Serpula) hyodysenteriae*, a gram-negative, anaerobic spirochete

---

The preceding chapters were devoted to anaerobic bacteria of importance in human medicine. Anaerobes are also of significance to veterinarians and veterinary microbiologists, constituting a major portion of the indigenous microflora of animals other than humans (hereafter simply referred to as "animals") and contributing significantly to animal diseases. Anaerobe-associated diseases occur frequently in animals, are associated with distinctive clinical characteristics, and require specific treatment often differing from that required for infectious diseases not involving anaerobes.[6] The specific anaerobes found in the indigenous microflora of animals, as well as those most often involved in infectious processes and intoxications, vary somewhat between animal species. Although many of the articles cited in this chapter refer to pigmented and nonpigmented species of *Bacteroides*, most pigmented species of *Bacteroides* have been reclassified as *Porphyromonas* or *Prevotella* spp.

Endogenous Anaerobes of Animals

As they do in humans, anaerobes inhabit many anatomical sites of animals as part of the indigenous microflora. Large numbers of them are found, for example, in the lower gastrointestinal tract, the oropharynx, and perhaps the vagina. The vast majority of anaerobes colonizing animals are the same species found as members of the indigenous microflora of humans (Chapter 3). However, certain anaerobes are either unique to animals or found in much higher numbers in animals than humans; examples include *Actinomyces bovis, A. hordeovulneris, A. suis, Bacteroides nodosus, Clostridium chauvoei, C. spiroforme, Eubacterium suis, Fusobacterium russii, Peptostreptococcus indolicus, Prevotella ruminicola* (formerly *B. ruminicola*), and *Treponema (Serpula) hyodysentariae.* Certain anaerobes that are common human pathogens are less frequently isolated from animals (e.g., *Actinomyces israelii* and perhaps *Bacteroides fragilis*). Table 13-1 names some of the anaerobes encountered in veterinary clinical specimens, the anatomic sites they inhabit as indigenous microflora (when applicable), and the diseases they cause.

Table 13-1. Anaerobes Encountered in Veterinary Clinical Specimens as Pathogens or Indigenous Microflora Contaminants

Anaerobe	Anatomic Site	Disease(s)	Comments
Actinomyces spp.	Oral cavity, GI tract		
A. bovis		Bovine actino-mycosis (lumpy jaw of cattle), mastitis in sows	Pus often contains sulfur granules; also recovered with *Brucella abortus* or *B. suis* from poll evil and fistulous withers (bursitic conditions in horses)
A. hordeovulneris	Oral cavity of dogs	Most pyothorax in dogs, occasional bronchopneu-monia, extensive subcutaneous pyo-granulomatous actinomycosis, osteomyelitis; organism is isolated sporadically from similar conditions in cats	
A. suis		Primary chronic granulomatous and suppurative mastitis in sows	Pus frequently contains sulfur granules

Table 13-1. Anaerobes Encountered in Veterinary Clinical Specimens as Pathogens or Indigenous Microflora Contaminants (*continued*)

Anaerobe	Anatomic Site	Disease(s)	Comments
Bacteroides spp.	Oral cavity, GI and GU tracts	Involved in a variety of purulent infectious processes of soft tissues (e.g., bovine liver abscesses, bite wounds, foot diseases)	Infectious processes are frequently poly-microbial
B. nodosus	Feet of sheep and cattle	Foot rot in sheep and cattle	Extracellular proteases; pili; foot rot produces a characteristic, unpleasant odor; may also involve *F. necrophorum* and *P. melaninogenica*
Clostridium spp.	Exogenous (soil) or GI tract		
C. botulinum	Exogenous	Botulism	An intoxication resulting from ingestion of toxin (neurotoxin); anti-toxin is available for treatment
type A		Botulism in humans, horses, cattle, and chickens	
type B		Botulism in humans, horses, and cattle	
type C_1		Botulism in water-fowl, farm chickens, turkeys, pheasants, herbivores, and dogs	
type C_2		Botulism in farm mink, sheep, cattle, and horses	
type D		Botulism in cattle	

Table 13-1. Anaerobes Encountered in Veterinary Clinical Specimens as Pathogens or Indigenous Microflora Contaminants (*continued*)

Anaerobe	Anatomic Site	Disease(s)	Comments
C. chauvoei	Exogenous and intestines of animals	Blackleg in cattle, wound infections in cattle and sheep	Spores transported hematogenously to muscle; produces alpha toxin (a lethal necrotizing protein), beta toxin (a DNase), gamma toxin (hyaluronidase), and delta toxin (oxygen-labile hemolysin); a vaccine containing *C. chauvoei* and *C. septicum* is available for immunization of cattle and sheep
C. colinum		Ulcerative enteritis of quail (quail disease), chickens, turkeys, pheasants, and other birds	
C. difficile	GI tract	Can cause fatal enterocolitis in guinea pigs, hamsters, rabbits, dogs, and foals	Usually follows antibiotic therapy with subsequent overgrowth of *C. difficile* in the colon; produces toxin A (an entero-toxin) and toxin B (a cytotoxin)
C. haemolyticum (*C. novyi* type D)	Exogenous and intestines of animals	Bacillary hemo-globinuria (red water disease) of cattle; port wine-colored urine; isolation of organism from liver infarct	Produces beta toxin (lecithinase; necrotizing, lethal; hemolytic), eta toxin (degrades tropomyosin and myosin), theta toxin (lipase)

Table 13-1. Anaerobes Encountered in Veterinary Clinical Specimens as Pathogens or Indigenous Microflora Contaminants (*continued*)

Anaerobe	Anatomic Site	Disease(s)	Comments
C. novyi (types A, B, and C)	Exogenous and intestines of animals	Gas gangrene in cattle, sheep, and other animals; causes big head in young rams; type B causes black disease (infectious necrotic hepatitis) in sheep, where organisms multiply in areas of liver necrosis resulting from migration of liver flukes (flat-worms)	Types A and B produce alpha toxin (lethal, necrotizing); type B produces beta toxin (lethal, necrotizing, hemo-lytic, lecithinase); type A produces gamma (necrotiz-ing, hemolytic, lecithinase), delta (oxygen-labile hemolysin), and epsilon (lipase) toxins; type B produces zeta (hemolysin) and eta (tropomyo-sinase) toxins; type C produces none of these toxins
C. perfringens	Exogenous or intestines	Causes entero-toxemic disease in sheep, goats, horses, cattle, dogs, and swine; also wound infec-tions (gas gan-grene) in dogs and horses	
type A		Enterotoxemia of lambs, cattle, goats, horses, and dogs; necrotic enteritis in fowl	Produces alpha toxin (lethal, phos-pholipase C, leci-thinase), kappa toxin (collagenase, gelatinase), neura-minidase, and enterotoxin

Table 13-1. Anaerobes Encountered in Veterinary Clinical Specimens as Pathogens or Indigenous Microflora Contaminants (*continued*)

Anaerobe	Anatomic Site	Disease(s)	Comments
type B		Lamb dysentery; sheep and goat dysentery in Europe and the Middle East; guinea pig enterotoxemia	Produces alpha toxin, beta toxin (lethal, necrotizing), epsilon toxin (lethal, permease), theta toxin (oxygen labile hemolysin, cytolysin), kappa toxin, lambda toxin (protease), mu toxin (hyaluronidase), nu toxin (DNase), and neuraminidase
type C		Enterotoxemia in lambs and pigs; necrotic enteritis in fowl	Produces alpha and beta toxins, delta toxin (hemolysin), theta, kappa, and nu toxins, neuraminidase, and enterotoxin
type D		Enterotoxemia of sheep; pulpy kidney disease of lambs	Produces alpha, epsilon, theta, kappa, and lambda toxins, neuraminidase, and enterotoxin
type E		Enterotoxemia in calves; lamb dysentery; guinea pig enterotoxemia; rabbit "iota" enterotoxemia	Produces alpha toxin, iota toxin (lethal, dermonecrotic), theta, kappa, and lambda toxins, and neuraminidase
C. septicum	Exogenous and intestines of animals	Wound infections in cattle, sheep, horses, and swine; causes malignant edema in cattle	Produces alpha toxin (lethal, necrotizing), beta toxin (DNase), gamma toxin (hyaluronidase), and delta toxin (oxygen-labile hemolysin)

Table 13-1. Anaerobes Encountered in Veterinary Clinical Specimens as Pathogens or Indigenous Microflora Contaminants (*continued*)

Anaerobe	Anatomic Site	Disease(s)	Comments
C. sordellii		Muscle and liver lesions in cattle and sheep, which can lead to sudden deaths	Produces alpha toxin (phospholipase C), lethal beta toxins (HT, equivalent to *C. difficile* toxin A; LT, equivalent to *C. difficile* toxin B), and an oxygen-labile hemolysin.
C. spiroforme	Uncertain	Enterotoxemia and explosive diarrheal disease of young rabbits (muco-hemorrhagic enteritis)	
C. tetani	Exogenous	Wound infection from soil, especially soil contaminated with feces; most animals are susceptible; cats and birds are relatively resistant; may cause "lockjaw"	Produces tetanospasmin (neurotoxin) and tetanolysin (oxygen-labile hemolysin); tetanus can be treated with antitoxin; a vaccine is available
Eubacterium spp.	Uncertain	Fish meningitis; porcine cystitis	
Fusobacterium spp.	Oral cavity, GI tract	Stomatitis, bite wounds	
F. necrophorum		Bovine liver abscesses following rumen acidosis and ruminal epithelial ulceration; calf diphtheria; necrotic rhinitis (bull-nose) of swine; foot rot in sheep and cattle (with *Bacteroides nodosus* and *P. melaninogenica*, respectively)	Produces endotoxin, hemolysin, leukocidin, cytoplasmic toxin

Table 13-1. Anaerobes Encountered in Veterinary Clinical Specimens as Pathogens or Indigenous Microflora Contaminants (*continued*)

Anaerobe	Anatomic Site	Disease(s)	Comments
Peptococcus niger	Uncertain	Opportunistic pathogen	
Peptostreptococcus spp.	Uncertain	Opportunistic pathogens	
Propionibacterium spp.	GI tract	Opportunistic pathogens	
Treponema (Serpula) hyodysenteriae	GI tract	Swine dysentery	Hemolysin

Source: Based on references 12, 14, 15, and 24.

The Role of Anaerobes in Veterinary Diseases

Anaerobes participate in a wide variety of infectious processes/diseases of animals. These conditions usually (1) involve anaerobes of the animal's indigenous microflora, (2) are polymicrobial, and (3) develop in proximity to mucous membrane surfaces and/or as a result of a "break" in a primary defense barrier.[8] Figure 13-1 and Table 13-2 illustrate the wide diversity of diseases that may involve anaerobes. Figure 13-1 depicts, by anatomic site, the infectious processes/diseases most apt to involve anaerobes in dogs. Table 13-2 contains information on the role of anaerobes in such conditions in a broader range of animal species.

Clostridial infections/intoxications are quite common in animals and of considerable economic importance. As described in Chapter 2, the natural habitats of *Clostridium* spp. are soil and the gastrointestinal (GI) tracts of humans and animals. Animals acquire exogenous clostridia via wound contamination or ingestion. Botulism usually follows ingestion of preformed *C. botulinum* toxins (botulins). Of the eight types/subtypes (A, B, C_1, C_2, D, E, F, G) of *C. botulinum*, types A, B, and E cause most cases of human botulism; types B, C_1, C_2, and D cause most animal cases. Many cases of animal botulism occur in wild water fowl, fewer cases occur in chickens, and even fewer occur in cattle and horses. Dogs, cats, and swine are relatively resistant to botulins.[12]

As is true for many other clostridial species, *C. perfringens* is widely distributed in soil and the GI tracts of animals. Six types have been identified: A, B, C, D, E, and F. Of these, three (B, C, and D) are of major clinical importance in animals. Type A causes most cases of clostridial food poisoning in humans. Types B and C cause severe enteritis, dysentery, toxemia, and high mortality in young lambs, calves, pigs, and foals. Type C causes enterotoxemia in these same animals. Type D causes an enterotoxemia of sheep, less frequently of goats, and rarely of cattle. Type A is suspected of causing hemorrhagic enteritis in young and mature cattle, horses, and sheep.[12] *Clostridium difficile* may be a cause of enterocolitis in foals.[23]

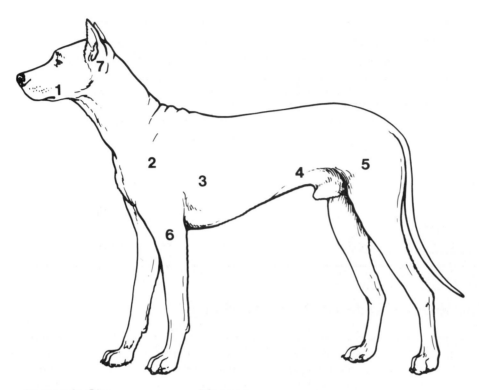

Anatomic Site	Disease
1. Oropharynx	Oropharyngeal abscess Retrobulbar abscess
2. Skin	Bite wound Traumatic puncture wound Foreign body–associated infectious processes
3. Respiratory tract	Pyothorax Lung abscess Aspiration pneumonia Chronic sinusitis
4. Abdomen	Intestinal perforation Liver abscess Cholangitis and cholecystitis Postsurgical infections
5. Reproductive tract	Pyometra Vaginal abscess
6. Musculoskeletal system	Osteomyelitis Myonecrosis (gas gangrene)
7. Central nervous system	Brain abscess Otitis media-interna

Figure 13-1. Infectious processes/diseases of dogs most apt to involve anaerobes (from reference 8, used with permission of Drs. S.W. Dow and R.L. Jones)

Table 13-2. The Role of Anaerobes in Infectious Processes/Diseases of Animals

Infectious Process/Disease	Comments
Bacillary hemoglobinuria (cattle)	*Clostridium hemolyticum*
Bacteremia	*Bacteroides* spp., including *B. fragilis*, have been isolated from blood cultures of dogs, *Fusobacterium* in cattle
Big head (rams)	*Clostridium novyi* type A
Blackleg (even-toed ungulates, rarely pigs)	*Clostridium chauvoei, C. septicum, C. sordellii* (rare)
Botulism	*Clostridium botulinum* types A-E
Brain abscesses of dogs and cats	*Bacteroides, Fusobacterium,* and *Peptostepto-coccus* spp. have been isolated
Braxy (sheep)	*Clostridium septicum*
Foot rot in cattle	A complex interaction among *Prevotella melaninogenica, Fusobacterium necrophorum,* and possibly *Actinomyces pyogenes*
Contagious foot rot in sheep	A polymicrobial infectious process involving synergism between *Bacteroides nodosus* and *Fusobacterium necrophorum*
Enterotoxemias (lambs, foals, calves, adult sheep, and goats)	*Clostridium perfringens* type B
Foot abscess (sheep)	*Fusobacterium necrophorum* and *Actinomyces pyogenes*
Intraabdominal infectious processes/diseases	
Infectious necrotic hepatitis/Black disease (sheep, cattle)	*Clostridium novyii* type B
Liver abscesses in cattle	*Fusobacterium necrophorum* has been reported to be the primary pathogen; predisposing factor is ruminal acidosis, which disrupts epithelial integrity, leading to portal bacteremia
Liver abscesses in dogs	*Clostridium perfringens* has been isolated in pure culture in several cases
Lamb dysentery	*Clostridium perfringens* type B
Myonecrosis (gas gangrene)	*Clostridium* spp.; *C. novyii* type B and *C. perfringens* in cattle
Malignant edema (cattle, sheep, pigs, horses)	*C. septicum*
Myositis/wound infections	*Bacteroides* spp., *Clostridium chauvoei, C. septicum, C. perfringens, Fusobacterium* spp., anaerobic cocci

Table 13-2. The Role of Anaerobes in Infectious Processes/Diseases of Animals (*continued*)

Infectious Process/Disease	Comments
Necrotic enteritis (lambs, calves, piglets)	*Clostridium perfringens* type C
Oropharyngeal and cutaneous infections	
Infected bite wounds in small animals	Anaerobes commonly isolated include *Prevotella melaninogenica, Fusobacterium* spp., and anaerobic cocci
Subcutaneous abscesses in cats	*Bacteroides* and *Fusobacterium* spp. are frequently isolated; *Prevotella melaninogenica, F. russii,* and *Porphyromonas asaccharolytica* are considered important pathogens in feline pyogenic infections
Osteomyelitis	
Chronic osteomyelitis in animals	*Bacteroides, Clostridium,* and *Fusobacterium* spp. were the anaerobes most frequently isolated
Osteomyelitis in dogs	*Actinomyces* spp. have been isolated
Pleuropulmonary infections	
Pyothorax in cats	*Actinomyces, Bacteroides, Clostridium,* and *Fusobacterium* spp. are commonly isolated
Pyothorax in dogs	*Actinomyces, Bacteroides,* and *Fusobacterium* spp. are frequently isolated
Pulpy kidney, enterotoxemia (lamb, adult sheep, cattle, goats)	*Clostridium perfringens* type D
Struck (adult sheep)	*Clostridium perfringens* type C
Tetanus	*Clostridium tetani*

Source: Based on references 8 and 25.

Actinomyces hordeovulneris, frequently misidentified as *A. viscosus,* is apparently a component of the oral flora of dogs. This facultative anaerobe is frequently isolated from mixed infections involving oral flora, such as most pyothorax, occasional bronchopneumonia, extensive subcutaneous pyogranulomatous actinomycosis, and osteomyelitis.[5a] *A. hordeovulneris* is less frequently isolated from similar disease processes in cats. This organism spontaneously produces L-phase variants *in vitro,* which may explain the unusually high rate of therapeutic failures with penicillin derivatives, should such L-phase variants also occur *in vivo.*[24]

Bacteroides nodosus is an anaerobic, gram-negative bacillus that colonizes the interdigital epithelial tissue of sheep and cattle feet. Extensive rainfall and wet pastures predispose to a disease known as foot rot, an economically important, contagious disease of sheep. Serious cases result in lameness, poor wool production, and susceptibility to other infectious diseases and may lead to annual losses of $50,000 or more for individual ranchers due to added labor, feed, and antibiotics and lower lamb prices.[22] In foot rot, *B. nodosus* interacts synergistically with *Fusobacterium necrophorum* and possibly *Actinomyces pyogenes*. Virulent strains of *B. nodosus* produce high levels of stable, extracellular proteases and possess abundant pili (fimbriae), whereas *F. necrophorum* possesses biologically active cell wall lipopolysaccharide (LPS) endotoxin, a hemolysin, a leukocidin, a leukotoxin, and a cytoplasmic toxin.[14] Vaccines based on whole *B. nodosus* cells or purified pili induce antibodies to the surface pili and provide some immunoprotection.[22]

Enterotoxigenic isolates of *Bacteroides fragilis* are associated with diarrheal disease in calves, foals, lambs, piglets and humans and are enteropathogenic in laboratory rabbits and gnotobiotic pigs. Myers et al. have recently described a rabbit model of value in screening isolates of *B. fragilis* for enterovirulence and studying the pathogenesis of disease caused by these organisms.[26] In this assay, enterotoxin-producing isolates elicit fluid accumulation in the intestinal tract and exfoliation of epithelial cells in the proximal colon of two-week-old rabbits.

Treponema (Serpula) hyodysenteriae is a gram-negative, anaerobic spirochete that causes swine dysentery. This common, mucohemorrhagic, diarrheal disease has been reported in every major pig-producing country, resulting in up to 90% morbidity and 30% mortality in infected herds.[21] Although its pathogenic mechanisms are not well understood, *T. hyodysenteriae* is known to produce a hemolysin. Even though it is an anaerobe, *T. hyodysenteriae* can survive for several months in pig feces or farm effluent at low environmental temperatures.[14]

As in human infectious diseases involving anaerobes, certain indications can alert veterinarians to anaerobe involvement, including the anatomic location of the infectious process, specific characteristics of the wound or exudate (including odor), response to previous therapy, and Gram-stained preparations of the exudate.[9] An investigation of pleuropneumonia in horses, for example, revealed that 62% of the horses from which anaerobes were isolated had foul-smelling specimens (pleural fluids and/or tracheobronchial aspirates) and/or breath (particularly after coughing).[27] Indications of anaerobe involvement in human disease are described in Chapter 4.

Frequency of Anaerobe Isolation from Veterinary Clinical Specimens

In an investigation of 3,133 specimens obtained from domestic animals, 1,927 (62%) were bacteriologically positive. Of these, 823 (26%) contained anaerobes.[16] Thus, anaerobes were recovered from 42% of the bacteriologically positive specimens. A mean of 1.9 different species of anaerobes was recovered from each anaerobe-positive specimen. The majority of the specimens had been obtained from dogs, cats, horses, cattle, sheep, and goats. The most commonly identified anaerobe was *Fusobacterium necrophorum*, representing 18% of the total isolated. *Bacteroides* spp. accounted for 44% of the total, with pigmented species (8%) isolated more frequently than either *B. fragilis* (7%) or *P. ruminicola* (7%). Fusobacteria, *Peptostreptococcus* spp., and clostridia represented 21%, 12%, and 10% of the total, respectively. The most frequently identified *Peptostreptococcus* species was *P. anaerobius*, and the most frequently identified *Clostridium* species was *C. perfringens*. With the exception of *P. ruminicola*, the anaerobes most commonly isolated from the veterinary specimens included in this particular study were the same species that are frequently isolated from human specimens. Figure 13-2 illustrates the relative frequencies of isolation of anaerobes in this investigation.

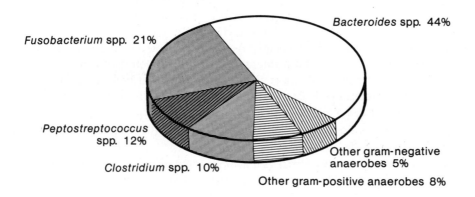

Figure 13-2. Relative frequencies of isolation of anaerobes from domestic animals (based on reference 16)

A more recent report provides data on the specific species of *Bacteroides* isolated most frequently from veterinary clinical specimens.[19] Of 1,148 *Bacteroides* spp. isolated at the Veterinary Medical Teaching Hospital, University of California, Davis, from 1984 to 1987, 742 (65%) were identified to species level. Of these, approximately 53% were nonpigmented and bile resistant, 28% were pigmented and bile sensitive, and 19% were nonpigmented and bile-sensitive. Ninety-one percent of the nonpigmented/bile-resistant isolates were members of the *B. fragilis* group. Thus, almost half (48%) of the *Bacteroides* isolates identified to the species level were members of the *B. fragilis* group. These results are depicted diagrammatically in Figure 13-3.

In a study involving dogs and cats, anaerobes were isolated from 78 (13%) of 599 specimens.[10] In a subgroup of 153 specimens that the authors considered "properly submitted" specimens, 93 (61%) were bacteriologically positive, and 53 (35%) contained anaerobes. Thus, anaerobes were recovered from 57% of the bacteriologically positive subgroup. A mean of 1.7 different species was recovered from each anaerobe positive specimen. Overall, *Clostridium perfringens* was the most commonly isolated anaerobe, and *Fusobacterium nucleatum* was the most common anaerobe recovered from the subgroup of 153 "properly submitted" specimens. *Bacteroides* spp. and *Fusobacterium* spp. each accounted for 30% of the isolates, followed by *Clostridium* spp. (14%), *Peptostreptococcus* spp. (11%), and *Actinomyces* spp. (9%). With the exception of *Fusobacterium russii,* the anaerobes recovered most often from the dog and cat specimens in this study are those frequently isolated from human specimens. Figure 13-4 illustrates the relative frequencies of isolation of anaerobes in this study.

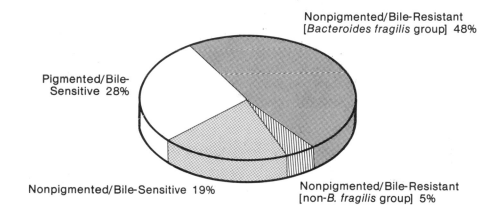

Figure 13-3. *Bacteroides* spp. isolated from veterinary specimens and identified to species level (based on reference 19)

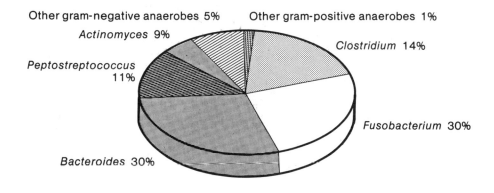

Figure 13-4. Relative frequencies of isolation of anaerobes from dogs and cats (based on reference 10)

In an investigation of pleuropneumonia in horses, anaerobes were recovered from specimens (pleural fluid and transbronchial aspirates) from 21 (46%) of the 46 horses.[27] The genus of anaerobe most frequently found was *Bacteroides,* representing 65% of the total recovered. The most common species of *Bacteroides* were *B. oralis* (35% of the *Bacteroides* isolates and 22.5% of the total anaerobes isolated) and *Prevotella melaninogenica* (formerly *B. melaninogenicus*); 19% of the *Bacteroides* and 12.5% of the total. *Clostridium* was the next most frequently encountered genus (20% of the total isolates), with *C. beijerinkii* the most common species (50% of the clostridial isolates and 10% of the total). Together, *Bacteroides* and *Clostridium* spp. represented 85% of the isolates, with all other genera representing only 15%.

The complete absence of *Fusobacterium* isolates in this study is curious, perhaps a reflection of the media and/or methodology used. Fusobacteria are especially fastidious anaerobes. They are frequently misidentified as *Bacteroides* spp. when identification systems that contain only human isolates in their data bases are used.[24]

Selection, Collection, and Transport of Specimens

Recommendations for selecting, collecting, and transporting veterinary specimens are essentially the same as those for human clinical specimens (Chapter 5). In order to generate clinically relevant information, veterinary microbiologists require specimens that contain few (preferably no) endogenous anaerobes. Specimens must be collected in a manner that minimizes their contamination by indigenous microflora that might be present at or near the collection site. Specimens must be transported to the laboratory as rapidly as possible in transport containers designed to minimize specimen exposure to oxygen. They should not be refrigerated or allowed to dry out en route.

As with human clinical specimens, aspirates and carefully collected tissue specimens are preferable to those collected by swab, which are notoriously contaminated with indigenous microflora and subject to drying en route to the laboratory. Voided urine, feces, and sputum are inappropriate specimens for **routine** anaerobic culture due to their contamination with endogenous anaerobes of the urethra, colon, and oral cavity, respectively. Urine collected by cystocentesis, lower respiratory tract specimens collected by transtracheal or transthoracic aspiration, and aspirated, usually sterile body fluids (e.g., pleural, peritoneal, pericardial, synovial, or cerebrospinal fluids and blood) are acceptable clinical materials for anaerobic culture.[11] Fecal specimens are acceptable when *T. hyodysenteriae, C. difficile,* or *C. perfringens* are suspected. Acceptable and unacceptable veterinary clinical specimens for anaerobic culture are listed in Table 13-3. Detailed recommendations regarding selection, collection, and transport of specimens can be found elsewhere.[9, 11]

Specimens submitted for anaerobic culture should be clearly labeled with the identity of the animal, its species, and the specimen source. If applicable, they should be accompanied by a brief history of the animal's condition and prior therapy.[11]

Table 13-3. Acceptable and Unacceptable Veterinary Clinical Specimens for Anaerobic Culture

Category of Specimen	Specimen Type	Comments
Acceptable		
	Transtracheal aspirates	
	Centesis samples from surgically prepared sites and usually sterile body sites (e.g., urinary bladder, blood, thoracic/pleural cavity, peritoneal cavity, pericardial cavity, cerebrospinal fluid, joints)	
	Fistulous tracts	Skin should first be decontaminated. A syringe should be used to obtain the specimen.
	Abscesses	
	Deep wound and aspirates from other soft tissues	
	Endometrial swabs	Must be obtained via a guarded swab device (e.g., Accu-CulShure® collection instrument).
	Surgical specimens obtained from usually sterile body sites	The deeper the better. Wounds should be debrided before swabbing for culture.
Unacceptable		
	Saliva or nasopharyngeal swabs	Except for tooth root abscess cultures.
	Gingival swabs	
	Bronchoscopy cultures	
	Vaginal or cervical swabs	
	Skin or superficial wounds	
	Gastric washes	
	Urine (free catch or catheter)	
	Feces, intestinal tract	Except for clostridial and treponemal cultures.

Source: Reference 25. Used with permission of Elsevier Science Publishing Co., New York.

Specimen Processing

For the most part, veterinary specimens submitted for anaerobic bacteriology are processed exactly like human clinical specimens (Chapter 6). They should be processed rapidly and should not be refrigerated. Processing specimens within an anaerobic chamber minimizes their exposure to oxygen. The use of holding systems is encouraged in laboratories not equipped with anaerobic chambers, following the precautions described in Chapter 6. A Gram-stained smear of the specimen should always be examined for the same reasons discussed in Chapter 6.

Hirsh et al. report an increase in the number of different species of anaerobes isolated from veterinary clinical specimens following substitution of commercially available PRAS plated media (Chapter 6) for "homemade" media stored under anaerobic conditions and replacement of anaerobic jars with a gloveless anaerobic chamber.[16] An eighteenfold increase in *Fusobacterium necrophorum* isolates and a sevenfold increase in *Bacteroides fragilis* isolates were noted, establishing these organisms as the most common and fourth most common isolates, respectively.

At the Veterinary Medical Teaching Hospital, University of California, Davis, specimens are inoculated onto PRAS blood agar plates containing Brucella agar base, hemin, and vitamin K_1 (Anaerobe Systems, San Jose, CA).[19] Inoculated plates are held under flowing, oxygen-free CO_2 until placed in an incubator within a gloveless anaerobic chamber (Anaerobe Systems) containing palladium catalyst pellets and a mixture of 10% CO_2, 10% H_2, and 80% N_2.

Processing and Identifying Anaerobic Isolates

Because many of the anaerobes recovered from veterinary specimens are identical or nearly identical to those isolated from human clinical specimens, much of the information in Chapters 7, 8, and 9 is applicable to veterinary specimens. For example, the Veterinary Medical Teaching Hospital, University of California, Davis, uses the following observations/tests/materials to identify anaerobic isolates:[19, 20] Gram stain observations, colony morphology, aerotolerance testing, PRAS chopped meat carbohydrate broth, PRAS horse serum and formate-fumarate supplements, PRAS peptone yeast glucose (PYG) and PYG formate-fumarate broths, GLC, tubes of PRAS biochemical media (Carr-Scarborough Microbiologicals, Inc, Stone Mountain, GA), catalase test, spot indole test, 20% bile, fluorescence, SPS disk, and special potency antimicrobial disks. Antisera and fluorescein-labeled antisera against *Clostridium chauvoei, C. novyi* types A and B, *C. septicum,* and *C. sordellii* are available from Burroughs Wellcome Company (Research Triangle Park, North Carolina) for use in the identification of these particular clostridia by passive protection tests (using guinea pigs) or direct staining of smears, respectively.

Many veterinary laboratories use "conventional" tubed biochemical systems to identify anaerobic isolates because these are the most reliable systems at present. Although use of miniaturized biochemical systems or miniature, preformed enzyme–based minisystems would greatly speed up and simplify anaerobe identifications, unfortunately, these systems do not as yet include unique veterinary species or strains in their data bases. Thus, such veterinary isolates are often misidentified by these systems.[1] These minisystems are, of course, entirely appropriate for use in identifying the numerous veterinary isolates that are identical to human isolates.

Gas-liquid chromatography has been used to demonstrate the presence of obligate anaerobes in aspirates from swine abscesses, udder secretions of cattle, and purulent specimens from a variety of animals, including dogs, cats, rodents, and ruminants.[3-5] In the latter study, a Packard-Becker 433 gas chromatograph with a flame ionization detector and a wide-bore, coated, fused silica column was used in conjunction with a digital processor. Although lactic acid is a reliable marker for bacterial infections, it is found in **both** aerobic and anaerobic infections. Acetic, propionic, and succinic acids are produced by obligate anaerobes, but they may be produced by other bacteria as well. However, obligate anaerobes (especially certain veterinary isolates, like *Fusobacterium necrophorum* and *Peptostreptococcus indolicus*) produce greater quantities of propionic acid than facultative anaerobes. Thus, if propionic acid is used to indicate the presence of obligate anaerobes, you must remember that it is **not** a specific marker.

These investigators feel that C_4 to C_6 volatile fatty acids (butyric, isobutyric, valeric, isovaleric, caproic, and isocaproic acids) can be used as specific markers for the presence of obligate anaerobes in veterinary clinical specimens. Further, they state that detection of C_4 to C_6 volatile fatty acids in such specimens might represent a more sensitive and reliable method of demonstrating the presence of obligate anaerobes than culturing. Unfortunately, perhaps due to the polymicrobial nature of the infectious processes, the chromatographic profiles were not sufficiently distinct to permit species or genus identifications.

Oligodeoxynucleotide probes have recently been used to detect *Treponema (Serpula) hyodysenteriae* in pig feces and to identify pure cultures of this organism.[21] These DNA probes bind to complementary regions of 16S rRNA molecules, which can be present in up to 10,000 copies per bacterial cell.

Treating Anaerobe-Associated Diseases of Animals

Surgical Debridement and Drainage

With few exceptions, the mainstay of treatment of anaerobe-associated infectious processes includes aggressive and complete surgical debridement and drainage.[7] Debridement is removing any foreign material and devitalized or contaminated tissue until the surrounding healthy tissue is exposed. Thoroughly removing infected, devitalized tissue and draining purulent exudates offer the following advantages:[7, 24]

- Improved tissue oxygenation, making the local environment less hospitable for anaerobic growth

- Elimination of necrotic tissues that provide favorable conditions for proliferation of anaerobes

- Removal of endogenous pyrogens and decreased absorption of microbial toxins

- Increased tissue blood flow, which accelerates delivery of white blood cells and antimicrobial agents

- The end results are eradication of local infection and reduction of the likelihood of local and hematogenous spread

Antimicrobial Therapy

Antimicrobial therapy is usually combined with surgical treatment, and initial agent selection is most often made empirically.[7] In general, the antimicrobial agents used to treat anaerobe-associated infectious processes in animals are the same as those used to treat such infectious processes in humans. Penicillins are the most widely used agents (especially for clostridial infections), but chloramphenicol, clindamycin, and metronidazole are recommended when penicillin-resistant anaerobes are encountered or increased tissue penetration is desired.[6]

Utilizing an agar dilution technique, investigators studied the susceptibility of anaerobic isolates from domestic animals.[16] Penicillin, ampicillin, and/or cephalothin resistance was encountered in 167 *Bacteroides* isolates, 68% of which were identified as *B. fragilis*. Approximately one-third of the penicillin-resistant species of *Bacteroides* was also resistant to tetracycline. A small percentage (3%) of *Peptostreptococcus anaerobius* isolates was resistant to penicillin. None of the 86 strains of fusobacteria tested was found to be resistant to any antimicrobial agents, and all anaerobes tested were susceptible to chloramphenicol, metronidazole, and clindamycin. Penicillin resistance has been reported in most *Bacteroides* spp., including two important veterinary pathogens—*Prevotella melaninogenica* and *Porphyromonas asaccharolytica* (formerly *B. asaccharolyticus*).

Penicillin-resistant *Fusobacterium* strains have been recovered from cattle.[24] The combination of amoxicillin and potassium clavulanate has proven to be effective against strains of *B. fragilis* isolated from dogs and for prophylaxis and treatment of bite wound infections.[7, 17]

Chloramphenicol seems to be an attractive alternative to penicillin in that it is active against virtually all anaerobes and is relatively inexpensive. Drug-dependent, hematologic toxicity occurs, however, in dogs and cats treated with chloramphenicol.[6] Clindamycin, used successfully to treat a variety of anaerobic infections in dogs, is also recommended as initial treatment for lung abscesses in dogs.[2, 6, 7] Although clindamycin therapy can result in gastrointestinal disturbances, pseudomembranous colitis appears to be a relatively rare complication in dogs treated with this drug.[6]

Metronidazole is a valuable drug for treating anaerobic infections in animals because of its activity against most (but not all) anaerobes and its effective tissue penetration. This drug has been successfully used to treat anaerobic infections in dogs, although there exists some concern about adverse reactions associated with higher doses.[6] Metronidazole has relatively poor activity against strains of *Actinomyces* and *Propionibacterium*.

In general, cephalosporins, aminoglycosides, tetracycline, erythromycin, lincomycin, and sulfonamides are not recommended for treating anaerobic infections in animals.[6] Cefoxitin has been used effectively in dogs to treat polymicrobial infectious processes involving obligate anaerobes and members of the family *Enterobacteriaceae*.[7]

Table 13-4 contains updated information on the susceptibility/resistance of anaerobes isolated from veterinary clinical specimens, determined by using a broth-disk elution technique.

Table 13-4. Percentage of Anaerobes Isolated from Veterinary Clinical Specimens Susceptible to Various Antimicrobial Agents

Anaerobe	Chloramphenicol	Clindamycin	Metronidazole	Penicillin G	Tetracycline
Bacteroides fragilis	99	94	98	16	77
B. fragilis group (including *B. fragilis*)	99	95	98.5	34	81
Nonpigmented *Bacteroides* (other than *B. fragilis* group)	100	98	99	83	96
Pigmented *Bacteroides**	100	99	100	89	96
Fusobacterium necrophorum	100	100	99	96	99
Fusobacterium spp. (other than *F. necrophorum*)	99	99	99	95	99
Peptostreptococcus anaerobius	100	99	99	96	100
Anaerobic cocci (other than *P. anaerobius*)	100	99	99	97	98
Clostridium perfringens	99	93	98	93	70
Clostridium spp. (other than *C. perfringens*)	99	85	98	90	77

Note: Boxed percentages are particularly noteworthy.

* Most pigmented *Bacteroides* spp. have been reclassified as *Porphyromonas* and *Prevotella* spp.

Source: Based on reference 18.

Susceptibility Testing

Veterinary anaerobic isolates are tested for susceptibility/resistance to anti-microbial agents in exactly the same manner as human anaerobic isolates. Therefore, the technical information and warnings contained in Chapter 10 are equally applicable to veterinary isolates.

The broth-disk elution (BDE) method of susceptibility testing was used to study a large number of anaerobes isolated from veterinary specimens at the Veterinary Medical Teaching Hospital, University of California, Davis.[18] Although the BDE method is not presently endorsed by the NCCLS for testing human anaerobic isolates, the authors feel it is well suited to the small veterinary clinical microbiology laboratory. Unfortunately, they had to use concentrations of antimicrobial agents that represent optimal, achievable blood levels (breakpoints) for humans (as recommended by the NCCLS). As yet, no such standards exist for optimal blood levels of these agents in animals. Antimicrobial agents and concentrations used in the study were as follows: chloramphenicol, 18 μg/ml; clindamycin, 4 μg/ml; metronidazole, 16 μg/ml; penicillin G, 8 U/ml; tetracycline, 6 μg/ml. A turbid 24- to 48-hour suspension of the isolate in PRAS chopped meat medium was used as inoculum; 0.1 ml was added to each tube of antimicrobial agent-containing PRAS brain heart infusion (BHI) broth and to a control tube of BHI broth containing no antimicrobial agents. Tubes were incubated aerobically for 24 hours, after which the growth in each antimicrobial agent–containing tube was compared to the amount of growth in the control tube. Results appear in Table 13-4.

Chapter in Review

- Anaerobes are important to veterinarians and veterinary microbiologists because they constitute a major portion of the indigenous microflora of animals and contribute significantly to animal diseases. The specific endogenous anaerobes of animals, as well as those most often involved in infectious processes and intoxications, vary somewhat between animal species. Many are identical to those found as constituents of the indigenous microflora of humans and known to cause infectious processes/intoxications in humans.

- Anaerobes inhabit many anatomical sites of animals as part of the indigenous microflora of those sites. For example, large numbers of anaerobes are found in the lower gastrointestinal tract, oropharynx, and perhaps the vagina of animals.

- Anaerobes that are either unique to animals or found in much higher numbers in them than in humans include *Actinomyces bovis, A. hordeovulnaris, A. suis, Bacteroides nodosus, Clostridium chauvoei, Fusobacterium russii, Peptostreptococcus indolicus, Prevotella ruminicola,* and *Treponema (Serpula) hyodysentariae.*

- Anaerobes and/or their toxins cause or contribute to a wide variety of infectious processes/diseases and intoxications in animals. Examples include big head in rams, bite wounds, black disease in sheep and other animals, black leg in cattle and sheep, botulism, brain abscesses, enterotoxemia, foot rot in sheep and cattle, gas gangrene, liver abscesses, lockjaw, lumpy jaw in cattle, malignant edema in cattle, red water disease in cattle, swine dysentery, tetanus, and a wide variety of pleuropulmonary and necrotizing soft tissue infectious processes.

- Of the eight types/subtypes (A, B, C_1, C_2, D, E, F, G) of *Clostridium botulinum*, types A, B, and E cause most cases of human botulism. Types B, C_1, C_2, and D cause most animal cases. Of the six types of *C. perfringens* (A, B, C, D, E, F), only three (B, C, and D) are of major clinical importance in animals. Enterotoxigenic type A causes human food poisoning. *Clostridium difficile* may be a cause of enterocolitis in foals.[23]

- *Bacteroides nodosus* is an anaerobic, gram-negative bacillus that interacts synergistically with *Fusobacterium necrophorum* and possibly *Actinomyces pyogenes* to cause foot rot, an economically important, contagious disease of sheep. *Treponema (Serpula) hyodysenteriae* is a gram-negative, anaerobic spirochete that causes swine dysentery.

- The anaerobes most commonly encountered in veterinary clinical specimens include *Bacteroides* spp. (especially the renamed pigmented species and *B. fragilis*), *Fusobacterium* spp. (especially *F. necrophorum, F. nucleatum*, and *F. russii*), *Clostridium* spp. (especially *C. perfringens*), and *Peptostreptococcus* spp. (especially *P. anaerobius*). With the exception of *F. russii*, these organisms are also frequently isolated from human clinical specimens.

- Recommendations for selection, collection, and transport of veterinary specimens are essentially the same as those for human clinical specimens. In order to obtain clinically relevant information, specimens must contain few (preferably no) endogenous anaerobes. They must be collected in a manner that minimizes their contamination by indigenous microflora that might be present at or near the collection site. Specimens must be transported to the laboratory as rapidly as possible in transport containers designed to minimize their exposure to oxygen and should not be refrigerated or allowed to dry out en route.

- In general, veterinary specimens submitted for anaerobic bacteriology are processed exactly like human clinical specimens. They should be processed rapidly and should not be refrigerated. Because many of the anaerobes recovered from veterinary specimens are identical to those isolated from human clinical specimens, much of the information in this book about processing of isolates, presumptive identifications, and definitive identifications is applicable to veterinary specimens.

- Treatment of anaerobe-associated infectious processes usually involves aggressive and complete surgical debridement, drainage, and antimicrobial therapy. The initial selection of an antimicrobial agent is most often made empirically. The antimicrobial agents used to treat anaerobic infectious processes in animals are usually the same as those used to treat such infectious processes in humans.

- Currently, anaerobic isolates from veterinary clinical specimens are tested for susceptibility/resistance to antimicrobial agents using the same techniques used to test anaerobic isolates from human clinical materials.

Self-Assessment Exercises

1. Name three anaerobes of significance in veterinary medicine that are of little or no significance in human medicine.

   a. _____

   b. _____

   c. _____

2. Name an anaerobe associated with each of the following diseases and an/the animal species in which each disease occurs: big head, black disease, black leg, lockjaw, lumpy jaw, malignant edema, overeating disease (enterotoxemia), swine dysentery, red water disease.

   _____

   _____

   _____

   _____

   _____

   _____

   _____

   _____

3. What are the three types of *Clostridium botulinum* of greatest importance in veterinary medicine?

   a. _____

   b. _____

   c. _____

4. Name the three types/subtypes of *Clostridium perfringens* of greatest importance in veterinary medicine.

   a. _____

   b. _____

   c. _____

5. Identify the essential elements in selection, collection, and transport of veterinary clinical specimens for anaerobic bacteriology.

   _____

   _____

   _____

6. List the three major components in the treatment of anaerobe-associated infectious processes in animals.

   a. _____

   b. _____

   c. _____

References

1. Adney, W. S., and R. L. Jones. 1985. Evaluation of theRapID–ANA system for identification of anaerobic bacteria of veterinary origin. J. Clin. Microbiol. 22:980–983.
2. Berg, J. N., C. M. Scanlan, G. M. Buening, et al. 1984. Clinical models for anaerobic bacterial infection of dogs and their use in testing the efficacy of clindamycin and lincomycin. Am. J. Vet. Res. 45:1299–1306.

3. Bogaard, A. E. J. M. van den, M. J. Hazen, and J. H. J. Maes. 1983. The detection of obligate anaerobic bacteria in swine abscesses. Vet. Microbiol. 8:389–396.

4. Bogaard, A. E. J. M. van den, M. J. Hazen, and U. Vecht. 1987. The detection of obligate anaerobic bacteria in udder secretions of dry cattle with mastitis during summer: a comparison between gas-liquid chromatography and bacteriological culturing methods. Vet. Microbiol. 14:173–182.

5. Bogaard, A. E. J. M. van den, M. J. H. Ing, and J. H. Maes. 1989. Rapid presumptive diagnosis of anaerobic infections in animals by gas-liquid chromatography. Am. J. Vet. Res. 50:1454–1459.

5a. Buchanan, A. M., Scott, J. L., Gerencser, M. A., Beaman, B. L., Jang, S. S., and Biberstein, E. L. 1984. *Actinomyces hordeovulneris* sp. nov., an agent of canine actinomycosis. Int. J. Syst. Bacteriol. 34:439–443.

6. Dow, S. W. 1988. Management of anaerobic infections. Vet. Clin. North Amer.: Sm. Anim. Prac. 18:1167–1182.

7. Dow, S. W. 1989. Anaerobic infections in dogs and cats, p. 1082–1085. *In* R. W. Kirk (ed.), Current veterinary therapy X: small animal practice. W. B. Saunders, Philadelphia.

8. Dow, S. W., and R. L. Jones. 1987. Anaerobic infections. Part I. Pathogenesis and clinical significance. Compend. Contin. Educ. Pract. Vet. 9:711–720.

9. Dow, S. W., and R. L. Jones. 1987. Anaerobic infections. Part II. Diagnosis and treatment. Compend. Contin. Educ. Pract. Vet. 9:827–839.

10. Dow, S. W., R. L. Jones, and W. S. Adney. 1986. Anaerobic bacterial infections and response to treatment in dogs and cats: 36 cases (1983–1985). J. Am. Vet. Med. Assoc. 189:930–934.

11. Dow, S. W., R. L. Jones, and R. A. W. Rosychuk. 1989. Bacteriologic specimens: selection, collection, and transport for optimum results. Compend. Contin. Educ. Pract. Vet. 11:686–701.

12. Fraser, C. M. (ed.). 1986. The Merck veterinary manual, 6th ed. Merck, Rahway, NJ.

13. George, W. L. 1989. Actinomycosis, p. 529–539. *In* S. M. Finegold and W. L. George (eds.), Anaerobic infections in humans. Academic Press, San Diego.

14. Gyles, C. L., and C. O. Thoen. 1986. Pathogenesis of bacterial infections in animals. Iowa State University Press, Ames.

15. Hatheway, C. L. 1990. Toxigenic clostridia. Clin. Microbiol. Rev. 3:66–98.

16. Hirsh, D. C., M. C. Indiveri, S. S. Jang, and E. L. Biberstein. 1985. Changes in prevalence and susceptibility of obligate anaerobes in clinical veterinary practice. J. Amer. Vet. Med. Assoc. 186:1086–1089.

17. Indiveri, M. C., and D. C. Hirsh. 1985. Clavulanic acid-potentiated activity of amoxicillin against *Bacteroides fragilis*. Am. J. Vet. Res. 46:2207–2209.

18. Jang, S. S., and D. C. Hirsh. Broth–disk elution determination of antimicrobial susceptibility of selected anaerobes isolated from animals. J. Vet. Diag. Invest., in press.

19. Jang, S. S., and D. C. Hirsh. Identity of *Bacteroides* isolates and previously named *Bacteroides* spp. in clinical specimens of animal origin. Am. J. Vet. Res., in press.

20. Jang, S. S., E. L. Biberstein, and D. C. Hirsh. 1988. A diagnostic manual of veterinary clinical bacteriology and mycology. Veterinary Medical Teaching Hospital, University of California, Davis.

21. Jensen, N. S., T. A. Casey, and T. B. Stanton. 1990. Detection and identification of *Treponema hyodysenteriae* by using oligodeoxynucleotide probes complementary to 16S rRNA. J. Clin. Microbiol. 28:2717–2721.

22. John, G. H., J. O. Carlson, C. V. Kimberling, and R. P. Ellis. 1990. Polymerase chain reaction amplification of the constant and variable regions of the *Bacteroides nodosus* fimbrial gene. 28:2456–2461.

23. Jones, R. L. 1989. Review article: diagnostic procedures for isolation and characterization of *Clostridium difficile* associated with enterocolitis in foals. J. Vet. Diagn. Invest. 1:84–86.

24. Jones, R. L. (Colorado State University). 1990. Personal communication.

25. McDonough, P., and A. B. Onderdonk. 1990. Veterinary clinical microbiology: part II. Clin. Microbiol. Newsl. 12:161–167.

26. Myers, L. L., D. S. Shoop, and J. E. Collins. 1990. Rabbit model to evaluate enterovirulence of *Bacteroides fragilis*. J. Clin. Microbiol. 28:1658–1660.

27. Sweeney, C. R., T. J. Divers, and C. E. Benson. 1985. Anaerobic bacteria in 21 horses with pleuropneumonia. J. Am. Vet. Med. Assoc. 187:721–724.

CHAPTER 14

FUTURE DIRECTIONS

Education
Cost Containment and Quality Assurance
Media
Presumptive Identifications
Definitive Identifications
Susceptibility Testing

Microbiologists one day will probably be able to perform some sort of simple chemical extraction procedure on a clinical specimen, inject the extract into an instrument (perhaps a computer-linked gas-liquid chromatograph), and have the instrument not only identify microorganisms (including anaerobes) in the specimen but also reveal their relative numbers. Perhaps the printout will also include updated susceptibility information the clinician could use as a basis for empirical antimicrobial therapy. Although certain microorganisms have actually been identified in research settings using similar approaches, the technology is not yet routinely available for the clinical laboratory. Until "kits" for this purpose become commercially available, clinical microbiology laboratories are unlikely to adopt this approach.

More likely, the immediate future of clinical anaerobic bacteriology will see increased emphasis on education, cost containment, overall quality assurance, specimen quality, improved media, and presumptive identifications; expansion of minisystem data bases to incorporate anaerobes less commonly isolated from human clinical specimens and unique veterinary anaerobes; instruments capable of automatically reading minisystem results; identifications based upon immunologic or genetic probe principles; identifications based upon rapid analysis of cellular contents; and improved susceptibility test systems.

Education

Education is the key to future improvements in clinical anaerobic bacteriology. According to Cox, "Education is the **missing link** in anaerobic bacteriology."[3] Finegold and George state that anaerobe-associated infectious processes "undoubtedly have been, and probably still are, the most commonly overlooked of bacterial infections," due primarily to a "lack of awareness or interest on the part of clinicians and laboratory workers."[6] Others have emphasized the "need for education of clinicians and microbiologists as to the importance of these organisms and optimum procedures for their recovery and identification."[11]

There exists a definite need to devote additional time to this extremely important aspect of medical microbiology in our education programs and to standardize the instruction. A 1988 survey of anaerobic bacteriology education in medical technology programs revealed marked differences in the amount of time devoted to this subject.[5] About 25% of the 320 responding programs devoted two hours or less to the subject in their microbiology lecture courses, and 50% devoted three hours or less. Only 52% conducted a microbiology laboratory course separate and distinct from the practicum or actual clinical experience in microbiology. Of the programs that conducted such a student laboratory course, 16% did **not** teach anaerobic bacteriology as part of that course. The amount of time dedicated to anaerobes in the practicum (clinical rotation) varied greatly, but about 27% of the responding programs devoted only two days or less to the subject, and 52% devoted four days or less.

Sixty-six percent of the responding programs expressed interest in a concise but comprehensive textbook on clinical anaerobic bacteriology. Fifty-seven percent expressed interest in a self-study course on the subject—one that students could use on their own to learn the topic. Sixty-three percent expressed interest in 35 mm color slide sets for use in teaching anaerobic bacteriology, and 31% expressed interest in prepackaged, commercially available materials that could be used to conduct a student "wet" laboratory session on this subject.

Of course, medical technologists are not the only health care professionals in need of education in anaerobic bacteriology. Physicians, dentists, veterinarians, nurses, physician's assistants, medical laboratory technicians, veterinary laboratory technologists/technicians, clinical microbiologists, and others all need to be educated in this important area.

Perhaps one day a single institution will exist where individuals from a variety of disciplines can receive **specialized** and **standardized** instruction in anaerobic bacteriology. Although certain differences exist between human and veterinary anaerobic bacteriology and between medical and dental anaerobic bacteriology, there are sufficient similarities to warrant a central facility for such training. It is likely that such a facility would require federal and/or corporate (perhaps pharmaceutical company) sponsorship.

In the meantime, excellent one- to three-day workshops are currently available through state health department laboratories and education-oriented companies such as Anaerobe Systems (San Jose, CA). Even after the establishment of a centralized anaerobic bacteriology training facility, there will always be a need to bring the knowledge directly to practitioners in the form of local/regional workshops and workshops conducted in conjunction with national society meetings.

Cost Containment and Quality Assurance

There will be ever-increasing emphasis on cost containment. The future will bring increased emphasis at the manufacturer level on rapid testing—instruments and kits designed to diagnose the problem as quickly as possible, cure the patient, and get him or her out of the hospital.

The **quality** of specimens submitted to anaerobic bacteriology laboratories simply **must** improve. Shrinking laboratory budgets demand that microbiologists end time-consuming, expensive workups of poor quality specimens. Hospital and laboratory administrators will begin to put "teeth" into policies that demand high-quality specimens. As if the old adage, "garbage in—garbage out," was not sufficient justification to demand good specimens, administrators will soon realize that the continued workup of poor quality specimens is a tremendous waste of money and potentially dangerous to patients.

There probably will be an increasing number of commercially available devices that enable collection of clinical specimens in a manner that will greatly reduce or even eliminate their contamination with indigenous microflora—products such as the Accu-CulShure® collection instrument described in Chapter 5.

The immediate future will also bring tighter rules regarding the **number** of isolates worked up from a single specimen, the number of **repeat** cultures performed, and the **frequency** of *Clostridium difficile* culturing, beta-lactamase testing, and susceptibility testing of anaerobes. There will be increased emphasis on the "output" phase of quality assurance as it pertains to the anaerobic bacteriology laboratory (Chapter 11). Microbiologists will increasingly interact with physicians and nursing service personnel to answer questions such as these: Are anaerobic cultures being requested appropriately? Are anaerobic bacteriology results being received in sufficient time to affect patient treatment? Are physicians using the results of tests they have requested?

Media

If microbiologists are themselves truly convinced of the importance of isolating anaerobes from clinical specimens, they must use media in which they have confidence. In an effort to reduce costs, too many microbiologists are currently using media of mediocre quality for the isolation of anaerobes. In actuality, the use of such media **increases** costs. Repeat cultures are expensive, as are the extra days that a patient remains undiagnosed in the hospital.

Some microbiologists are under the impression that media that have been stored in air can be made suitable for the culture of anaerobes by placing the media under anaerobic conditions prior to use. As mentioned in Chapter 6, additional studies are needed to determine the extent to which this practice can actually remove toxic substances such as superoxide anions and hydrogen peroxide from the surface of ordinary and so-called "reducible" media that have been stored in the presence of oxygen (i.e., in air, in gas-permeable wrappers). Studies are also needed to determine if such media will support growth of the more oxygen-sensitive anaerobes (e.g., fusobacteria, gram-positive cocci, and *Porphyromonas* spp.)

More stringent controls on media production at the manufacturer level will also be instituted. The ability of a particular medium to isolate *Bacteroides fragilis* and *Clostridium perfringens* from a clinical specimen or support the growth of QC strains of these particular organisms is **insufficient** evidence that the media will support growth of more fastidious anaerobes (e.g., *Fusobacterium*, *Peptostreptococcus*, and *Porphyromonas* spp.). Thus, manufacturer-level QC of media, as well as in-house QC, should include **the most fastidious** of pathogenic anaerobes—**not** the most aerotolerant.

Presumptive Identifications

With ever-increasing emphasis on speed and cost containment in microbiology laboratories will probably come increased emphasis on presumptive identifications via the use of disks, simple procedures, and easy-to-follow flow charts. Such procedures must be rapid, inexpensive, and reasonably accurate. Perhaps most important, there is a need for a rapid (same day) method of differentiating between obligate anaerobes and facultative anaerobes—something that will replace the overnight aerotolerance test. Surely, there must be an intrinsic compound (perhaps an enzyme) unique to one or the other group that would differentiate between them. We urge our research colleagues to discover such a compound and develop practical methods for its use in clinical microbiology laboratories.

Definitive Identifications

The currently available miniaturized identification systems ("minisystems") contain in their data bases many of the anaerobes **routinely** isolated from clinical specimens, and these systems perform reasonably well when inoculated with such isolates (Appendix C). However, minisystems that will identify a greater variety of isolates, including those **less frequently** encountered in human clinical specimens, are needed. Many of these less common anaerobes are misidentified by existing minisystems. Also, incorporation of unique veterinary isolates into minisystem data bases will enable veterinary microbiology laboratories to use these systems without fear of misidentifications. Automated reading of minisystems will reduce the amount of time required for interpretation of reactions and will eliminate the subjectivity associated with interpreting the various color changes.

Investigations might also demonstrate that **anaerobic** inoculation and incubation of the four-hour, preexisting enzyme–based identification systems might improve their accuracy. Because the enzymes being tested usually function under anaerobic conditions, they may very well function less effectively in the presence of oxygen. Although such changes in technique might be met with initial resentment on the part of manufacturers and users alike, the extra time and effort would be a small price to pay for decreased rates of nonidentifications or misidentifications with these minisystems.

Additional commercial identification kits based upon immunologic principles will probably become available. These might include fluorescence, latex agglutination, and enzyme-linked procedures employing monoclonal antibodies. They might be of value in identifying anaerobes in clinical specimens as well as in pure culture.

Although a variety of immunofluorescence and latex agglutination procedures for detection/identification of *Bacteroides* spp. have been commercially available over the years, they have had specificity and/or sensitivity problems, and have been withdrawn from the market.[13]

Genetic probes for identification of anaerobes have been developed (e.g., those described by Kuritza et al. for clinically important *Bacteroides* spp.; by Dix et al. for *B. forsythus, Fusobacterium nucleatum, Porphyromonas gingivalis, Prevotella intermedia,* and *Wolinella recta;* by Wren et al. for toxigenic *Clostridium difficile;* and by van Damme-Jongsten et al. for enterotoxigenic strains of *C. perfringens*).[4, 9, 12, 14] Such probes will undoubtedly become commercially available in the near future and will most likely detect bacterial rRNA or multicopy plasmids using complementary strands of DNA or synthetic oligonucleotides as probes. Laboratories of the future will most likely employ polymerase chain reaction (PCR) technology to identify anaerobes, as has been reported for toxigenic strains of *Clostridium difficile.*[8] These newer methodologies will undoubtedly enable microbiologists to identify anaerobes directly in clinical specimens, in many cases eliminating the need to isolate organisms in pure culture.

The MIDI-MIS system (Chapter 9) will probably increase in popularity, not only because of its ability to identify over 375 anaerobes by cellular fatty acid analysis, but also because of its versatility (i.e., its additional "libraries" of *Enterobacteriaceae,* staphylococci, streptococci, mycobacteria, and yeasts). It is an objective system, requiring relatively little technologist time. Although the initial cost of the instrumentation is high (~$38,000), the software packages ("libraries") are relatively inexpensive, and the cost of expendables per identification is quite low (~$1.30 at present). Multivariate analysis of cellular fatty acids was successfully used to assess the taxonomic relationships between *Wolinella* spp., *Bacteroides gracilis, B. ureolyticus,* and *Campylobacter fetus* subsp. *venerealis* and to increase the number of reliable characteristics for classification and identification of organisms within the genera *Wolinella, Campylobacter, Bacteroides, Prevotella,* and *Porphyromonas.*[1]

Using GLC and related techniques, investigators have been attempting for many years to **directly** identify anaerobes and other organisms in clinical specimens. Definite progress is being made toward that goal, and we expect to see standardized, commercially available systems in the near future. Johnson et al., for example, have described a method for identifying *Clostridium difficile* in stool specimens using culture-enhanced GLC.[7] In this technique, GLC is used to detect four distinctive peaks produced by *C. difficile* following incubation of the stool specimen in a cefoxitin-containing selective broth. The method is 99.6% sensitive and 99.0% specific when compared to culture techniques or cytotoxin testing. Assays of this type (i.e., those performed directly on clinical specimens) could probably be modified to enable use of the MIDI-MIS instrumentation.

Susceptibility Testing

The area of clinical anaerobic bacteriology in most critical need of improvement is susceptibility testing. No single technique currently provides reliable information for all anaerobes. There is no universally agreed upon "gold standard." Anaerobe susceptibility test systems of the future will probably require **anaerobic** inoculation and incubation, making them somewhat more tedious and time-consuming to perform. This would, however, be a small price to pay for reliable results. Using *Clostridium* *difficile* and *Prevotella melaninogenica*, Mortelmans and Cox recently demonstrated that relatively short exposure times to atmospheric oxygen dramatically affect anaerobe survival.[10] They speculate that exposure of the anaerobes to oxygen while susceptibility tests are being set up on the open bench may at least in part explain reported difficulties in obtaining reliable and reproducible results. Citron et al. tested the growth of fastidious anaerobes in microtiter trays containing six different broths and presented additional data supporting such a hypothesis.[2]

Automation of anaerobe susceptibility testing is also likely to occur. Automation of the broth-disk elution system, for example, would make a simple system even easier, and refinement and strict standardization of the technique should increase its accuracy to the point where it could once again be endorsed by the NCCLS.

Although additional studies and strict standardization of the procedure are required, the Epsilometer (or "E test") technique is likely to gain in popularity (Chapter 10). It is relatively easy, inexpensive, and rapid; is a practical method for testing small numbers of clinically significant anaerobic isolates; and although it is **not** a disk-diffusion technique, microbiologists are apt to feel comfortable with a method that bears some familiarity to a system they've used for many years (namely, the "Kirby-Bauer" system). The use of **anaerobic** conditions at **all** stages of this technique would probably improve its accuracy, including storage of test strips, streaking of PRAS plated media, and addition of test strips to plates.

In conclusion, we envision a future in which microbiologists of the world are united in their common goals: comprehensive and standardized education, increased quality and speed of microbiological services, reduced health care costs, and superior patient care—especially as these worthy goals impact clinical anaerobic bacteriology.

References

1. Brondz, I., and I. Olsen. 1991. Multivariate analyses of cellular fatty acids in *Bacteroides, Prevotella, Porphyromonas, Wolinella,* and *Campylobacter* spp. J. Clin. Microbiol. 29:183–189.
2. Citron, D. M., M. E. Cox, M.I. Ostovari, and E.J.C. Goldstein. 1989. Growth of fastidious anaerobes in six different broths in microtiter trays under strictly anaerobic conditions vs. the National Committee on Clinical Laboratory Standards (NCCLS) recommended procedure. Abstract 39, Third European Congress on Anaerobic Bacteria and Infections, Munich, FRG.
3. Cox, M. (Anaerobe Systems). 1988. Personal communication.
4. Dix, K., S. M. Watanabe, S. McArdle, D. I. Lee, C. Randolph, B. Moncla, and D. E. Schwartz. 1990. Species-specific oligodeoxynucleotide probes for the identification of periodontal bacteria. J. Clin. Microbiol. 28:319–323.
5. Duben-Engelkirk, J., and P. G. Engelkirk. 1990. Analysis of anaerobic bacteriology education in medical technology programs, p. 155–166. *In* Proceedings of the second annual southwestern allied health symposium, New Orleans.
6. Finegold, S. M., and W. L. George (eds.). 1989. Anaerobic infections in humans. Academic Press, San Diego.
7. Johnson, L. L., L. V. McFarland, P. Dearing, V. Raisys, and F. D. Schoenknecht. 1989. Identification of *Clostridium difficile* in stool specimens by culture-enhanced gas-liquid chromatography. J. Clin. Microbiol. 27:2218–2221.

8. Kato, N., C-Y. Ou, H. Kato, S. L. Bartley, V. K. Brown, V. R. Dowell, Jr., and K. Ueno. 1991. Identification of toxigenic *Clostridium difficile* by the polymerase chain reaction. J. Clin. Microbiol. 29:33–37.

9. Kuritza, A. P., C. E. Getty, P. Shaughnessy, R. Hesse, and A. A. Salyers. 1986. DNA probes for identification of clinically important *Bacteroides* species. J. Clin. Microbiol. 23:343–349.

10. Mortelmans, K., and M. Cox. 1990. A quantitative survival study of anaerobic bacteria exposed to atmospheric oxygen. Abstract C-274, Annual Meeting of the American Society for Microbiology, Anaheim.

11. Sutter, V. L., D. M. Citron, M. A. C. Edelstein, and S. M. Finegold. 1985. Wadsworth anaerobic bacteriology manual, 4th ed. Star Publishing, Belmont, CA.

12. Van Damme-Jongsten, M., J. Rodhouse, R. J. Gilbert, and S. Notermans. 1990. Synthetic DNA probes for detection of enterotoxigenic *Clostridium perfringens* strains isolated from outbreaks of food poisoning. J. Clin. Microbiol. 28:131–133.

13. Weissfeld, A. S., and A. C. Sonnenwirth. 1981. Rapid detection and identification of *Bacteroides fragilis* and *Bacteroides melaninogenicus* by immunofluorescence. J. Clin. Microbiol. 13:798–800.

14. Wren, B. W., C. L. Clayton, N. B. Castledine, and S. Tabaqchali. 1990. Identification of toxigenic *Clostridium difficile* strains by using a toxin A gene-specific probe. J. Clin. Microbiol. 28:1808–1812.

APPENDIX A

EQUIPMENT, SERVICES AND SUPPLIES

The materials and services listed in this appendix will enable a laboratory to provide clinical anaerobic bacteriology services. Some of them are optional. Items routinely found in all clinical microbiology laboratories (e.g., inoculating loops, incubators, microscopes, glass microscope slides, cover slips) are not included. This appendix contains the names of representative commercial sources and is not intended to be a complete listing of all possible sources. Additional information concerning commercial sources of these materials and services can be found in Appendix B.

Anaerobe identification systems
    Biochemical-based minisystems
        Analytab Products (API 20 A®)
        Becton Dickinson Microbiology Systems (Minitek® and SCEPTOR®)
    High-resolution gas-liquid chromatography
        Microbial ID, Inc. (MIDI-MIS®)
    Preexisting enzyme-based minisystems
        Analytab Products (AN-IDENT®)
        Austin Biological Laboratories (ABL), Inc.
        Baxter Healthcare Corp., MicroScan Division
        Innovative Diagnostic Systems, Inc. (RapID ANA II™)
        Vitek Systems (ANI Card® and RapID ANA II™)
    VPI® anaerobic culture tube system
        Bellco Glass, Inc.

Anaerobic bags/pouches and accessories
    Becton Dickinson Microbiology Systems (GasPak Pouch® and Type A Bio Bag®)
    Difco Laboratories (Anaerobic Pouch System, Catalyst-Free®)

Anaerobic blood culture bottles
    Becton Dickinson Microbiology Systems (BBL® prepared culture bottles and BACTEC®
        bottles)
    Difco Laboratories (Bacto®-blood culture bottles)

Anaerobic chambers
    Coy Laboratory Products, Inc.
    Don Whitley Scientific Ltd. (Gloveless models - Europe)
    Forma Scientific, Inc.
    Lab-Line Instruments, Inc.

Pacific Diagnostics (Gloveless models - Australia and New Zealand)
Shel-Lab (Anaerobe Systems' BACTRON® gloveless models)

Anaerobic holding chamber
Carr-Scarborough Microbiologicals, Inc.

Anaerobic jars and jar accessories
Anaerobe Systems
Becton Dickinson Microbiology Systems (including GasPak® envelopes, GasPak Plus®
    envelopes, and Anaerosphere® gas generation system)
Difco Laboratories
Don Whitley Scientific Ltd.
Oxoid U.S.A., Inc.
Scott Laboratories, Inc. (ANAPAK® system)

Anaerobic media
Dehydrated media
Becton Dickinson Microbiology Systems
Difco Laboratories (including Bacto® Anaerobe Broth MIC for microdilution
    broth susceptibility testing)
Lab M Ltd.
Oxoid U.S.A., Inc.
Tubed and plated media
Anaerobe Systems (PRAS plated and tubed media)
Becton Dickinson Microbiology Systems
Carr-Scarborough Microbiologicals, Inc.
Cultech Diagnostics (VAPRUF® prereduced anaerobic media)
Pacific Diagnostics (PRAS plated and tubed media—Australia and New Zealand)
Remel (including PRAS tubed media)
Scott Laboratories, Inc. (including PRAS tubed media)

Anaerobic transport media/containers (for aspirates and swabs)
Anaerobe Systems
Becton Dickinson Microbiology Systems (Port-A-Cul® tubes and swabs and Culturette®
    swabs)
Carr-Scarborough Microbiologicals, Inc. (AnaTrans® and Anaswab®)
Medical Wire & Equipment Co. (TRANSWAB® and TRANSTUBE®)
Scott Laboratories, Inc. (ANAPORT® and ANASWAB®)

Antimicrobial disks and/or other disks useful in identifying anaerobes
Anaerobe Systems
Becton Dickinson Microbiology Systems (Sensi-Discs® and Taxo® discs)
Difco Laboratories (Dispens-O-Discs®)
Medical Wire & Equipment Co. (Microrings® AN and AC)
Oxoid U.S.A., Inc.
Remel
Scott Laboratories, Inc.

Beta-lactamase testing
Becton Dickinson Microbiology Systems (Cefinase® disks)

***Clostridium difficile* products**
Baxter Healthcare Corp., Bartels Division (*C. difficile* Cytotoxicity Assay®)
Becton Dickinson Microbiology Systems (CDT™ latex agglutination test)

Difco Laboratories (Immuno-CUBE™ test system for *Clostridium difficile* called C.diff-CUBE™)

Meridian Diagnostics, Inc. (Meritec™-*C. difficile* latex agglutination test and Premier™ *C. difficile* Toxin A enzyme immunoassay procedure)

Educational materials (textbooks, 35-mm slide sets, etc.)

Star Publishing Company

Gas-liquid chromatograph and accessories

Dodeca

Hewlett-Packard Company

Newsletter (*Anaerobe Abstracts*; published quarterly)

Unicornucopia

Proficiency surveys

College of American Pathologists (CAP)

QC organisms (anaerobes)

American Type Culture Collection

Anaerobe Systems

Becton Dickinson Microbiology Systems (Quali-Disc® lyophilized QC cultures)

Difco Laboratories

Remel (BACTi DISKS®)

Scott Laboratories, Inc. (Q-Check Culti-loops®)

Reagents

Anaerobe Systems

Becton Dickinson Microbiology Systems

Difco Laboratories (Spot-Test® reagents)

Remel

Scott Laboratories, Inc.

Reference laboratories

Acculab

Anaerobe Reference Laboratory, Centers for Disease Control

Microbiology Reference Laboratory

Microbiology Specialists, Inc.

Susceptibility testing standards

The National Committee for Clinical Laboratory Standards (NCCLS)

Susceptibility testing systems

AB Biodisk (Epsilometer® or E test®)

Baxter Healthcare Corp., MicroScan Division

Becton Dickinson Microbiology Systems (SCEPTOR®)

Innovative Diagnostic Systems, Inc. (MicroTech panels)

Microtech Medical Systems, Inc.

MSI/Micro-Media Systems

Spiral System Instruments, Inc.

Workshops

Anaerobe Systems

APPENDIX B

COMMERCIAL SOURCES

This is not intended to be a complete listing of commercial sources. Please contact Star Publishing Company to have a company name, address, phone number, FAX number, and/or product(s) added, deleted, or corrected in future editions.

AB Biodisk, Pyramidvagen 7, S-17136 Solna, Sweden, Phone +46-8 730 07 60, Telex 13578 Gnostic S, FAX +46-8 83 81 58

AB Biodisk North America, Inc., 11260 Overland Avenue, Suite 6A, Culver City, CA, USA 90230, Phone (213) 558-4274, FAX (213) 836-4839

Acculab, 700 Barksdale Road, Newark, DE, USA 19711, Phone (302) 292-8888, FAX (302) 292-8468

American Type Culture Collection, 12301 Parklawn Drive, Rockville, MD, USA 20852-1776, Phone (800) 638-6597 or (301) 881-2600, FAX (301) 231-5826

Anaerobe Reference Laboratory, Anaerobic Bacteria Branch, Centers for Disease Control, 1600 Clifton Road, Bldg. 5, Rm. 112G10, Atlanta, GA, USA 30333, Phone (404) 639-3654

Anaerobe Systems, 2200 Zanker Road, Suite C, San Jose, CA, USA 95131, Phone (800) 443-3108 or (408) 432-9103, FAX (408) 432-9481

Analytab Products (API), 200 Express Street, Plainview, NY, USA 11803, Phone (800) 645-7034 or (516) 349-4000, FAX (516) 349-8719

API Systems SA, La Balme les Grottes F-38390, Montalieu-Vercieu, France, Phone 74.90.66.77, Telex 340467, FAX 74.90.66.80

Austin Biological Laboratories (ABL), Inc., 6620-A Manor Road, Austin, TX, USA 78723-4731, Phone (800) 666-3342 or (512) 928-1304, FAX (512) 928-3607

Baxter Healthcare Corp., Bartels Division, 2950 Northup Way, Bellevue, WA, USA 98004, Phone (800) BARTELS or (206) 828-4983, FAX (06) 828-6262

Baxter Healthcare Corp., Microscan Division, 1584 Enterprise Blvd., West Sacramento, CA, USA 95691, Phone (800) 631-7216 or (916) 372-1900, FAX (916) 372-2081

Becton Dickinson Europe, Diagnostic Systems Division, 5 chemin des Sources, Meylan, France F-38240, Phone 76.90.80.35, Telex 980640 F, FAX 76.90.19.65

Becton Dickinson Microbiology Systems, P.O. Box 243, Cockeysville, MD, USA 21030, Phone (301) 771-0100, FAX (301) 584-7121

Bellco Glass, Inc., P.O. Box B, 340 Edrudo Road, Vineland, NJ, USA 08360-3493, Phone (800) 257-7043 or (609) 691-1075, FAX (609) 691-3247

Carr-Scarborough Microbiologicals, Inc., P.O. Box 1328, Stone Mountain, GA, USA 30086, Phone (800) 241-0998 or (404) 987-9300, FAX (404) 987-9345

College of American Pathologists (CAP), 5202 Old Orchard Road, Skokie, IL, USA 60077-1034

Coy Laboratory Products, Inc., 22 Metty Drive, Ann Arbor, MI, USA 48103, Phone (313) 663-1320, FAX (313) 663-4453

Cultech Diagnostics, P.O. Box 134, Owensville, MO, USA 65066, Phone (800) 325-0167 or (314) 437-4117, FAX (314) 437-5263

Difco Laboratories, P.O. Box 331058, Detroit, MI, USA 48232-7058, Phone (800) 521-0851 or (313) 961-0800, FAX (313) 591-3530

Dodeca, P.O. Box 1383, Fremont, CA, USA 94538, Phone (415) 490-4385

Don Whitley Scientific Ltd., 14 Otley Road, Shipley, West Yorkshire, England, BD17 5BR, Phone (0274) 595728, Telex 517550, FAX (0274) 531197

Forma Scientific, Inc., Box 649, Mill Creek Road, Marietta, OH, USA 45750, Phone (614) 373-4763, Telex 298205, FAX (614) 373-8466

Hewlett-Packard Company, Analytical Product Group, P.O. Box 10301, Palo Alto, CA, USA 94303, Phone (415) 857-5731

Hewlett Packard Ltd., Miller House, The Ring, Bracknell, Berks RG12 1XN, UK, Phone 0344 424898, FAX 0344 860015

Innovative Diagnostic Systems, Inc., 3404 Oakcliff Road, Suite C-1, Atlanta, GA, USA 30340, Phone (800) 225-5443 or (404) 457-2691, FAX (404) 458-3771

Lab-Line Instruments, Inc., 15th & Bloomingdale Avenues, Melrose Park, IL, USA 60160-1491, Phone (800) 323-0257 or (708) 450-2600, FAX (708) 450-0943

Lab M Ltd., Topley House, POB 19, Bury, Lancs BL9 6AU, UK, Phone 061 797 5729, FAX 061 762 9322

Medical Wire & Equipment Co. (USA), Hamilton Business Park, Unit 9B, Franklin Road, Dover, NJ, USA 07801, Phone (800) 321-3244 or (201) 361-0225, FAX (201) 361-0703

Medical Wire and Equipment Co. (Bath) Ltd., Corsham, Wiltshire SN13 9RT, UK, Phone 0225-810361, Telex 449366 Medcor G, FAX 0225-810153

Meridian Diagnostics, Inc., 3471 River Hills Drive, Cincinnati, OH, USA 45244, Phone (800) 543-1980 or (513) 271-3700, FAX (513) 271-0124

Meridian Diagnostics Europe, Srl., Via Strobino 4, 20025 Legnano, Milano, Italy, Phone (331) 544178, FAX (331) 543294

Microbial ID, Inc., 115 Barksdale Professional Center, Newark, DE, USA 19711, Phone (302) 737-4297, FAX (302) 737-7781

Microbiology Reference Laboratory, 10703 Progress Way, Cypress, CA USA 90630, Phone (800) 445-0185 or (714) 220-1900, FAX (714) 220-9213

Microbiology Specialists Inc., 8911 Interchange, Houston, TX, USA 77054, Phone (713) 663-6888, FAX (713) 663-7722

MicroTech Medical Systems, Inc., 401 Laredo Street, Unit I, Aurora, CO, USA 80011, Phone (800) 642-8378

MSI/Micro-Media Systems, 2330 Denison, Cleveland, OH, USA 44109, Phone (800) 423-6496 or (216) 398-2700, FAX (216) 351-3777

National Committee for Clinical Laboratory Standards (NCCLS), 771 E. Lancaster Avenue, Villanova, PA, USA 19085, Phone (215) 525-2435

Oxoid USA, Inc., 9200 Rumsey Road, Columbia, MD, USA 21045, Phone (800) 638-7638 or (301) 997-2216, FAX (301) 995-1523

Pacific Diagnostics, P.O. Box 658, Archerfield, Queensland 4108, Australia, Phone (617) 273-7111, FAX (617) 273-7122

Remel, 12076 Santa Fe Drive, Lenexa, KS, USA 66215-3594, Phone (800) 255-6730 or (913) 888-0939

Scott Laboratories, Inc. (now Adams Scientific, Inc.), 771 Main Street, West Warwick, RI, USA 02893, Phone (800) 556-6480 or (401) 828-5250, FAX (401) 828-5613

Shel-Lab, Sheldon Manufacturing, Inc., 300 N. 26th Avenue, Cornelius, OR, USA 97113, Phone (503) 640-3000, FAX (503) 640-1366

Spiral System Instruments, Inc., 7830 Old Georgetown Road, Bethesda, MD, USA 20814, Phone (301) 657-1620, FAX (301) 652- 5036

Star Publishing Company, 940 Emmett Avenue, Suite 3, Belmont, CA, USA 94002, Phone (800) 852-6226 or (415) 591-3505, FAX (415) 591-3898

Technology for Medicine, Inc., 270 Marble Avenue, Pleasantville, NY, USA 10570-2982, Phone (800) 431-1720 or (914) 747-3020, FAX (914) 747-0498

Unicornucopia, c/o P.G. Engelkirk, 12803 Kittybrook Lane, Houston, TX, USA 77071, Phone (713) 728-4132

Vitek Systems, 595 Anglum Drive, Hazelwood, MO, USA 63042, Phone (800) MD-VITEK or (314) 731-8500, FAX (314) 731-8800

APPENDIX C

PUBLISHED EVALUATIONS OF MINISYSTEMS FOR ANAEROBE IDENTIFICATION

This is an abbreviated listing of evaluations published since 1984. Please bear in mind that minisystems continuously evolve. Thus, presently available products may have been modified since these evaluations were performed.

AN-IDENT
J. Clin. Microbiol. 22:32-35, 1985 (includes PRAS II & RapID-ANA)
J. Clin. Microbiol. 22:52-55, 1985 (includes API 20A & RapID-ANA)
J. Clin. Microbiol. 22:333-335, 1985 (includes API-ZYM also; concerns certain oral gram-negative species)
Diagn. Microbiol. Infect. Dis. 5:9-15, 1986
Am. J. Clin. Pathol. 85:716-718, 1986 (includes Minitek Anaerobe II & RapID-ANA; concerns *C. difficile*)
J. Clin. Microbiol. 26:144-146, 1988 (includes API 20A, API-ZYM, Minitek Anaerobe II, & RapID-ANA; concerns *C. difficile*)
J. Clin. Microbiol. 29:231-235, 1991

API 20 A
J. Clin. Microbiol. 19:915-916, 1984 (*C. difficile*)
J. Clin. Microbiol. 21:122-126, 1985 (includes PRAS II & RapID-ANA)
J. Clin. Microbiol. 22:52-55, 1985 (includes AN-IDENT & RapID-ANA)
J. Clin. Microbiol. 26:144-146, 1988 (includes AN-IDENT, API-ZYM, Minitek Anaerobe II, & RapID-ANA; concerns *C. difficile*)

API-ZYM
Eur. J. Clin. Microbiol. 3:294-300, 1984
J. Clin. Microbiol. 22:333-335, 1985 (includes AN-IDENT; concerns certain oral gram-negative species)
J. Clin. Microbiol. 26:144-146, 1988 (includes AN-IDENT, API 20A, Minitek Anaerobe II, & RapID-ANA; concerns *C. difficile*)
Anaerobes Today, pp. 229-230, 1988 (concerns clostridia)

ATB 32 A Anaerobe ID System
J. Clin. Microbiol. 26:1063-1065, 1988 (concerns pigmented *Bacteroides* spp.)
Anaerobes Today, pp. 254-255, 1988 (concerns *Fusobacterium* & *Bacteroides* spp.)
J. Clin. Microbiol. 27:2509-2513, 1989
J. Clin. Microbiol. 28:1519-1524, 1990

MicroScan Anaerobe Combo Panel
J. Clin. Microbiol. 20:81-83, 1984 (includes PRAS II)

MicroScan Rapid Anaerobe Identification System
J. Clin. Microbiol. 28:1135-1138, 1990 (includes discussion of automated reading of panels)

Minitek
J. Clin. Microbiol. 21:645-646, 1985
Am. J. Clin. Pathol. 85:716-718, 1986 (includes AN-IDENT & RapID-ANA; concerns *C. difficile*)
Diagn. Microbiol. Infect. Dis. 7:69-72, 1987 (includes RapID-ANA)
J. Clin. Microbiol. 26:144-146, 1988 (includes AN-IDENT, API 20A, API-ZYM, & RapID-ANA; concerns *C. difficile*)

PRAS II
J. Clin. Microbiol. 20:81-83, 1984 (includes MicroScan Anaerobe Combo Panel)
J. Clin. Microbiol. 21:122-126, 1985 (includes API 20A & RapID-ANA)
J. Clin. Microbiol. 22:32-35, 1985 (includes AN-IDENT & RapID-ANA)

RapID-ANA and RapID-ANA II
J. Clin. Microbiol. 21:122-126, 1985 (includes API 20A & PRAS II)
J. Clin. Microbiol. 21:894-898, 1985
J. Clin. Microbiol. 22:32-35, 1985 (includes AN-IDENT & PRAS II)
J. Clin. Microbiol. 22:52-55, 1985 (includes AN-IDENT & API 20A)
J. Clin. Microbiol. 22:980-983, 1985 (concerns veterinary isolates)
J. Med. Technol. 3:45-48, 1986
J. Clin. Microbiol. 2:289-293, 1986 (concerns bile-susceptible *Bacteroides*)
Am. J. Clin. Pathol. 85:716-718, 1986 (includes AN-IDENT & Minitek; concerns *C. difficile*)
Diagn. Microbiol. Infect. Dis. 7:69-72, 1987 (includes Minitek)
J. Clin. Microbiol. 25:491-493, 1987 (concerns certain *Bacteroides* spp.)
J. Clin. Microbiol. 26:144-146, 1988 (includes AN-IDENT, API 20A, API-ZYM, & Minitek Anaerobe II; concerns *C. difficile*)
J. Clin. Microbiol. 26:2226-2228, 1988 (concerns oral spirochetes)
Pathology 20:256-259, 1988
J. Clin. Microbiol. 29:457-462, 1991
J. Clin. Microbiol. 29:874-878, 1991

Vitek ANI Card
J. Clin. Microbiol. 26:225-230, 1988

APPENDIX **D**

LABORATORY PROCEDURES

This appendix contains information about the following laboratory procedures:

Aerotolerance Testing

Simplified Method[24]

Purpose: To differentiate anaerobes from bacteria with other atmospheric requirements

Materials required:
Pure culture of isolate
Brucella blood agar (BRU) plate (or acceptable alternative blood agar [BA] medium)
Chocolate agar (CA) plate
Sterile inoculating loop
Anaerobic incubation system (chamber, jar, or bag/pouch)
CO_2 incubator

Procedure: Inoculate a portion of a colony on the pure culture/subculture (PC/SC) plate to a BRU/BA plate, streaking in such a manner as to obtain well-isolated colonies. Inoculate a portion of the **same** colony to one-quarter of a chocolate agar (CA) plate; four different isolates can be tested on one CA plate. Incubate the BRU/BA plate anaerobically and the CA plate in the CO_2 incubator for 24-48 hours at 35°C. Incubate both plates for up to seven days if no growth appears in 24-48 hours; examine daily.

Interpretation of results: An organism that grows only on the anaerobically incubated BRU/BA plate is considered an **anaerobe.** An organism that grows equally well on both plates is considered a **facultative anaerobe.** An organism that grows well on the anaerobically incubated BRU/BA plate but only slightly on the CO_2-incubated CA plate is considered an **aerotolerant anaerobe.** An organism that grows only on the CO_2-incubated CA plate is considered an **obligate aerobe.** (Note: This procedure is not quite as precise as the technique described in the next section.)

Relationship to Oxygen and Carbon Dioxide (CDC Method) [9]

Purpose: To group bacteria on the basis of their relationship to oxygen and carbon dioxide

Materials required:
Plates of CDC anaerobe blood agar or acceptable BA alternative
Lombard-Dowell broth (or acceptable alternative)
Pure culture of anaerobic isolate
Sterile 0.001 ml calibrated loop
No. 3 McFarland nephelometer standard
Candle extinction jar
Aerobic (non-CO_2) incubator
CO_2 incubator
Anaerobic incubation system (chamber, jar, or bag/pouch)

Procedure:

- Remove plates of CDC anaerobe blood agar (BA) or acceptable alternative media from refrigerator, and allow them to warm to room temperature.

- Label plates appropriately for incubation in the various atmospheres. Label one plate "AN" for anaerobic incubation, one "MI" for microaerophilic incubation, one "CJ" for candle extinction jar incubation, and one "AIR" for aerobic incubation.

- Use colonies from a 48-hour-old pure BA culture of the isolate to be tested or a 24- to 48-hour-old enriched thioglycollate broth or chopped meat broth culture of the organism. Prepare a turbid suspension of the organism (equivalent to a No. 3 McFarland standard or greater) in about 2 ml of Lombard Dowell (LD) broth.

- Using a 0.001 ml calibrated loop, inoculate one-quarter of each of the four labeled BA plates with one loopful of cell suspension or broth culture. Streak in such a manner as to obtain well-isolated colonies. Four isolates can be tested on each plate.

- Incubate plates in appropriate atmospheres. The AN plate is incubated in an anaerobic system (chamber, jar, bag/pouch, or evacuation-replacement jar containing 10% hydrogen, 5% carbon dioxide, 85% nitrogen). The MI plate is incubated in an atmosphere of 5% oxygen, 10% carbon dioxide, 85% nitrogen. The CJ plate is incubated in a candle extinction jar. The AIR plate is incubated in an aerobic, non-CO_2 incubator. All plates are incubated at 35°C for 24-48 hours. The approximate amounts of oxygen and carbon dioxide in each of the atmospheres are shown in Table D-1.

Table D-1. Percentages of Oxygen and Carbon Dioxide in Various Atmospheres

Atmosphere	Oxygen (%)	Carbon Dioxide (%)
Air	21	< 0.5
Candle extinction jar	12-17	3-5
Microaerophilic environment	5	10
Anaerobic environment	0	5

Source: Based on reference 9.

- Observe and record the relative amount of growth on the four plates, using the coding system shown in Table D-2. If growth on CDC anaerobe blood agar is insufficient, use chocolate agar (CA) to determine the organism's relationship to oxygen. Some organisms (e.g., *Haemophilus influenzae* and *H. paraphrophilus*) may appear as obligate anaerobes when blood agar plates are used but will appear as facultative anaerobes when CA plates are used.

Table D-2. Coding System for Extent of Growth

Degree of Growth	No. of Colonies	Size of Colonies
0	No growth	n/a
1	Sparse (< 30)	Tiny (< 1 mm)
2	Sparse	Small to medium (1-5 mm)
3	Sparse	Large (> 5 mm)
4	Moderate (30-300)	Tiny
5	Moderate	Small to medium
6	Moderate	Large
7	Abundant (> 300)	Tiny
8	Abundant	Small to medium
9	Abundant	Large

Source: Based on reference 9.

- Determine the group in relationship to oxygen, using the following definitions:

Obligate anaerobe (AN)	Growth in anaerobic system; no growth in 5% oxygen, 10% carbon dioxide, 85% nitrogen, a candle extinction jar, or air
Microaerotolerant anaerobe (MA)	Growth in an anaerobic system and in 5% oxygen, 10% carbon dioxide, 85% nitrogen; no growth in a candle extinction jar or air
Aerotolerant anaerobe (AT)	Best growth in an anaerobic system; moderate growth in 5% oxygen, 10% carbon dioxide, 85% nitrogen; less growth in a candle extinction jar and air
Facultative anaerobe (FA)	Approximately the same degree of growth in an anaerobic system; 5% oxygen, 10% carbon dioxide, 85% nitrogen; a candle extinction jar; and air
Microaerophilic aerobe (MI)	Best growth in 5% oxygen, 10% carbon dioxide, 85% nitrogen; less growth in a candle extinction jar; usually no growth in air or an anaerobic system
Obligate aerobe (OA)	Best growth in air and in candle extinction jar; less growth in 5% oxygen, 10% carbon dioxide, 85% nitrogen; little, if any, growth in an anaerobic system

- If growth occurs in the candle extinction jar, but there is no growth in air, the organism is considered a capnophile.

Bile[24]

Purpose: To determine the effect of bile on growth of an isolate

Materials required:
Pure culture of the isolate
Tube of PRAS peptone-yeast-glucose (PYG)
Tube of PRAS PYG containing 2% commercial dehydrated oxgall (equivalent to 20% bile)
Sterile inoculating loop

Procedure: Inoculate the tubes, and incubate them anaerobically until growth is observed in the control tube. THIO containing 20% bile and 0.1% sodium deoxycholate and THIO lacking bile and deoxycholate can be used in place of PRAS PYG media.

Interpretation of results: Compare the growth in the tubes, and record results as follows:

Inhibition	Less growth in the bile-containing tube than in the control tube (in this case, the organism is said to be bile-sensitive)
No inhibition	Same amount of growth in the bile-containing tube as in the control tube (in this case, the organism is said to be bile tolerant or bile resistant)
Enhancement	More growth in the bile-containing tube than in the control tube

Note: BBE agar or bile disks can also be used to determine an organism's resistance or sensitivity to bile.

Catalase Test[9]

Purpose: To determine if an isolate produces catalase, an enzyme that catalyzes the conversion of hydrogen peroxide to water and oxygen

Materials required:
    Pure culture of the isolate
    Glass microscope slide
    Sterile inoculating loop or wooden applicator stick
    Hydrogen peroxide, 3%

Procedure: Using an inoculating loop or wooden applicator stick, remove some of the growth from the PC/SC plate, and rub it onto a small area of a glass microscope slide. Be careful not to transfer any of the agar because sheep red blood cells contain catalase, which could cause a false-positive test result. Add a drop of 3% hydrogen peroxide to the organism on the slide, and watch for the production of bubbles of oxygen gas.

Interpretation of results: The presence of gas (oxygen) bubbles within 30 seconds indicates a positive test; the absence of bubbles constitutes a negative test. Most anaerobes are catalase-negative. Notable exceptions are *Actinomyces viscosus*, certain members of the *B. fragilis* group (*B. distasonis, B. fragilis, B. ovatus, B. thetaiotaomicron*), some *Propionibacterium* spp., and *Staphylococcus saccharolyticus*.

Note: A method using 15% H_2O_2 has been described, which is said to be more sensitive.[24]

Colony Observations

The characteristics shown in Table D-3 should be recorded whenever colonies are described. Thus, these categories should be included on the worksheet.

Table D-3. Colony Information To Be Recorded on Worksheet

Characteristic	Examples/Comments
Diameter	Tiny (equal to or less than 1 mm), small (1-3 mm in diameter), medium (3-5 mm in diameter), large (greater than 5 mm)
Form	Punctiform, circular, filamentous, irregular, rhizoid, spindle (Figure 7-3)
Elevation	Flat, raised, convex, pulvinate, umbonate, umbilicate (Figure 7-3)
Margin	Entire, undulate, lobate, erose, filamentous, curled (Figure 7-3)
Color	Accurately describe the color
Surface	Glistening (shiny), dull, smooth
Density	Opaque, translucent, transparent
Consistancy	Butyrous (butterlike), viscous, membranous, brittle
Internal structure	Granular, "ground glass," mosaic, concentric rings

Table D-3. Colony Information To Be Recorded on Worksheet (*continued*)

Characteristic	Examples/Comments
Hemolysis	Alpha-hemolytic, beta-hemolytic, nonhemolytic (gamma-hemolytic), double zone hemolysis
Other	Swarming, greening of the medium, browning of the medium, pitting, odor, spider colonies, breadcrumb colonies, molar tooth colonies, fried egg colonies, fluorescence

Culture Media

Because most modern laboratories obtain their plated and broth media from commercial sources, specific media formulations and preparation techniques are not described in this book. Such information is contained in the product inserts that accompany the commercial media. Readers are strongly encouraged to review the section on media in Chapter 6 as well as published evaluations of commercially available media. Table D-4 provides the names and purposes of many of the media in use. Such media are available from Anaerobe Systems, Becton Dickinson Microbiology Systems, Carr-Scarborough, Remel, and Scott Laboratories, among others. The table does not differentiate, between ordinary, "reducible," and PRAS media, but readers are strongly urged to investigate those differences. Microbiologists wishing to prepare their own media should consult other publications.[7, 10, 14, 24]

Table D-4. Anaerobic Bacteriology Media and Their Purposes

Media	Purpose
Liquid media	
Carbohydrate broths (e.g., arabinose, glucose, glycerol, lactose, maltose, mannitol, mannose, rhamnose, salicin, starch, sucrose, trehalose, xylose). Note: These and certain other media are available either as PRAS or non-PRAS.*	To determine whether a particular organism is capable of fermenting specific carbohydrates (Acid is produced if the carbohydrate is fermented.)
Chopped meat medium	Supports growth of most clinically encountered anaerobes; useful as a holding medium or for preservation of clostridial cultures by freezing; can be used for sporulation of, proteolysis of, or toxin production by clostridia
Chopped meat glucose medium	Enrichment broth that supports growth of most clinically encountered anaerobes; useful as a holding medium or for sporulation of or toxin production by clostridia

Table D-4. Anaerobic Bacteriology Media and Their Purposes (*continued*)

Media	Purpose
Chopped meat starch medium	To isolate *Clostridium botulinum* from mixed microbial populations by a heat or ethanol spore selection technique; excellent holding medium for clostridia or for demonstrating proteolysis and indole production by clostridia
Enriched thioglycollate (THIO) medium	Supports growth of most clinically encountered anaerobes; contains hemin, vitamin K_1, and L-cystine.
Esculin broth	To detect esculin hydrolysis
Hydrogen sulfide (H_2S) semisolid medium	To detect hydrogen sulfide (H_2S) production
Indole-nitrite medium	To detect indole production and the ability to reduce nitrates
Iron milk medium	Differential medium used to determine how an organism reacts with the ingredients of whole milk; possibilities include gas production, blackening, coagulation, and digestion of milk proteins; especially useful in characterization of *Clostridium* spp.
Lombard-Dowell (LD) broth medium	To prepare bacterial suspensions for use in inoculating other media; also used as a basal medium for preparing differential media
LD broth with glucose (LD Glucose)	To study metabolic products by GLC
LD broth with lactate (LD Lactate)	To detect the ability of an organism to utilize lactic acid
LD broth with threonine (LD Threonine)	To study the ability of an organism to utilize the amino acid threonine to produce increased amounts of propionic acid
Motility medium	To detect motility; may also be used to ship anaerobic isolates to another laboratory because it contains no added fermentable carbohydrate
Peptone-yeast extract (PY) broth	A control medium in the analysis of metabolic products by GLC; uninoculated PY broth shows only trace amounts, if any, of volatile and nonvolatile acids when tested by GLC

Table D-4. Anaerobic Bacteriology Media and Their Purposes (*continued*)

Media	Purpose
Peptone-yeast extract-glucose (PYG) broth	A control medium in the analysis of metabolic products by GLC; uninoculated PYG broth shows only trace amounts, if any, of volatile and nonvolatile acids when tested by GLC
PY broths containing glucose or carbohydrates other than glucose (e.g., arabinose, arginine, fructose, galactose, lactose, maltose, mannitol, mannose, salicin, trehalose, etc.)	To determine whether a particular organism is capable of fermenting specific carbohydrates; acid is produced if the carbohydrate is fermented
Thiogel medium	To detect the ability of an organism to hydrolyze gelatin
Urea semisolid medium	To detect the ability of an organism to hydrolyze urea; especially useful in differentiating *Clostridium bifermentans* (urease-negative) from *C. sordellii* (urease-positive)

Agar slants

CDC anaerobe blood agar (BA) slants	Growth obtained is used for lyophilization of cultures and to prepare cell suspensions for inoculation of differential media
Chopped meat (CM) agar slants	To demonstrate spore production by *Clostridium* spp. that do not sporulate readily
Heart infusion agar (HIA) slants	Growth obtained is used to detect catalase activity and for lyophilization of cultures
LD agar slants	Growth obtained is used to detect catalase activity, for lyophilization of cultures, and to prepare cell suspensions for inoculation of differential media

Plated media

Bacteroides bile esculin (BBE) agar	A selective medium included in the primary setup to obtain rapid (sometimes 24-hour) presumptive identification of the *Bacteroides fragilis* group; as a result of esculin hydrolysis, members of the bile-tolerant *B. fragilis* group cause a brown discoloration of the normally pale yellow medium

Table D-4. Anaerobic Bacteriology Media and Their Purposes (*continued*)

Media	Purpose
Cycloserine-cefoxitin-fructose (CCFA) agar	A selective medium for *Clostridium difficile,* which causes the normally pinkish-colored medium to turn yellow in the vicinity of growth; CCFA is used to isolate *C. difficile* from watery stool specimens in cases of antibiotic-associated colitis (pseudomembranous colitis) and antibiotic-associated diarrhea
CDC anaerobe blood agar (BA)	An enriched, nonselective medium containing hemin, vitamin K_1, and L-cystine; supports the growth of most aerobic, facultatively anaerobic, micro-aerophilic, and obligately anaerobic bacteria
CDC anaerobe blood agar with kanamycin and vancomycin (KVA)	An enriched, selective medium; supports the growth of *Bacteroides* spp.
CDC anaerobe blood agar with paromomycin and vancomycin (PVA)	An enriched, selective medium containing laked blood; supports the growth of most anaerobic, gram-negative bacilli
CDC anaerobe blood agar with phenethylalcohol (PEA)	An enriched, selective medium; supports the growth of most obligate anaerobes but inhibits growth of gram-negative, facultatively anaerobic bacteria
CDC modified McClung Toabe egg yolk agar (EYA)	A differential medium to aid in isolation and identification of clostridia; allows detection of lecithinase, lipase, and pro-teolyic activities
CDC stiff anaerobe blood agar (Stiff BA)	An enriched medium used to decrease the swarming of clostridia
Egg yolk agar (EYA)	Primarily used to observe lecithinase, lipase, and proteolysis reactions of *Clostridium* spp.; also of value in pre-sumptive identification of other lipase-positive organisms (e.g., *Prevotella intermedia, Fusobacterium necrophorum,* and some strains of *Prevotella loescheii*)
LD agar	Used in one of the quadrants of Presumpto Plate 1 for detecting degree of growth, indole production, and catalase activity

Table D-4. Anaerobic Bacteriology Media and Their Purposes (*continued*)

Media	Purpose
LD bile agar	Used in one of the quadrants of Presumpto Plate 1; contains 2.0% oxgall; inhibits bile-sensitive organisms but allows growth of bile-tolerant organisms; some strains of *Bacteroides fragilis* produce a characteristic insoluble precipitate in LD bile agar
LD deoxyribonucleic acid agar	Used in one of the quadrants of Presumpto Plate 2 to detect the ability of an organism to hydrolyze (depolymerize) deoxyribonucleic acid (DNA) by means of deoxyribonuclease (DNase) activity
LD egg yolk agar (EYA)	Used in one of the quadrants of Presumpto Plate 1 to detect lecithinase, lipase, and proteolytic activities of bacteria
LD esculin agar	Used in one of the quadrants of Presumpto Plate 1 to detect esculin hydrolysis, H_2S production, and catalase activity
LD gelatin agar	Used in one of the quadrants of Presumpto Plate 3 to detect gelatin hydrolysis
LD glucose agar	Used in one of the quadrants of Presumpto Plate 2 to detect fermentation of glucose and stimulation of growth by glucose
LD lactose agar	Used in one of the quadrants of Presumpto Plate 3 to detect fermentation of lactose
LD mannitol agar	Used in one of the quadrants of Presumpto Plate 3 to detect fermentation of mannitol
LD milk agar	Used in one of the quadrants of Presumpto Plate 2 to test the ability of an organism to hydrolyze milk proteins
LD neomycin egg yolk agar (NEYA)	A selective and differential medium for selective isolation and presumptive identification of certain neomycin-resistant anaerobes (e.g., certain *Clostridium* spp.); allows detection of lecithinase, lipase, and proteolytic activities
LD Presumpto Plates	Contain various media described in this table

Table D-4. Anaerobic Bacteriology Media and Their Purposes (*continued*)

Media	Purpose
LD rhamnose agar	Used in one of the quadrants of Presumpto Plate 3 to detect fermentation of mannitol
LD starch agar	Used in one of the quadrants of Presumpto Plate 2 to detect the ability of an organism to hydrolyze starch
Laked blood-kanamycin-vancomycin (LKV) medium (sometimes referred to as KVLB)	A selective medium included in the primary setup for isolation of *Bacteroides* and *Prevotella* spp.; of greater importance, laked blood causes a more rapid brown-black pigmentation of colonies of *Prevotella* (previously, pigmented species of *Bacteroides*) than does nonlaked blood; most strains of *Porphyromonas* fail to grow on LKV due to their susceptibility to vancomycin

* VPI formulated anaerobe tube media are prereduced and anaerobically sterilized (PRAS), whereas CDC formulated media are not.

Source: Based on references 10 and 24.

PRAS Media

Only a brief summary of the preparation procedure for prereduced, anaerobically sterilized media is presented here. For complete step-by-step instructions, see the *VPI Anaerobe Laboratory Manual, 4th ed.*[14]

- Add weighed out, dry ingredients to a flask. Add water, salts solution, and resazurin (redox indicator). Use an appropriate size flask that will minimize air space (e.g., a 750 ml flask for 500 ml of medium or a 1000 ml flask for 750 ml of medium). Place a chimney onto the flask to prevent the medium from boiling over. Add a few glass beads or a magnetic stirrer to the flask to obtain an even boil.

- Boil the medium until the resazurin turns from pink (oxidized) to colorless (reduced), which usually takes five to ten minutes. Oxygen is driven off in this step, and the ingredients are partially reduced, which minimizes the oxidation of the reducing agent that will later be added. Oxidized reducing agents are toxic to some fastidious anaerobes.

- Remove the flask from the heat, and replace the chimney with a two-hole stopper fitted with a cannula that will deliver a stream of oxygen-free CO_2 to the medium. This prevents access of air while the medium is cooled to room temperature by placing the flask into an ice bath.

- Add a weighed quantity of cysteine hydrochloride (a reducing agent) to further lower the Eh of the medium. Then adjust the pH to the desired level (7.0-7.1 for most media).

- Switch from oxygen-free CO_2 to oxygen-free nitrogen. Dispense the medium into tubes that are being flushed with oxygen-free nitrogen. Stopper the tubes with rubber stoppers, and autoclave. Exercise caution while inserting rubber stoppers and autoclaving because the tubes can break.

- Cool the tubes, and protect them from sunlight during storage. Sunlight inactivates resazurin. Media are satisfactory for use until they become oxidized, in which case the resazurin turns the media pink.

Desulfoviridin Test[6, 26]

Purpose: To demonstrate the production of desulfoviridin, an enzyme (sulfite reductase) produced by certain sulfate-reducing organisms

Materials required:
Pure culture of organism to be tested
Sterile, distilled water
Reagent = 2.0 N NaOH
365 nm UV light source

Procedure:

- Prepare a turbid (equivalent to or greater than a McFarland #3) cell suspension in 0.5 ml distilled water

- Add two drops of reagent

- Immediately expose the cell suspension to long-wave ultraviolet light in a darkened room

Interpretation of results: The appearance of a characteristic red fluorescence (due to the sirohydrochlorin chromatophore of desulfoviridin) immediately after addition of the reagent constitutes a positive test (suggestive of *Desulfovibrio* spp. [motile, curved, or straight cells], *Desulfomonas pigra* [nonmotile, straight cells], or *Bilophila wadsworthia* [nonmotile, pleomorphic cells]).

Disks

Bile Disk[25]

Purpose: A bile disk can be used to determine the ability of a gram-negative bacillus to grow in the presence of relatively high concentrations of bile. Members of the *Bacteroides fragilis* group are bile resistant, as are certain other gram-negative bacilli.

Materials required:
BRU plate (or acceptable BA substitute)
Bile disk

Procedure: Add the bile disk to the heavily inoculated first quadrant of the PC/SC plate of the gram-negative bacillus. Press the disk firmly to the surface of the PC/SC plate to ensure uniform diffusion of the bile into the medium. Incubate anaerobically for 24-48 hours.

Interpretation of results: If there is a zone of inhibition of growth around the disk (see manufacturer's package insert for specific interpretive criteria), the organism is bile sensitive. Growth up to the disk indicates that the organism is bile resistant. Other indicators of bile resistance include good growth on the BBE plate and growth in 20% bile broth.

Nitrate Disk[28]

Purpose: The nitrate disk is used to determine an organism's ability to reduce nitrate to nitrite. The test is a "miniaturized" version of the tube-type nitrate reduction test used in the aerobic bacteriology laboratory. Most anaerobic, gram-negative bacilli are nitrate-negative, but *Bacteroides ureolyticus* is a notable exception. The nitrate test is also of value in the presumptive identification of gram-positive and gram-negative cocci. A small, nitrate-positive, anaerobic, gram-positive bacillus with growth that is stimulated by arginine can be presumptively identified as *Eubacterium lentum*. Tiny, nitrate-positive, anaerobic, gram-negative cocci can be presumptively identified as *Veillonella* spp.

Materials required:
    BRU plate (or acceptable BA substitute)
    Nitrate disk

Procedure: Press the nitrate disk firmly to the heavily inoculated first quadrant of the PC/SC plate of the isolate. Incubate anaerobically for 24-48 hours.

Interpretation of results: Remove the disk from the surface of the growth, place it into a sterile petri dish, and add the usual nitrate reagents "A" and "B" to the disk. A red color indicates the presence of nitrite, and thus a positive test. If no red color develops, sprinkle zinc dust on top of the disk. As in the standard procedure, production of a red color at this point indicates a negative test, and absence of a red color constitutes a positive test (i.e., that nitrate has been reduced **beyond** nitrite).

Special Potency Antimicrobial Disks[23, 24]

Purposes: The primary purpose of these three disks is to verify that a gram-negative bacillus is **truly** gram-negative, as opposed to one of the clostridia (e.g., *C. ramosum, C. clostridioforme*) that routinely stains pink with the Gram stain procedure. Beyond that, in conjunction with other information (such as colony morphology), the disks can differentiate *Fusobacterium* spp. from other true gram-negative bacilli.

Materials required:
    BRU plate (or acceptable BA alternative)
    Vancomycin disk (5 μg)*
    Kanamycin disk (1 mg)*
    Colistin disk (10 μg)*

    (* It is critical to use disks of the prescribed potency.)

Procedure: Add disks to the heavily inoculated first quadrant of the PC/SC plate of the pink-staining isolate, and press them firmly to the surface of the PC/SC plate to ensure uniform diffusion of the agents into the medium. Incubate anaerobically for 24-48 hours.

Interpretation of results: A zone of no growth >10 mm in diameter around the disk indicates that the organism is susceptible to the antimicrobial agent. A zone size of 10 mm or less is interpreted as resistant. Analyze the disk results in a stepwise manner, as shown in Table D-5 and Flow Charts 8-6 and 8-7. Remember that these disks give no information whatsoever about antimicrobial agents that might be used for therapy.

Table D-5. Interpretation of Special Potency Antimicrobial Disk Results

	Vancomycin (5 μg)	Colistin (10 μg)	Kanamycin (1 mg)
Gram-positive organisms (and vancomycin-sensitive *Porphyromonas* spp.)	S		
Most gram-negative organisms	R		
Fusobacterium spp. (and *Bacteroides ureolyticus* group)	R	S	S
Most *Bacteroides* and *Prevotella* spp.	R	V	R
Bacteroides fragilis group	R	R	R

Source: Based on reference 24.

SPS Disk[27]

Purpose: To determine whether a gram-positive coccus is susceptible or resistant to sodium polyanethol sulfonate (SPS). This test is especially valuable for presumptive identification of *Peptostreptococcus anaerobius,* which is susceptible to SPS.

Materials required:
    BRU plate (or acceptable BA substitute)
    SPS disk

Procedure: Add the SPS disk to the heavily inoculated first quadrant of the PC/SC plate of the gram-positive coccus, and press it firmly to the surface of the PC/SC plate to ensure uniform diffusion of the agent into the medium. Incubate anaerobically for 24-48 hours.

Interpretation of results: Examine the plate for a zone of no growth around the disk, which indicates that the organism is susceptible to SPS. (See manufacturer's package insert for exact interpretive guidelines.) An anaerobic, gram-positive coccus that is susceptible to SPS may be presumptively identified as *P. anaerobius.*

Fluorescence

Purpose: To determine whether colonies of a particular organism fluoresce under long-wave (365-366 nm), ultraviolet (UV) light, a characteristic of value in making presumptive identifications of several anaerobes

Materials required:
    PC/SC plate containing colonies of the anaerobe to be tested
    Long-wave, UV light source (There is a variety of commercial sources of such lights; a
        Wood's lamp can be used.)

Procedure:

- Turn off bench light or room lights
- Remove cover from PC/SC plate
- Turn on UV light source, and hold it over the colonies
- Observe for fluorescence

Interpretation of results: Examples of anaerobes that fluoresce under long-wave, UV light are contained in Table D-6.

Table D-6. Fluorescence of Anaerobes Under Long-Wave UV Light

Species	Color	References/Comments
Bacteroides levii	Pink-orange or red-orange at 24-48 hours; red-brown at 7 days; none at 14 days	22
Clostridium difficile	Chartreuse	24
Fusobacterium necrophorum	Chartreuse	24
Fusobacterium nucleatum	Chartreuse	24
Prevotella bivia	Light orange to pink (coral)	24
Prevotella disiens	Light orange to pink (coral)	24
Prevotella intermedia	Strain variation: red-orange or brilliant red at 24-48 hours; pink-orange or red-orange at 7 days; orange at 14 days	22
Prevotella melaninogenica	Strain variation: yellow, red-orange, or brilliant red at 24-48 hours; yellow or orange at 7 days; yellow, orange, or red-brown at 14 days	22
Porphyromonas asaccharolytica	Strain variation: yellow or red at 24-48 hours; red-orange or red-brown at 7 days; none at 14 days	22
Veillonella spp.	Red	2,4; Media-dependent; weaker fluorescence than *Prevotella* and *Porphyromonas* spp.; fades rapidly on exposure of colonies to air

Gram Stain Observations

Certain anaerobes have a characteristic cell morphology that can alert the microbiologist to the presence of a particular organism. Use the characteristics and terms listed in Table D-7 to describe organisms seen on Gram stain.

Table D-7. Gram Stain Information To Be Recorded on Worksheet

Organism Characteristic	Examples/Comments
Gram reaction	Gram-positive, gram-negative, or gram-variable
Cell shape	Cocci (spherical, oval, elongated), coccobacilli, bacilli, or filaments
Cell size	Comment if especially large or small
Arrangement	Diplococci, tetrads, chains (long? short?), clusters, masses, diphtheroids
Spores	Present or absent (If present, are they terminal or subterminal? Do they cause swelling of the cells? Are free spores present?)
Others	Banded, beaded, bifurcated (forked), branching, clubbed, comma shaped, curved, fusiform, pleomorphic, spiral shaped, straight, swellings, vacuoles

Growth Stimulation Tests[24]

Purpose: To determine if an isolate's growth is enhanced by a supplement that is a potential growth stimulator (Examples of such supplements are arginine and formate-fumarate.)

> **Materials required:**
> Pure culture of the isolate
> PRAS PYG broth tube lacking the potential growth stimulator
> PRAS PYG broth tube containing the potential growth stimulator
> Inoculating loop

Procedure: In these very simple tests, the amount of growth of an organism in a broth tube containing a potential growth stimulator is compared to the amount of growth in a broth tube lacking that supplement. The two tubes are inoculated equally and incubated anaerobically for 48-72 hours.

If the organism is a small, gram-positive, nondiphtheroid rod, inoculate it into a tube containing PRAS PYG broth and a tube containing PRAS PYG broth plus **arginine.**

Perform the **formate-fumarate growth stimulation test** on gram-negative bacilli that produce small, translucent to transparent colonies (that may or may not be pitting) and yield the typical *Fusobacterium* special potency antimicrobial disk results (resistant to vancomycin and susceptible to both kanamycin and colistin). The test is performed in PRAS PYG broth tubes.

Interpretation of results: The amount of growth in one tube is compared to that in the other. If there is more growth in the tube containing the supplement, you can conclude that growth of that particular organism is stimulated by that substance.

Greater growth in the tube containing arginine represents a positive arginine stimulation test. Growth of *Eubacterium lentum* is stimulated by arginine.

Stimulation of growth by formate-fumarate is characteristic of the pitting, anaerobic, gram-negative bacilli (i.e., the *Bacteroides ureolyticus* group).

Lecithinase and Lipase Reactions[24]

Purpose: To help identify many species of *Clostridium* (see Table D-8), as well as lipase-positive strains of *Prevotella intermedia*, *P. loescheii,* and *Fusobacterium necrophorum*

Table D-8. Lecithinase and Lipase Reactions of Selected *Clostridium* spp.

Clostridium Species	Lecithinase	Lipase
C. barati	+	−
C. bifermentans	+	−
C. limosum	+	−
C. novyi type A	+	+
C. perfringens	+	−
C. sordellii	+	−
C. sporogenes	−+	+
C. subterminale	−+	−

Materials required:
   Egg yolk agar (EYA) plate
   Inoculating loop

Procedure: Streak the egg yolk agar plate in a manner that will produce well-isolated colonies. Incubate anaerobically for 24-48 hours. Remove the cover of the petri dish to observe the reactions.

Interpretation of results: The **lecithinase reaction** occurs within the EYA medium and appears as a relatively wide, opaque or "cloudy" zone around colonies. It is **not** a surface phenomenon.

The **lipase reaction** occurs on the surfaces of colonies, immediately beneath colonies, and on the EYA surface in a relatively narrow zone around the colonies. It is described as being iridescent and multicolored, giving the appearance of "oil on water" or "mother of pearl." The lipase reaction is more evident when natural light is used (e.g., near a window), and the plate must be tilted so that the light reflects off the surface and into the eyes.

Motility Tests

CDC method[9]

Purpose: To determine whether or not an isolate is motile and, if motile, the type of motility

Materials required:
Well-isolated colony on CDC anaerobe blood agar
Sterile, distilled water
Sterile Pasteur pipette (or acceptable alternative)
Sterile inoculating needle or loop
Small (22 x 22 mm) glass coverslip
Clean microscope slide
Microscope (light, phase-contrast, or darkfield)

Procedure: Using a Pasteur pipette, place a small (~3 μl) drop of water onto a clean microscope slide. Using an inoculating needle or loop, touch a well isolated colony (preferably while viewing under a stereomicroscope), and prepare a light suspension of cells in the drop of water. Carefully fold a small coverslip over the drop so that no air bubbles are trapped. Examine immediately, using the low-power, high-power, and oil immersion objectives. A phase contrast condenser permits detection of spores as well as motility.

Interpretation of results: Motile organisms will be in motion; nonmotile organisms will not. If you detect motion, record if it is "tumbling," "darting," or straight directional. Motile, gram-negative, anaerobic bacilli include *Wolinella* spp. and *Campylobacter concisus,* among many others. *Mobiluncus* spp. and many, but not all, *Clostridium* spp. are motile. You can determine the arrangement of flagella using one of the flagella staining procedures described later in this appendix.

Hanging Drop Method[24]

Purpose: To determine whether or not an isolate is motile and, if motile, the type of motility

Materials required:
A young (four- to six-hour) broth culture of the isolate
Sterile Pasteur pipette (or acceptable alternative)
Glass coverslip
Depression microscope slide
Microscope (light, phase-contrast, or darkfield)

Procedure: Using a Pasteur pipette, transfer a drop of a young broth culture to a glass coverslip. Invert the coverslip over the cavity in a depression slide. Examine the "hanging drop" microscopically for motile organisms using the high-dry objective.

Interpretation of results: If you detect motion, record whether it is "tumbling," "darting," or straight directional motility. You can determine the arrangement of flagella using one of the flagella staining procedures described later in this appendix.

Nagler Test[24]

Purpose: Although no longer widely used, the Nagler test has been employed to presumptively identify *C. perfringens.* The test would be set up when the organism on the PC/SC plate is suspected of being *C. perfringens* (i.e., an anaerobic, boxcar-shaped, gram-positive bacillus; the fact that the organism is lecithinase-positive may also be known at this point if an EYA plate was set up at the same time as the PC/SC plate).

Materials required:
Pure culture of the isolate
Egg yolk agar plate
Cotton swab
C. perfringens type A antitoxin (hereafter referred to as the reagent)

Procedure: Using a cotton swab, coat half the plate with the reagent. Several minutes later, streak the suspected *C. perfringens* straight across the EYA plate, from the untreated to the treated side. Then incubate the plate anaerobically for 24-48 hours.

Interpretation of results: Examine the plate for lecithinase activity. Lecithinase activity surrounding the growth on the untreated side of the plate but no lecithinase activity on the treated side indicates a **positive Nagler test.** Although the Nagler test was originally developed for the presumptive identification of *C. perfringens,* it is now known that any of four species of *Clostridium* (*C. perfringens, C. barati, C. bifermentans, C. sordellii*) can produce a positive test. For this reason, the Nagler test is not as useful or popular as it once was for presumptive identification of *C. perfringens.* A double zone of hemolysis and a positive reverse CAMP test result are of much greater value for presumptive identification of this organism.

Lecithinase activity surrounding the growth on both sides of the plate or on neither side of the plate indicates a **negative Nagler test.**

Oxidase Test[9]

Purpose: To determine if the anaerobic isolate produces cytochrome oxidase (Relatively few anaerobes produce this enzyme. Oxidase-positive anaerobes include *Anaerobiospirillum succiniciproducens, Bacteroides ureolyticus, Campylobacter mucosalis, C. sputorum* subsp. *sputorum, Wolinella curva,* and *W. recta.*)

Materials required:
BRU/BA pure culture of the isolate to be tested
Whatman no. 1 filter paper
Kovac oxidase reagent
Disposable plastic or platinum wire inoculating loop (do not use a Nichrome loop)

Procedure: Place a piece of Whatman no. 1 filter paper in a clean container (e.g., a petri dish lid). Saturate a small section of it with the reagent. Transfer part of a colony to and rub it into the wet area of the paper.

Interpretation of results: If the spot turns blue within ten seconds, the test is positive. If the spot does not turn blue or does so after ten seconds, the test is negative.

Presumpto Plates

Many of the observations and simple tests necessary for presumptive identifications have been incorporated into specialized types of plated media, collectively referred to as Presumpto plates. Three types of Presumpto plates are available: Presumpto 1, Presumpto 2, and Presumpto 3 plates. Each is a "quad plate" (meaning that it is divided into four separate quadrants), thus providing a total of 12 separate quadrants when all three plates are used. As shown in Table 8-5, the three Presumpto plate system enables a user to determine 18 different characteristics about the anaerobe that has been inoculated onto the plates. Presumpto plates are available from several commercial sources, including Carr-Scarborough Microbiologicals, Remel, and Scott Laboratories.

Additional information can be found in a CDC publication.[8] Presumpto plates were introduced prior to the development and commercial availability of four-hour, preexisting enzyme-based identification systems (minisystems). Systems providing **same day** definitive identifications are more practical than those providing only PIDs after an additional 24-48 hours of incubation.

Reverse CAMP Test[3, 12]

Purpose: To presumptively identify *C. perfringens* (Greater than 95% of *C. perfringens* isolates produce a positive result. Only very rarely will an organism other than *C. perfringens* be reverse CAMP test positive.)

Materials required:
Pure culture of the isolate suspected of being *C. perfringens*
BRU plate (or acceptable alternative)
Inoculating loop
Stock culture of beta-hemolytic Group B streptococci (*Streptococcus agalactiae*)

Procedure: Using an inoculating loop, streak the Group B streptococcus down the center line of a BRU plate. Streak the unknown organism perpendicular to but not touching the streptococcus streak. Several unknowns can be tested on a single BRU plate. Then incubate the plate anaerobically for 24-48 hours.

Interpretation of results: Examine the plate for a zone of enhanced hemolysis (usually in the shape of an arrowhead) at the junction of the streptococcus streak and the unknown organism. An arrowhead-shaped zone of enhanced hemolysis constitutes a positive test. Absence of such a zone indicates a negative test.

The reverse CAMP test was described prior to the development and commercial availability of four-hour, preexisting enzyme-based identification systems (minisystems). As previously mentioned, it is more logical to use systems that provide **same day** definitive identifications than those that provide only PIDs after an additional 24-48 hours of incubation.

Spores[14, 24]

There is no one medium that can be used to demonstrate spore production by all species or strains of sporeforming anaerobes. Chopped meat, chopped meat glucose, and egg yolk media often support sporulation very well. When present, spores can be observed following Gram or special spore staining methods. On Gram stain, vacuoles are often mistaken for spores, but vacuoles remain unstained when spore staining methods are used.

Sporulation can often be demonstrated using the heat test. Inoculate two tubes of starch broth, being careful not to touch the walls of the tubes above the medium with the inoculum. Stopper the tubes. Heat one of them in a 70-80°C water bath for ten minutes, making certain that the level of water in the bath is above the level of the starch medium. Start timing when a thermometer placed in a tube containing a volume of water equal to the volume of starch medium reads 79-80°C. Incubate both the heated and unheated tubes anaerobically at 37°C and examine daily for growth for up to five days. Only sporeforming organisms will survive the heat test. Thus, growth in both tubes constitutes a positive test for spores.

An alternate procedure involves the use of alcohol.[16] Gently mix together equal volumes of 95% ethanol and a one-week-old thioglycollate or chopped meat broth culture. Allow the mixture to remain at room temperature for 30 minutes, subculture to BRU/BA, incubate anaerobically for two days, and then examine for growth. Only sporeforming organisms will survive the alcohol treatment.

Spot Indole and Indole Derivatives Tests[1, 9, 17]

Purpose: To determine if an isolate is capable of producing indole or an indole derivative from the amino acid tryptophan (Most anaerobes are indole negative. Notable exceptions include *Bacteroides ovatus, B. splanchnicus, B. thetaiotaomicron, B. uniformis,* a few *Clostridium* spp. [e.g., *C. bifermentans, C. sordellii,* and *C. sphenoides*], most *Fusobacterium* spp. [except *F. mortiferum*], *Peptostreptococcus asaccharolyticus, Porphyromonas asaccharolytica, Prevotella intermedia,* and *Propionibacterium acnes.*)

Materials required:
> BRU/BA pure culture of the isolate (preferably a 48-72-hour culture)
> 1% (w/v) p-dimethylaminocinnamaldehyde in 10% (v/v) concentrated hydrochloric acid
> Sterile inoculating loop or wooden applicator stick
> Whatman no. 1 filter paper
> Clean container (e.g., a petri dish lid)

Procedure: Place a piece of filter paper in a clean container. Saturate a small section of the filter paper with 1% p-dimethylaminocinnamaldehyde. Using an inoculating loop or wooden applicator stick, remove some of the growth from the PC/SC plate, and rub it onto the saturated area.

Interpretation of results: Rapid development (usually within one minute) of a blue-to-green spot indicates a positive test (i.e., production of indole from tryptophan). A pink-to-orange spot indicates a negative test. The appearance of a violet or "hot pink" spot within one minute indicates the presence of an indole derivative (e.g., 3-indole propionic acid produced by *Clostridium sporogenes* and some species of *C. botulinum*).[5] Other indole derivatives produced by some clostridial species are skatole and 3-indole butyric acid.

Alternatively, place a sterile, blank disk on an area of heavy growth on the PC/SC plate, and press it firmly to the surface. After several minutes, add one drop of 1% p-dimethylaminocinnamaldehyde to the disk. The disk will become blue to green if the organism produces indole. If the organism is indole negative, it will become pink to orange. This method cannot be used if the plate contains more than one organism because indole diffuses in the agar. All organisms might appear indole positive if one of the organisms on the plate is producing indole.

Staining Procedures

Gram Staining Procedure for Anaerobic Isolates or Clinical Specimens[9]

Purpose: To determine the Gram reaction and cellular morphology of anaerobic isolates or organisms present in clinical specimens

Materials required:
> Clinical specimen or anaerobic isolate on CDC anaerobe blood agar (or acceptable alternative BA medium)
> Sterile Pasteur pipette (or acceptable alternative)
> Sterile distilled water
> Clean microscope slide
> Absolute (100%) methanol
> Crystal violet solution (available commercially)
> Gram's iodine solution (available commercially)
> Safranin (or basic fuchsin) solution (available commercially)
> 95% ethanol
> Microscope
> Immersion oil

Procedure:

- **For anaerobic isolate:** Using a Pasteur pipette, place a small drop of sterile water on a clean microscope slide. Touch a well-isolated colony (preferably with the aid of a stereomicroscope) with a sterile inoculating needle or loop, and prepare a light suspension of the organism in the drop of water.

- **For clinical specimen:** Prepare a thin smear of the specimen on a clean glass microscope slide.

- Allow the smear to air dry. Do not heat to hasten drying.

- Flood the air-dried smear with absolute methanol, and allow the methanol to remain on the smear for one minute.[19]

- Tilt the slide so that the methanol drains off, and allow the smear to air dry (usually requires five to ten minutes to dry completely).

- Flood the smear with crystal violet solution, and allow the solution to remain on the smear for one minute.

- Wash the crystal violet off with a gentle stream of water (preferably sterile and distilled). Drain to remove excess water.

- Flood the smear with Gram's iodine mordant (preferably fortified with PVP[18]). Allow the iodine to remain on the smear for one minute.

- Tilt the slide, and allow the decolorizing solution (preferably 95% alcohol) to flow gently over the smear until no blue color is seen in the liquid running off the end of the slide (about 30 seconds). Do not overdecolorize!

- Immediately wash off the decolorizing solution with a gentle stream of water (preferably sterile and distilled).

- Flood the smear with safranin or basic fuchsin solution, and allow the solution to remain on the smear for one minute. Moore has recommended use of aqueous basic fuchsin (0.8% w/v) or dilute carbolfuschin (1:20 v/v aqueous dilution of Ziehl's carbolfuchsin).[21]

- Wash the safranin or basic fuchsin off with a gentle stream of water (preferably sterile and distilled).

- Allow the stained smear to drain dry before examining it, or gently blot the stained smear with a fresh piece of blotting paper.

Interpretation of results: Gram-positive organisms appear blue/purple; gram-negative organisms appear pink/red. Record information regarding the Gram reaction (positive or negative), cell morphology, and the presence and location of spores (if applicable). Refer to "Gram Stain Observations" earlier in this appendix.

Flagella Stains

Wet-Mount Method[13]

Purpose: To determine the number and location of flagella on motile isolates (This simple technique provides temporary preparations.)

Materials required:
Pure culture of organism to be tested (Blood agar plates work well; broth cultures sometimes produce undesirable background precipitation.)
Applicator stick (or wire loop)

Sterile water
Pasteur pipette
Clean, glass microscope slide
Clean, glass coverslip
Ryu stain (See the following section for recipe. The final stain can be stored at room temperature in a syringe fitted with a 0.22 μm pore-size porous membrane between syringe and needle. During storage, the needle should be tightly capped to prevent drying of the stain.)
Light microscope

Procedure:

- Touch the margin of a colony with an applicator stick or wire loop, and then touch the stick into a drop of sterile water on a clean microscope slide. Avoid agitation of the stick or loop, which causes shearing of flagella from the cells. (Alternatively, touch a loopful of water to the colony margin, and then touch the loopful of motile cell-containing water to a drop of water on a slide. This provides a lighter cell suspension and ensures a higher proportion of flagellated cells.)

- Add a coverslip to the faintly turbid drop, and examine microscopically for motile cells.

- If you see motile cells, wait five to ten minutes to allow about half the cells to adhere to the glass.

- Using a Pasteur pipette, apply two drops of Ryu stain to the edge of the coverslip. Capillary action will draw the stain under the coverslip and mix it with the cell suspension. Let this stand five to ten minutes at room temperature. Place it in a moist chamber to delay observation or preserve the specimen overnight. Drying of the preparation causes undesirable precipitation of the stain.

- Examine the preparation microscopically for flagellated cells.

Interpretation of results: If you see flagellated cells, record the following, as appropriate:

- Cells are flagellated
- Single polar flagellum
- Tuft of polar flagella
- Peritrichous flagella
- Tuft of flagella, concave side of cell
- Single subpolar flagellum

Modified Ryu Method[9, 15]

Purpose: To determine the number and location of flagella on motile isolates (This technique provides permanent preparations.)

Materials required:
Pure culture of the isolate
Sterile Pasteur pipette (or acceptable alternative)
Filtered (0.45 μm porosity), distilled water
Clean microscope slide
Sterile inoculating needle or loop

Ryu flagella-staining solution
　　Solution I (mordant):
　　　　10 ml of 5% aqueous phenol
　　　　2 g tannic acid
　　　　10 ml saturated aqueous solution of aluminum potassium sulfate-12 hydrate
　　Solution II (stain):
　　　　saturated ethanolic solution of crystal violet (12 g in 100 ml of 95% ethanol)
　　Mix one part solution II with ten parts solution I. Filter through filter paper to remove coarse precipitate. This solution will remain stable for several weeks at room temperature.

Procedure:

- Using a Pasteur pipette, transfer two drops of filtered, distilled water to a clean microscope slide, keeping the drops separated.

- Carefully touch a well-isolated colony with an inoculating needle, and then touch the center of each drop of water for a few seconds **without** mixing.

- Allow the preparations to air dry at room temperature. Do not heat to hasten the drying process. Fixation of the cells is not required.

- Flood the slide with Ryu flagella-staining solution, and allow three to five minutes for staining to occur. (Staining time may vary between different lots of staining solution.)

- Wash the slide thoroughly (front and back) with running tap water. Drain. Allow to dry at room temperature.

- Examine for the presence and arrangement of flagella, using the oil immersion objective. Start at the periphery of the preparation, where flagellated cells are most likely to be found.

Interpretation of results: If you see flagellated cells, record the following, as appropriate:

- Cells are flagellated

- Single polar flagellum

- Tuft of polar flagella

- Peritrichous flagella

- Tuft of flagella, concave side of cell

- Single subpolar flagellum

Spore Stains

Schaeffer-Fulton Modification of the Wirtz Method[7]

　Purpose: To stain endospores

　Materials required:
　　Air-dried smear of organism to be stained (Use of acid-cleaned slides will prevent fading of the green spores.)
　　5% aqueous malachite green
　　Bunsen burner
　　Distilled water
　　0.5% aqueous safranin O
　　Blotting paper

Procedure:

- Fix the air-dried smear using absolute methanol (as in Gram staining procedure)
- Flood the slide 30-60 seconds with 5% aqueous malachite green
- Heat to steaming three or four times
- Rinse off excess stain with distilled water
- Flood the slide with 0.5% aqueous safranin O, and allow to remain in contact with the smear for 30 seconds
- Rinse the slide with distilled water, and blot dry

Interpretation of results: Cells will stain red, and spores will stain green.

Bartholomew and Mittwer "Cold" Method[7]

Purpose: To stain endospores

Materials required:
Air-dried smear of organism to be stained
Saturated aqueous malachite green (~7.6%)
Distilled water
0.25% aqueous safranin
Blotting paper

Procedure:

- Fix the air-dried smear using absolute methanol (as in Gram staining procedure)
- Flood the slide with saturated aqueous malachite green; allow to remain on slide for ten minutes
- Rinse the slide with distilled water for about ten seconds
- Flood the slide with aqueous safranin
- Rinse the slide with distilled water, and blot dry

Interpretation of results: Cells will stain red, and spores will stain green.

Stock Cultures

Recommendations for preparing and storing stock cultures differ somewhat from one authority to another. Sutter et al. state they may be prepared from either liquid (supplemented THIO or chopped meat broth) or solid media.[24] In either case, cultures should be young (24-48 hours old) and actively growing. Add 0.5 ml of a liquid culture to an equal volume of sterile skim milk (20% powdered skim milk in distilled water) in a screw-cap, half dram vial. Freeze and store the stock culture at -70°C. When stock cultures are prepared from solid media, thoroughly suspend colonies in the skim milk prior to freezing. Always plate out a portion of the culture to check its purity.

Gavan recommends lyophilization (freeze drying) of stock cultures but suggests ultrafreezing or refrigeration when lyophilization facilities are unavailable.[11] For ultrafreezing, suspend a loop full of a log phase culture in 0.5-1.0 ml of sterile, defibrinated sheep blood in a screw-cap vial. Quick freeze the suspension in a dry ice ethanol bath, liquid nitrogen, or an ultrafreezer and store at -40°C or lower. Recover organisms after thawing cultures in a 37°C water bath for several minutes. Alternatively, anaerobes can be stored in chopped meat medium for two to three months at 4-8°C.

The CDC laboratory manual contains step-by-step instructions for lyophilization and ultrafreezing of anaerobic stock cultures.[7] Sterile, defibrinated rabbit blood is recommended for ultrafreezer storage of nonsporeforming organisms and brain storage medium for *Clostridium* spp. Suspensions prepared from actively growing broth cultures are quick frozen at -70°C and then stored at -20°C or -42°C for clostridia and -42°C or lower for nonsporeformers. Following thawing, either chopped meat dextrose medium (sporeformers and nonsporeformers) or THIO (nonsporeformers) is used to recultivate the organisms.

Urease Test[20]

Purpose: To determine if an isolate produces urease, an enzyme capable of catalyzing the hydrolysis of urea. The urease test is especially important in distinguishing *B. ureolyticus* (urease-positive) from other pitting, anaerobic, gram-negative bacilli (urease-negative). It is also a useful test for identifying *Actinomyces naeslundii*, *A. viscosus*, *Bifidobacterium infantis*, *Clostridium sordellii*, *Peptococcus niger*, *Peptostreptococcus prevotii*, *P. tetradius*, and *Staphylococcus saccharolyticus*, all of which are urease-positive. Urease can be detected a variety of ways, including rapid tube, disk, and spot tests. Perhaps the easiest way to test for urease is by using Urea Differentiation Disks® (Difco), which is the method described here. The disk method is accurate and does not require preparation of reagents.

> **Materials required:**
> Pure culture of the isolate
> Small, sterile tube
> Sterile distilled water
> Inoculating loop
> Urea Differentiation Disks® (Difco)

Procedure: First suspend a loop full of organisms from the PC/SC plate in a small volume (0.4 ml) of sterile water. Then add a urea disk to the tube, and incubate the tube at 37°C for up to four hours.

Interpretation of results: A color change from pale yellow to dark pink represents a positive test (urea hydrolysis). Development of a pale orange color upon addition of the disk to the bacterial suspension is a negative reaction.

References

1. Bartley, S.L. (Centers for Disease Control). 1990. Personal communication.
2. Brazier and Riley. 1988. UV red fluorescence of *Veillonella* spp. J. Clin. Microbiol. 26:383–384.
3. Buchanan, A. G. 1982. Clinical laboratory evaluation of a reverse CAMP test for presumptive identification of *Clostridium perfringens*. J. Clin. Microbiol. 16:761–762.
4. Chow, A. W., V. Patten, and L. B. Guze. 1975. Rapid screening of *Veillonella* by ultraviolet fluorescence. J. Clin Microbiol. 2:546–548.
5. Dezfulian, M., and V. R. Dowell, Jr. 1981. Physiological characterization of *Clostridium botulinum* and development of practical isolation and identification procedures. Academia Press, New York
6. Dowell, V. R., Jr. 1988. Procedures for isolation and characterization of anaerobic bacteria. Unpublished document that accompanied a lecture presented at the IV National Symposium on Clinical Microbiology of the Sociedade Brasileira de Microbiologia, Rio de Janeiro, Brazil. Centers for Disease Control, Atlanta.
7. Dowell, V. R., Jr., and T. M. Hawkins. 1974. Laboratory methods in anaerobic bacteriology: CDC laboratory manual. Centers for Disease Control, Atlanta.
8. Dowell, V. R., Jr., and G. L. Jones. 1981. Procedures for use of differential agar media in the identification of anaerobic bacteria. Centers for Disease Control, Atlanta.

9. Dowell, V. R., Jr., and G. L. Lombard. 1984. Procedures for preliminary identification of bacteria. Centers for Disease Control, Atlanta.

10. Dowell, V. R., Jr., G. L. Lombard, F. S. Thompson, and A. Y. Armfield. 1977. Media for isolation, characterization, and identification of obligately anaerobic bacteria. Centers for Disease Control, Atlanta.

11. Gavan, T. L. 1991. Quality control in microbiology, p. 1281–1293. *In* J. B. Henry, D. A. Nelson, R. H. Tomar, and J. A. Washington (eds.), Clinical diagnosis and management by laboratory methods. W. B. Saunders Company, Philadelphia.

12. Hansen, M. V., and L. P. Elliott. 1980. New presumptive identification test for *Clostridium perfringens*: reverse CAMP test. J. Clin. Microbiol. 12:617–619.

13. Heimbrook, M. E., W. L. L. Wang, and G. Campbell. 1989 Staining bacterial flagella easily. J. Clin. Microbiol. 27:2612–2615.

14. Holdeman, L. V., E. P. Cato, and W. E. C. Moore (eds.). 1977 (with 1987 update). VPI anaerobe laboratory manual, 4th ed. Virginia Polytechnic Institute and State University, Blacksburg.

15. Kodaka, H., A. Y. Armfield, G. L. Lombard, and V. R. Dowell, Jr. 1982. Practical procedure for demonstrating bacterial flagella. J. Clin. Microbiol. 16:948–952.

16. Koransky, J. R., S. D. Allen, and V. R. Dowell, Jr. 1978. Use of ethanol for selective isolation of sporeforming microorganisms. Appl. Environ. Microbiol. 35:762–765.

17. Lombard, G. L., and V. R. Dowell, Jr. 1983. Comparison of three reagents for detecting indole production by anaerobic bacteria in microtest systems. J. Clin. Microbiol. 18:609–613.

18. Magee, C. M., G. Rodeheaver, M. T. Edgerton, and K. F. Edlich. 1975. A more reliable Gram staining technique for diagnosis of surgical infections. Amer. J. Surg. 130:341–346.

19. Mangels, J. I., M. E. Cox, and L. H. Lindberg. 1984. Methanol fixation: an alternative to heat fixation of smears before staining. Diag. Microbiol. Infect. Dis. 2:129–137.

20. Mills, C. K., B. Y. Grimes, and R. L. Gherna. 1987. Three rapid methods compared with a conventional method for detection of urease production in anaerobic bacteria. J. Clin. Microbiol. 25:2209–2210.

21. Moore, H. B. 1981. Rapid methods in microbiology: IV. Presumptive and rapid methods in anaerobic bacteriology. Amer. J. Med. Technol. 47:705–712.

22. Slots, J., and H. S. Reynolds. 1982. Long-wave UV light fluorescence for identification of black-pigmented *Bacteroides* spp. J. Clin. Microbiol. 16:1148–1151.

23. Sutter, V. L., and S. M. Finegold. 1971. Antibiotic disk susceptibility tests for rapid presumptive identification of gram-negative bacilli. Appl. Microbiol. 21:13–20.

24. Sutter, V. L., D. M. Citron, M. A. C. Edelstein, and S. M. Finegold. 1985. Wadsworth anaerobic bacteriology manual, 4th ed. Star Publishing, Belmont, CA.

25. Weinberg, L. G., L. L. Smith, and A. H. McTighe. 1983. Rapid identification of the *Bacteroides fragilis* group by bile disk and catalase tests. Lab. Med. 14:785–788.

26. Widdel, F., and N. Pfennig. 1984. Dissimilatory sulfate- or sulfur-reducing bacteria, p. 663–679. *In* N. R. Krieg and J. G. Holt (eds.), Bergey's manual of systematic bacteriology, vol 1. William and Wilkins Company, Baltimore.

27. Wideman, P. A., V. L. Vargo, D. Citronbaum, and S. M. Finegold. 1976. Evaluation of the sodium polyanethol sulfonate disk for the identification of *Peptostreptococcus anaerobius*. J. Clin. Microbiol. 4:330–333.

28. Wideman, P. A., D. M. Citronbaum, and V. L. Sutter. 1977. Simple disk test for detection of nitrate reduction by anaerobic bacteria. J. Clin. Microbiol. 5:315–319.

COMPENDIUM OF CLINICALLY ENCOUNTERED ANAEROBES

This appendix contains information about the following organisms:

Appendix E

Key to abbreviations and symbols used in this appendix:

AN = obligate anaerobe
MA = microaerotolerant anaerobe
AT = aerotolerant anaerobe
FA = facultative anaerobe
ST = subterminal
T = terminal
NF = no flagella
SSF = single subpolar flagellum
PF = peritrichous flagella
V or ( ) = variable
superscript $+$ = some strains positive
superscript $-$ = some strains negative
NG = no growth
A = acetic acid
P = propionic acid
IB = isobutyric acid
B = butyric acid
IV = isovaleric acid
L = lactic acid
S = succinic acid
C = caproic acid
PA = phenylacetic acid
* = considered a cardinal feature in the identification of this particular organism

Definitions

Primary tests: Consist primarily of "observations"; those attributes of an organism that can be learned from growth on primary isolation plates; attributes that do not require the use of reagents (other than Gram stain reagents); examples: growth/no growth on primary media (BRU/BA, PEA, BBE, KVLB), colony morphology, double zone hemolysis, pitting, pigmentation, fluorescence, gram reaction, cell morphology, and wet mount motility

Secondary tests: Tests that produce results **on the same day that the pure culture/subculture (PC/SC) plate becomes available;** include tests that were set up at the same time that the PC/SC plate was inoculated (e.g., special potency antimicrobial disks [Vanc, Kan, Col], Pen & Rif disks, bile disk, SPS disk, nitrate disk), as well as tests performed as soon as the PC/SC plate is removed from the incubator or observations that can be made at that time (e.g., aerotolerance test results, spot indole, catalase, oxidase, rapid urease disk test, lecithinase and lipase activities [if EYA plate was set up at same time as PC/SC plate], desulfoviridin, Ryu flagella stain)

Tertiary tests: Tests that **produce results 24-48 hours after they are inoculated with growth from the PC/SC plate** (e.g., tubed carbohydrate tests, H_2S in TSI, esculin hydrolysis, 20% bile, milk, starch, gelatin, DNase, lecithinase and lipase activities [if EYA was not inoculated at same time as PC/SC plate], growth stimulation tests, reverse CAMP test, Nagler test, GLC analysis of acids from PYG, cellular fatty acids by GLC, toxin assays) (If an organism cannot be identified by primary and secondary tests, a particular laboratory may opt to use one of the four-hour rapid enzyme panels [e.g., RapID ANA II® or AN-IDENT®] rather than "tertiary tests." Laboratories in smaller hospitals are particularly apt to go that route.)

Most diagrams illustrate the appearance of the organism when grown in PY broth (left) and PYG broth (right) and, if applicable, following flagella staining. Scale = $\underline{\quad 10\ \mu m \quad}$

Readers are reminded that biochemical test results may vary, depending upon the specific methodolgy being used. Biochemical data in this appendix were furnished by the Centers for Disease Control (CDC), Atlanta.

Clostridium spp.

Clostridium bifermentans

Colony morphology (on BRU/BA):

Diameter:	0.5-4.0 mm
Form:	circular
Elevation:	flat or raised
Margins:	irregular, lobate, or scalloped
Color:	gray
Surface:	shiny and smooth
Density:	translucent or opaque
Internal structure:	granular or slightly mottled
Hemolysis:	most strains beta-hemolytic

Cell morphology (in PYG broth):

0.6-1.9 x 1.6-11.0 μm; straight bacilli; motile, peritrichous; occur singly, in pairs, or in short chains; oval, subterminal spores that usually do not cause swelling of cells

Primary tests:

Relationship to oxygen:	MA,AN
Swarming on BRU/BA:	−
Double zone of hemolysis:	−
Chartreuse fluorescence:	−
Gram stain reaction:	+
Position of spores:	ST
Motility:	+

Secondary tests:

Flagella:	PF
Indole:	+
Indole derivatives:	−
Lecithinase:	+
Lipase:	−
Urease:	−

Tertiary tests:

Esculin hydrolysis:	−
Proteolysis in milk:	+
Gelatin hydrolysis:	−
DNase:	−
Glucose fermentation:	+
Lactose fermentation:	−
Mannitol fermentation:	−
Rhamnose fermentation:	−
Major acids in PYG:	A
(Often isoacids, but variable)	

Clostridium clostridiiforme
(Clostridium clostridioforme)

Colony morphology (on BRU/BA):

Diameter:	0.5-2.0 mm
Elevation:	convex to slightly peaked
Margin:	entire, sightly scalloped or erose
Color:	gray-white
Density:	translucent to opaque
Internal structure:	usually mottled or mosaic
Hemolysis:	nonhemolytic

Cell morphology (in PYG broth):

0.3-0.9 x 1.4-9.0 μm; straight, gram-variable bacilli with pointed ends; usually occur in pairs, but may also occur singly or in short chains; oval, subterminal spores that cause swelling of cells; often difficult to demonstrate spores; easily confused with a gram-negative bacillus

Primary tests:

Relationship to oxygen:	AN
Swarming on BRU/BA:	−
Double zone of hemolysis:	−
Chartreuse fluorescence:	−
Gram stain reaction:	−+
Position of spores:	ST
Motility:	−+

Secondary tests:

Flagella:	PF
Indole:	−
Indole derivatives:	−
Lecithinase:	−
Lipase:	−
Urease:	−

Tertiary tests:

Esculin hydrolysis:	+
Proteolysis in milk:	−
Gelatin hydrolysis:	−
DNase:	−
Glucose fermentation:	+
Lactose fermentation:	+−
Mannitol fermentation:	−
Rhamnose fermentation:	+−
Major acids in PYG:	A

Additional important information:

Heat shock or ethanol treatment useful to demonstrate that this organism is a spore-former

Clostridium difficile

Colony morphology (on BRU/BA):

Diameter:	2-5 mm
Form:	circular (occasionally rhizoid)
Elevation:	flat to low convex
Color:	grayish or whitish
Surface:	matte to glossy
Density:	opaque
Hemolysis:	nonhemolytic

Cell morphology (in PYG broth):

0.5-1.9 x 3.0-16.9 μm straight bacilli; may produce chains of 2-6 cells aligned end to end; oval, subterminal (rarely terminal) spores cause swelling of cells

Primary tests:

Relationship to oxygen:	AN
Swarming on BRU/BA:	−
Double zone of hemolysis:	−
Chartreuse fluorescence:	+*
Gram stain reaction:	+
Position of spores:	ST
Motility:	+

Secondary tests:

Flagella:	SSF*ᴾꟳ
Indole:	−
Indole derivatives:	−
Lecithinase:	−
Lipase:	−
Urease:	−

Tertiary tests:

Esculin hydrolysis:	+⁻
Proteolysis in milk:	−
Gelatin hydrolysis:	−⁺
DNase:	−⁺
Glucose fermentation:	+
Lactose fermentation:	−
Mannitol fermentation:	+
Rhamnose fermentation:	−
Major acids in PYG: A, IB, B, IV, V, IC	

Additional important information:

Characteristic appearance on CCFA (large, yellow, ground-glass colonies; yellowing of the medium around the colonies)

Characteristic "horse stable" odor

A common cause of pseudomembranous colitis and antibiotic-associated diarrhea

Produces toxins A (an enterotoxin and cytotoxin) and B (an extremely potent cytotoxin)

Clostridium ramosum

Colony morphology (on BRU/BA):

Diameter:	0.5-2 mm
Form:	circular to slightly irregular
Elevation:	convex or raised
Margin:	entire, scalloped, or erose
Color:	gray-white to colorless
Surface:	smooth
Density:	translucent to semiopaque
Internal structure:	mottled, mosaic, or granular
Hemolysis:	nonhemolytic

Cell morphology (in PYG broth):

0.5-0.9 x 2.0-12.8 μm straight bacilli; frequently appears gram-negative; occur singly, in pairs, in short chains (often in "V" arrangements), or irregular masses; spores are rarely seen, but when present, they are round, usually terminal, and cause swelling of cells

Primary tests:

Relationship to oxygen:	AN
Swarming on BRU/BA:	−
Double zone of hemolysis:	−
Chartreuse fluorescence:	−
Gram stain reaction:	−
Position of spores:	rare (T)*
Motility:	→*

Secondary tests:

Flagella:	−
Indole:	−
Indole derivatives:	−
Lecithinase:	−
Lipase:	−
Urease:	−

Tertiary tests:

Esculin hydrolysis:	+
Proteolysis in milk:	−
Gelatin hydrolysis:	−
DNase:	−
Glucose fermentation:	+
Lactose fermentation:	+
Mannitol fermentation:	+−
Rhamnose fermentation:	V
Major acids in PYG:	A

Additional important information:

Although this organism usually stains gram-negative, it is sensitive to vancomycin

One of very few anaerobes that is rifampin-resistant

Heat shock or ethanol treatment useful to demonstrate that this organism is a spore-former

Clostridium septicum

Colony morphology (on BRU/BA):

Diameter:	1-5 mm
Form:	circular
Elevation:	slightly raised
Margin:	variable (markedly irregular, rhizoid, swarming)
Color:	gray
Surface:	glossy
Density:	translucent
Hemolysis:	beta-hemolytic
Other:	usually forms an invisible film over the entire agar surface

Cell morphology (in PYG broth):

0.6-1.9 x 1.9-35.0 μm straight or curved bacilli; occur singly and in pairs; become gram-negative with age; stain unevenly; oval, sub-terminal spores that cause swelling of cells

Primary tests:

Relationship to oxygen:	MA, AN
Swarming on BRU/BA:	+*
Double zone of hemolysis:	−
Chartreuse fluorescence:	−
Gram stain reaction:	+
Position of spores:	ST
Motility:	+

Secondary tests:

Flagella:	PF
Indole:	−
Indole derivatives:	−
Lecithinase:	−
Lipase:	−
Urease:	−

Tertiary tests:

Esculin hydrolysis:	+
Proteolysis in milk:	−
Gelatin hydrolysis:	+
DNase:	+
Glucose fermentation:	+
Lactose fermentation:	+
Mannitol fermentation:	−
Rhamnose fermentation:	−
Major acids in PYG:	A, B

Additional important information:

Associated with gas gangrene and malignancy

Closely related to *C. chauvoei* (an animal pathogen); most reliable differentiation requires serological, pathological, or toxicological means

Appendix E

Clostridium sordellii

Colony morphology (on BRU/BA):

Diameter:	1-4 mm
Form:	circular to irregular
Elevation:	flat or raised
Margin:	scalloped, lobate, or entire
Color:	gray or chalklike
Surface:	dull or shiny
Density:	translucent to opaque
Internal structure:	granular or mottled
Hemolysis:	variable (slightly beta-hemolytic on rabbit blood agar)

Cell morphology (in PYG broth):

0.5-1.7 x 1.6-20.6 μm straight bacilli; occur singly and in pairs; oval, central to subterminal spores that cause slight swelling of cells; free spores often seen

Primary tests:

Relationship to oxygen:	MA, AN
Swarming on BRU/BA:	−
Double zone of hemolysis:	−
Chartreuse fluorescence:	−
Gram stain reaction:	+
Position of spores:	ST
Motility:	+

Secondary tests:

Flagella:	PF
Indole:	+
Indole derivatives:	−
Lecithinase:	+
Lipase:	−
Urease:	+*

Tertiary tests:

Esculin hydrolysis:	−
Proteolysis in milk:	+
Gelatin hydrolysis:	−
DNase:	−
Glucose fermentation:	+
Lactose fermentation:	−
Mannitol fermentation:	−
Rhamnose fermentation:	−
Major acids in PYG:	A

Additional important information:

Distinguished from *C. bifermentans* most readily by its ability to produce urease

Clostridium sphenoides

Colony morphology (on BRU/BA):

Diameter:	1-2 mm
Form:	circular
Elevation:	low convex
Margin:	entire or erose
Color:	gray
Surface:	glossy
Density:	translucent
Internal structure:	often mottled
Hemolysis:	nonhemolytic

Cell morphology (in PYG broth):

0.3-1.1 x 1.3-8.6 μm; straight bacilli with tapered or round ends; occur singly, in pairs, or short chains; cells usually stain gram-negative; spores are oval and subterminal (occasionally terminal), and cause swelling of cells

Primary tests:

Relationship to oxygen:	AN
Swarming on BRU/BA:	−
Double zone of hemolysis:	−
Chartreuse fluorescence:	−
Gram stain reaction:	−
Position of spores:	TST
Motility:	+

Secondary tests:

Flagella:	PF
Indole:	+
Indole derivatives:	−
Lecithinase:	−
Lipase:	−
Urease:	−

Tertiary tests:

Esculin hydrolysis:	−
Proteolysis in milk:	−
Gelatin hydrolysis:	V
DNase:	V
Glucose fermentation:	+
Lactose fermentation:	+
Mannitol fermentation:	+
Rhamnose fermentation:	+
Major acids in PYG:	A

Additional important information:

Usually a gram-negative bacillus with tapered ends
Heat shock or ethanol treatment useful to demonstrate that this organism is a spore-former

Clostridium sporogenes

Colony morphology (on BRU/BA):

Diameter:	2-6 mm
Form:	irregularly circular
Elevation & color:	raised, yellowish-gray center
Margin:	flattened periphery composed of tangled filaments ("Medusa head" colony); coarse rhizoid
Surface:	matte
Density:	opaque
Hemolysis:	usually beta-hemolytic
Other:	firmly adherent to the agar

Cell morphology (in PYG broth):

0.3-1.4 x 1.3-16.0 μm baccili; occur singly; oval, subterminal spores cause swelling of cells; sporulates readily on most media

Primary tests:

Relationship to oxygen:	AN
Swarming on BRU/BA:	−
Double zone of hemolysis:	−
Chartreuse fluorescence:	−
Gram stain reaction:	+
Position of spores:	ST
Motility:	+

Secondary tests:

Flagella:	PF
Indole:	−
Indole derivatives:	+*
Lecithinase:	−+
Lipase:	+*
Urease:	−

Tertiary tests:

Esculin hydrolysis:	+
Proteolysis in milk:	+
Gelatin hydrolysis:	−+
DNase:	−+
Glucose fermentation:	+
Lactose fermentation:	−
Mannitol fermentation:	−
Rhamnose fermentation:	−
Major acids in PYG:	A, IB, B, IV

Additional important information:

Cannot be differentiated from *Clostridium botulinum* without use of mouse toxin neutralization assay

Clostridium tertium

Colony morphology (on BRU/BA):

Diameter:	2-4 mm
Form:	circular
Elevation:	low convex
Margin:	slightly irregular
Color:	white to gray
Surface:	matte
Internal structure:	usually mottled or granular
Hemolysis:	variable (may be alpha, beta, or nonhemolytic)

Cell morphology (in PYG broth):

0.5-1.4 x 1.5-10.2 μm straight bacilli; occur singly and in pairs; large, oval, terminal (occasionally subterminal) spores that cause marked swelling of cells

Primary tests:

Relationship to oxygen:	FA, AT*
Swarming on BRU/BA:	−
Double zone of hemolysis:	−
Chartreuse fluorescence:	−
Gram stain reaction:	+−
Position of spores:	T*
Motility:	+

Secondary tests:

Flagella:	PF
Indole:	−
Indole derivatives:	−
Lecithinase:	−
Lipase:	−
Urease:	−

Tertiary tests:

Esculin hydrolysis:	+
Proteolysis in milk:	−
Gelatin hydrolysis:	−
DNase:	+
Glucose fermentation:	+
Lactose fermentation:	+
Mannitol fermentation:	+
Rhamnose fermentation:	−
Major acids in PYG:	A, B

Additional important information:

Can easily be confused with a nonsporeforming gram-negative bacillus
Can distinguish from a *Bacillus* species by spore formation only under anaerobic conditions
(*Bacillus* spp. produce spores only under aerobic conditions)

Clostridium tetani

Colony morphology (on BRU/BA):

Diameter:	4-6 mm
Elevation:	flat
Margin:	irregular and rhizoid
Color:	gray
Surface:	matte
Density:	translucent
Hemolysis:	usually a narrow zone of beta hemolysis
Other:	may swarm on moist plates

Cell morphology (in PYG broth):

0.5-1.7 x 2.1-18.1 μm bacilli; occur singly and in pairs; become gram-negative after 24 hrs incubation; spores are usually round and terminal, but may be oval and/or subterminal; spores may create a "drumstick" or "tennis racket" appearance

Primary tests:

Relationship to oxygen:	AN
Swarming on BRU/BA:	+*
Double zone of hemolysis:	−
Chartreuse fluorescence:	−
Gram stain reaction:	+
Position of spores:	T*
Motility:	+

Secondary tests:

Flagella:	PF
Indole:	+⁻
Indole derivatives:	−
Lecithinase:	−
Lipase:	−
Urease:	−

Tertiary tests:

Esculin hydrolysis:	−
Proteolysis in milk:	−
Gelatin hydrolysis:	+
DNase:	+
Glucose fermentation:	−
Lactose fermentation:	−
Mannitol fermentation:	−
Rhamnose fermentation:	−
Major acids in PYG:	A, P, B

Additional important information:

Note that *C. tetani* is asaccharolytic
Often very fastidious and difficult to culture
Often mixed with other organisms from wounds

Nonsporeforming, Anaerobic, Gram-Positive Bacilli

Actinomyces spp.

Colony morphology (on BRU/BA):

Young (24-48-hour-old) microcolonies are composed of branched filaments sometimes referred to as "spider colonies"; older (7-14 day-old) macrocolonies are 0.5-5.0 mm in diameter; rough or smooth; rough colonies (sometimes referred to as "breadcrumb," "molar tooth," or "raspberrylike" colonies) are dry to crumby in texture; smooth colonies are soft to mucoid; usually white to gray-white or creamy white, but some species produce pigmented colonies

Cell morphology (in THIO):

Variable; straight or slightly curved bacilli (0.2-1.0 μm in width); slender filaments (1.0 x 10-50 μm or longer), with true branching; filaments may be straight or wavy, with varying degrees of branching, and may have swollen, clubbed, or clavate ends; short rods (1.5-5.0 μm in length), with or without clubbed ends, that may occur singly, in pairs, in short chains, or in small clusters; longer branched bacilli (5.0-10.0 μm in length); gram-positive, but irregular staining gives rise to a beaded or barred appearance; nonmotile

---

Actinomyces israelii

Colony morphology (on BRU/BA):

24-48 hours: small "spider colonies" are common (branching filaments radiating from a central point); less common are tiny colonies having one or two branching filaments or larger colonies with many filaments but no central tangled mass

7-14 days: rough colonies with or without a central depression; circular or irregular with undulate, lobate, or erose edges; described as "molar tooth" or "breadcrumb" in appearance or "raspberrylike"; surface texture often granular with ground-glass appearance; convex to pulvinate to umbonate; opaque; cream or gray-white

Cell morphology (in THIO):
Intertwining, branching filaments resembling a microcolony; club-shaped rods in thioglycollate broth

Primary tests:		Tertiary tests:	
Relationship to oxygen:	FA, AT, MA, AN	Gelatin hydrolysis:	–
48 h colonies < 1 mm diam:	+	Proteolysis in milk:	–
Red pigmentation:	–	Glucose fermentation:	+
Beta hemolysis:	–	Lactose fermentation:	+⁻
Rough colonies:	+	Mannitol fermentation:	+⁻
Branched bacilli:	+	Rhamnose fermentation:	+⁻
		Major acids in PYG:	A, L, S

Secondary tests:

Catalase:	–
Indole:	–
Urease:	–

Actinomyces meyeri

Colony morphology (on BRU/BA):

As described for genus, except microcolonies are smooth and nonfilamentous; mature colonies are pinpoint to 1 mm; circular; flat to convex; translucent to opaque; white; shiny, smooth surface; entire margin; alpha or nonhemolytic

Cell morphology (in THIO):

As described for genus, except branching may be difficult to demonstrate; cells are usually short; long, straight or curved filaments without branching and chains of longer bacilli may be seen; terminal swellings occasionally present

Primary tests:		**Tertiary tests:**	
Relationship to oxygen:	FA, AT, MA, AN	Gelatin hydrolysis:	−
48 h colonies < 1 mm diam:	−+	Proteolysis in milk:	−
Red pigmentation:	−	Glucose fermentation:	+
Beta hemolysis:	−	Lactose fermentation:	−+
Rough colonies:	−	Mannitol fermentation:	−
Branched bacilli:	+	Rhamnose fermentation:	−
		Major acids in PYG:	A, L, S

Secondary tests:

Catalase:	−
Indole:	−
Urease:	−

Actinomyces naeslundii

Colony morphology (on BRU/BA):

24-48 hours: small colony with a dense central mass of diphtheroidal cells or filaments surrounded by long, branched filaments projecting in all directions

7-14 days: 1-2 mm; smooth; circular; low convex to umbonate; entire edge; may produce rough colonies, with a surface texture ranging from granular to "raspberrylike"

Cell morphology (in THIO):

Variable in appearance; intertwining branching filaments; club-shaped bacilli; thin filaments; bacilli with bifurcated ends; small bacilli with or without metachromatic staining

Primary tests:		**Tertiary tests:**	
Relationship to oxygen:	FA, AT, MA, AN	Gelatin hydrolysis:	−
48 h colonies < 1 mm diam:	−+	Proteolysis in milk:	−
Red pigmentation:	−	Glucose fermentation:	+
Beta hemolysis:	−	Lactose fermentation:	−+
Rough colonies:	−+	Mannitol fermentation:	−
Branched bacilli:	+	Rhamnose fermentation:	−
		Major acids in PYG:	A, L, S

Secondary tests:

Catalase:	−
Indole:	−
Urease:	+ (but often weak or very late)

Actinomyces odontolyticus

Colony morphology (on BRU/BA):

24-48 hours: small smooth to finely granular colony; entire to irregular edge; slightly raised to convex; soft and white

7-14 days: 1-2 mm; circular to irregular; entire or irregular edge; low convex to umbonate; smooth to finely granular; white to gray-white, soft, opaque; colonies usually become deep red after 2-10 days of incubation on blood agar

Cell morphology (in THIO):

Variable in appearance; club-shaped bacilli; bacilli with bifurcated ends; thin filaments; small bacilli with or without metachromatic staining

Primary tests:		**Tertiary tests:**	
Relationship to oxygen:	FA, AT, MA, AN	Gelatin hydrolysis:	—
48 h colonies < 1 mm diam:	—	Proteolysis in milk:	—
Red pigmentation:	+*	Glucose fermentation:	+
Beta hemolysis:	—	Lactose fermentation:	−+
Rough colonies:	—	Mannitol fermentation:	—
Branched bacilli:	+	Rhamnose fermentation:	—
		Major acids in PYG:	A, L, S

Secondary tests:	
Catalase:	—
Indole:	—
Urease:	—

Actinomyces pyogenes

Colony morphology (on BRU/BA):

As described for genus

Cell morphology (in THIO):

As described for genus

Primary tests:		**Tertiary tests:**	
Relationship to oxygen:	FA, AT	Gelatin hydrolysis:	+*
48 h colonies < 1 mm diam:	—	Proteolysis in milk:	+*
Red pigmentation:	—	Glucose fermentation:	+
Beta hemolysis:	+*	Lactose fermentation:	—
Rough colonies:	—	Mannitol fermentation:	—
Branched bacilli:	—	Rhamnose fermentation:	—
		Major acids in PYG:	A, L, S

Secondary tests:	
Catalase:	—
Indole:	—
Urease:	—

Actinomyces viscosus

Colony morphology (on BRU/BA):

As described for genus

Cell morphology (in THIO):

As described for genus

Primary tests:

Relationship to oxygen:	FA, AT, MA, AN
48 h colonies < 1 mm diam:	−
Red pigmentation:	−
Beta hemolysis:	−
Rough colonies:	−⁺
Branched bacilli:	+

Secondary tests:

Catalase:	+*
Indole:	−
Urease:	+*

Tertiary tests:

Gelatin hydrolysis:	−
Proteolysis in milk:	−
Glucose fermentation:	+
Lactose fermentation:	+⁻
Mannitol fermentation:	−
Rhamnose fermentation:	−
Major acids in PYG:	A, L, S

Bifidobacterium dentium

Colony morphology (on BRU/BA):
(for the genus *Bifidobacterium*)

Elevation:	convex
Margin:	entire
Color:	cream to white
Surface:	smooth, glistening
Consistency:	soft

Cell morphology (in PYG broth):
(for the genus *Bifidobacterium*)

Bacilli of various shapes; short, regular, thin cells with pointed ends; coccoidal regular cells; long cells with slight bends, protuberances, or a variety of branchings; pointed, slightly bifurcated, club-shaped or spatulated extremities; single or in chains; in starlike aggregates or in "V" or "palisade" arrangements

Primary tests:

Relationship to oxygen:	FA, AT, MA, AN
48 h colonies < 1 mm diam:	−
Red pigmentation:	−
Beta hemolysis:	−
Rough colonies:	−
Branched bacilli:	−

Secondary tests:

Catalase:	−
Indole:	−

Tertiary tests:

Gelatin hydrolysis:	−
Proteolysis in milk:	−
Glucose fermentation:	+
Lactose fermentation:	+*
Mannitol fermentation:	−
Rhamnose fermentation:	−
Major acids in PYG:	

A, L (> 1:2 ratio of A:L; the reverse is true for *Streptococcus* spp.)

Eubacterium alactolyticum

Colony morphology (on BRU/BA):

Diameter:	punctiform to 0.5 mm
Form:	circular
Elevation:	convex to pulvinate
Margin:	entire
Surface:	smooth, shiny
Density:	translucent to opaque

Cell morphology (in PYG broth):

0.3-0.6 x 1.6-7.5 μm; in pairs resembling flying birds, clumps, and "Chinese letter" configurations

Primary tests:		**Tertiary tests:**	
Relationship to oxygen:	AN	Gelatin hydrolysis:	—
48 h colonies < 1 mm diam:	+	Proteolysis in milk:	—
Red pigmentation:	—	Glucose fermentation:	V
Beta hemolysis:	—	Lactose fermentation:	—*
Rough colonies:	—	Mannitol fermentation:	—
Branched bacilli:	—	Rhamnose fermentation:	—
		Major acids in PYG: A, B*, C*	

Secondary tests:	
Catalase:	—
Indole:	—

Eubacterium lentum

Colony morphology (on BRU/BA):

Diameter:	0.5-2.0 mm
Form:	circular
Elevation:	raised to low convex
Margin:	entire to erose
Surface:	dull to shiny; smooth
Density:	translucent to semiopaque
Internal structure:	sometimes mottled when viewed by obliquely transmitted light

Cell morphology (in PYG broth):

0.2-0.4 x 0.2-2.0 μm; diphtheroidal; small, pleomorphic bacilli; occur singly or in pairs and short chains

Primary tests:		**Tertiary tests:**	
Relationship to oxygen:	AN	Gelatin hydrolysis:	—
48 h colonies < 1 mm diam:	—	Proteolysis in milk:	—
Red pigmentation:	—	Glucose fermentation:	—
Beta hemolysis:	—	Lactose fermentation:	—
Rough colonies:	—	Mannitol fermentation:	—
Branched bacilli:	—	Rhamnose fermentation:	—
		Major acids in PYG:	(A)

Secondary tests:	
Catalase:	V*
Indole:	—

Additional important information:

Produces H_2S in TSI

Appendix E

Eubacterium limosum

Colony morphology (on BRU/BA):

Diameter:	punctiform up to 2 mm
Form:	circular
Elevation:	convex
Margin:	entire
Consistency:	translucent to slightly opaque
Internal structure:	sometimes mottled when viewed by obliquely transmitted light

Cell morphology (in PYG broth):

0.6-0.9 x 1.6-4.8 μm; often seen to have swollen ends and bifurcations; occur singly, in pairs, and in small clumps

Primary tests:

Relationship to oxygen:	AN
48 h colonies < 1 mm diam:	—
Red pigmentation:	—
Beta hemolysis:	—
Rough colonies:	—
Branched bacilli:	—

Secondary tests:

Catalase.	—
Indole:	—

Tertiary tests:

Gelatin hydrolysis:	—
Proteolysis in milk:	—
Glucose fermentation:	+
Lactose fermentation:	—
Mannitol fermentation:	+*
Rhamnose fermentation:	—
Major acids in PYG:	A, B*

Propionibacterium acnes

Colony morphology (on BRU/BA):

Diameter:	punctiform to 0.5 mm
Form:	circular
Margin:	entire to pulvinate
Color:	white to gray
Surface:	glistening
Density:	translucent to opaque

Cell morphology (in PYG broth):

0.3-1.3 x 1.0-10 μm; small, pleomorphic bacilli with or without metachromatic granules; club-shaped; "Chinese letter" configurations are common

Primary tests:

Relationship to oxygen:	FA, AT, MA, AN
48 h colonies < 1 mm diam:	−
Red pigmentation:	−
Beta hemolysis:	−
Rough colonies:	−
Branched bacilli:	−

Secondary tests:

Catalase:	+*
Indole:	+*

Tertiary tests:

Gelatin hydrolysis:	+
Proteolysis in milk:	+
Glucose fermentation:	+
Lactose fermentation:	−
Mannitol fermentation:	−+
Rhamnose fermentation:	−
Major acids in PYG:	
A, P*, (L), (S)	

Additional important information:

An anaerobic, gram-positive diphtheroid that is catalase-positive and indole-positive can be presumptively identified as *P. acnes*

Propionibacterium propionicus (formerly *Arachnia propionica*)

Colony morphology (on BRU/BA):

24-48 hours: small, branching, filamentous colonies that resemble *A. israelii;* smooth colonies are rarely observed

7-14 days: up to 2 mm; white to gray-white; variable in appearance; may be rough ("molar tooth" or "breadcrumb"); may be smooth, convex, with undulate edges; crumbly to soft

Cell morphology (in THIO):

Extremely variable in appearance; cells from young (24-48 hour) broth cultures tend to be more filamentous than older cells; branched structures (5-20 μm or longer) may occur, as may swollen, spherical cells (up to 5.0 μm); young broth cultures may contain intact filamentous microcolonies; short, diphtheroid rods (0.2-0.3 x 3.0-5.0 μm) usually predominate in aged (7-14 days) broth cultures, but septate or nonseptate filaments may also be seen; cells may have distended or clubbed ends

Primary tests:

Relationship to oxygen:	FA, AT, MA, AN
48 h colonies < 1 mm diam:	+
Red pigmentation:	−
Beta hemolysis:	−
Rough colonies:	+
Branched bacilli:	+

Secondary tests:

Catalase:	−
Indole:	−

Tertiary tests:

Gelatin hydrolysis:	−
Proteolysis in milk:	−
Glucose fermentation:	+
Lactose fermentation:	+−
Mannitol fermentation:	−
Rhamnose fermentation:	−
Major acids in PYG:	A, P*, L

Anaerobic, Gram-Negative Bacilli

Bacteroides fragilis group

Cell morphology:	coccobacilli or straight bacilli of variable length; some filamentous forms
BBE agar:	good growth (colonies > 1 mm in diameter)*; cause browning of medium due to esculin hydrolysis

Primary tests:

Brown-black pigmentation:	—
Red fluorescence:	—
Pitting colonies:	—
Colonies < 1 mm in diameter:	—

Secondary tests:

Catalase:	V
Oxidase:	—
Indole:	V
Vancomycin (5 μg disk):	R
Kanamycin (1 mg disk):	R*
Colistin (10 μg disk):	R
Penicillin (2 U disk):	R*
Rifampin (15 μg disk):	S

Tertiary tests:

Growth in 20% bile:	+
Glucose:	+
Lactose:	+
Butyric acid in PYG:	—
Succinic acid in PYG:	+

Additional important information:

Usually resistant to all three special potency antimicrobial disks

---

Bacteroides caccae (a member of the *B. fragilis* group)

Colony morphology (on BRU/BA):

Diameter:	0.5-1 mm
Form:	circular
Elevation:	convex
Margin:	entire
Color:	gray
Surface:	shiny and smooth
Density:	translucent
Hemolysis:	rabbit blood may be slightly hemolyzed

Cell morphology (in PYG broth):

1.4-1.6 x 2.5-12.0 μm; occur singly or in pairs;
cells may appear vacuolated or beaded

Secondary tests:

Catalase:	—
Indole:	—

Tertiary tests:

DNase:	+
Arabinose:	+
Cellobiose:	+⁻
Mannitol:	—
Rhamnose:	V
Salicin:	+
Trehalose:	+

Bacteroides distasonis (a member of the *B. fragilis* group)

Colony morphology (on BRU/BA):

Diameter:	pinpoint to 0.5 mm
Form:	circular
Elevation:	convex
Margin:	entire
Color:	gray-white
Surface:	smooth
Density:	translucent to opaque
Hemolysis:	some strains alpha-hemolytic

Cell morphology (in PYG broth):

0.6-1.0 x 1.6-11.0 μm straight bacilli with rounded ends; occur singly and occasionally in pairs

Secondary tests:

Catalase:	+* −
Indole:	−

Tertiary tests:

DNase:	−
Arabinose:	−
Cellobiose:	−
Mannitol:	−
Rhamnose:	+
Salicin:	+
Trehalose:	+

Bacteroides fragilis (a member of the *B. fragilis* group)

Colony morphology (on BRU/BA):

Diameter:	1-3 mm
Form:	circular
Elevation:	low convex
Margin:	entire
Density:	translucent to semiopaque
Internal structure:	concentric rings when viewed by obliquely transmitted light
Hemolysis:	a few strains are beta-hemolytic

Cell morphology (in PYG broth):

0.8-1.3 x 1.6-8.0 μm; pale-staining, pleomorphic bacilli with rounded ends; occur singly and in pairs; swellings and vacuoles are often present

Secondary tests:

Catalase:	+* −
Indole:	−

Tertiary tests:

DNase:	−
Arabinose:	−
Cellobiose:	−
Mannitol:	−
Rhamnose:	−
Salicin:	−
Trehalose:	−

Appendix E

Bacteroides merdae (a member of the *B. fragilis* group)

Colony morphology (on BRU/BA):

Diameter:	0.5-1.0 mm
Form:	circular to slightly irregular
Elevation:	convex
Margin:	entire
Color:	white
Surface:	shiny and smooth
Hemolysis:	rabbit blood is slightly hemolyzed

Cell morphology (in PYG broth):

1.6 x 3.1-12 μm; occur singly, in pairs, or short chains

Secondary tests:		Tertiary tests:	
Catalase:	—	DNase:	—
Indole:	—	Arabinose:	—
		Cellobiose:	+
		Mannitol:	—
		Rhamnose:	+
		Salicin:	+
		Trehalose:	+

Bacteroides ovatus (a member of the *B. fragilis* group)

Colony morphology (on BRU/BA):

Diameter:	0.5-1.0 mm
Form:	circular
Elevation:	convex
Margin:	entire
Color:	pale buff
Density:	semiopaque
Consistency:	may have a mottled appearance
Hemolysis:	nonhemolytic

Cell morphology (in PYG broth):

0.6-0.8 x 1.6-5.0 μm; oval cells with rounded ends; occur singly and occasionally in pairs

Secondary tests:

Catalase:	+
Indole:	+

Tertiary tests:

DNase:	+
Arabinose:	+
Cellobiose:	+
Mannitol:	+
Rhamnose:	+
Salicin:	+
Trehalose:	+

Bacteroides stercoris (a member of the *B. fragilis* group)

Colony morphology (on BRU/BA):

Diameter:	0.5-1.0 mm
Form:	circular
Elevation:	convex
Margin:	entire
Surface:	shiny and smooth
Density:	transparent to translucent
Hemolysis:	beta-hemolytic

Cell morphology (in PYG broth):

1.6 x 2.4-12.6 μm; occur singly and in pairs; vacuolated cells are sometimes seen

Secondary tests:		**Tertiary tests:**	
Catalase:	−	DNase:	+
Indole:	+	Arabinose:	−+
		Cellobiose:	−
		Mannitol:	−
		Rhamnose:	+
		Salicin:	−
		Trehalose:	−

Bacteroides thetaiotaomicron (a member of the *B. fragilis* group)

Colony morphology (on BRU/BA):

Diameter:	puntiform
Form:	circular
Elevation:	convex
Margin:	entire
Color:	whitish
Surface:	shiny
Density:	semiopaque
Consistency:	soft
Hemolysis:	nonhemolytic

Cell morphology (in PYG broth):

0.7-1.1 x 1.3-8.0 μm; irregularly staining, pleomorphic bacilli with rounded ends; occur singly and in pairs

Secondary tests:		**Tertiary tests:**	
Catalase:	+*	DNase:	+
Indole:	+	Arabinose:	+
		Cellobiose:	+
		Mannitol:	−+
		Rhamnose:	+
		Salicin:	−+
		Trehalose:	+

Bacteroides uniformis (a member of the *B. fragilis* group)

Colony morphology (on BRU/BA):

Diameter:	0.5-2.0 mm
Form:	circular
Elevation:	low convex
Margin:	entire
Color:	gray to white
Density:	translucent to slightly opaque
Hemolysis:	usually nonhemolytic, but some strains produce a slight greening of the agar

Cell morphology (in PYG broth):

0.6-1.0 x 1.5-11.0 μm; occur singly and in pairs; an occasional filament may be seen; vacuoles are often present but do not cause extensive swelling of cells

Secondary tests:

Catalase:	—*
Indole:	+

Tertiary tests:

DNase:	+
Arabinose:	+
Cellobiose:	+
Mannitol:	—
Rhamnose:	—+
Salicin:	+−
Trehalose:	—

Bacteroides vulgatus (a member of the *B. fragilis* group)

Colony morphology (on BRU/BA):

Diameter:	1-2 mm
Form:	circular
Elevation:	convex
Margin:	entire
Color:	grayish
Density:	semiopaque
Hemolysis:	nonhemolytic

Cell morphology (in PYG broth):

0.5-0.8 x 1.5-8.0 μm; pleomorphic bacilli with swellings or vacuoles; occur singly and occasionally in pairs or short chains

Secondary tests:

Catalase:	—+
Indole:	—

Tertiary tests:

DNase:	—
Arabinose:	+
Cellobiose:	—
Mannitol:	—
Rhamnose:	+
Salicin:	—
Trehalose:	—

Other *Bacteroides* spp. (or former *Bacteroides* spp.)

Bacteroides gracilis

Colony morphology (on BRU/BA):

three colony types—
- (1) 1 mm; convex; translucent
- (2) up to 5 mm; agar-pitting or corroding
- (3) up to 5 mm; spreading

Cell morphology (in PYG broth):

0.4 x 4.0-6.0 μm; slim, straight rods with tapered and rounded ends; cells may demonstrate "twitching" movements, but cells have no flagella

Primary tests:

Colonies < 1 mm diam:	+*
Brown-black pigmentation:	−
Red fluorescence:	−
Pitting colonies:	−+

Secondary tests:

Catalase:	−
Oxidase:	−
Indole:	−
Motility:	−
Urease:	−

Tertiary tests:

Bile inhibition:	+
Esculin hydrolysis:	−
Lipase:	−
DNase:	−
Glucose fermentation:	−
Lactose fermentation:	−
Gelatin hydrolysis:	−
Proteolysis in milk:	−
H_2S in TSI:	+*
Rifampin (15 μg):	S
Acids in PYG:	
Butyric acid	−
Succinic acid	+
Phenylacetic acid	−

Prevotella intermedia (formerly *B. intermedius*)

Colony morphology (on BRU/BA):

Diameter:	0.5-2.0 mm
Form:	circular
Elevation:	low convex
Margin:	entire
Color:	after 48 hours of anaerobic incubation, colonies may be tan, gray, reddish-brown, or black; pigmentation occurs more rapidly on agar containing hemolyzed blood than on agar containing whole blood; on hemolyzed rabbit blood agar, about one-third of strains have dark brown to black colonies within 2 days, another third within 7 days, and most of the remainder within 14 days; an occasional strain requires 18-21 days for definitive pigmentation to be noted
Surface:	smooth
Density:	translucent; older colonies may be opaque
Hemolysis:	hemolytic
Other:	almost all strains fluoresce under shortwave UV light within 2-4 days, but only a brick-red fluorescence should be interpreted as indicative of probable pigmentation

Appendix E

Cell morphology (in PYG broth):

Most cells are 0.4-0.7 x 1.5-2.0 μm; coccobacilli common; some up to 12 μm in length

Primary tests:		Tertiary tests:	
Colonies < 1 mm diam:	−	Bile inhibition:	+
Brown-black pigmentation:	+*	Esculin hydrolysis:	−
Red fluorescence:	+	Lipase:	+*
Pitting colonies:	−	DNase:	+
		Glucose fermentation:	+
Secondary tests:		Lactose fermentation:	−
		Gelatin hydrolysis:	+*
Catalase:	−	Proteolysis in milk:	+*
Oxidase:	−	Rifampin (15 μg):	S
Indole:	+	Acids in PYG:	
Motility:	−	Butyric acid	−
Urease:	−	Succinic acid	+
		Phenylacetic acid	−

Bacteroides ureolyticus

Colony morphology (on BRU/BA):

Variable; some are 1.0 mm, circular, convex to slightly umbonate, entire to slightly undulating margins, and gray-white; others produce spreading or swarming growth extending for a few mm from the slightly raised center; the flat spreading growth appears to occupy a slight depression in the agar, so that the edge of the spreading growth is not detectably raised above the surface of the medium; the pitting can usually be seen best if the plate surface is observed at about a 30-45° angle

Cell morphology (in PYG broth):

0.5 x 1.5-4.0 μm; filaments exceeding 20 μm in length may occur; cells of some strains have polar tufts of long pili (not flagella), which produce a "twitching" motility

Primary tests:		Tertiary tests:	
Colonies < 1 mm diam:	+*	Bile inhibition:	+
Brown-black pigmentation:	−	Esculin hydrolysis:	−
Red fluorescence:	−	Lipase:	−
Pitting colonies:	+* −	DNase:	−
		Glucose fermentation:	−
Secondary Tests:		Lactose fermentation:	−
		Gelatin hydrolysis:	−
Catalase:	−	Proteolysis in milk:	−
Oxidase:	+*	H₂S in TSI:	+*
Indole:	−	Rifampin (15 μg):	S
Motility:	−	Acids in PYG:	
Urease:	+*	Butyric acid	−
		Succinic acid	+
		Phenylacetic acid	−

Porphyromonas spp.

P. asaccharolytica

Colony morphology (on BRU/BA):

Diameter:	0.5-1.0 mm
Form:	circular
Elevation:	convex
Color:	light gray after incubation for 48 hours; 6-14 days are required for black color formation
Density:	opaque
Hemolysis:	variable on rabbit blood

Cell morphology (in PYG broth):

0.8-1.5 x 1.0-3.5 μm; longer cells are occasionally seen; shorter (almost spherical) cells from solid medium

Primary tests:

Colonies < 1 mm diam:	−
Brown-black pigmentation:	+*
Red fluorescence:	+
Pitting colonies:	−

Secondary tests:

Catalase:	−
Oxidase:	−
Indole:	+
Motility:	−
Urease:	−

Tertiary tests:

Bile inhibition:	+
Esculin hydrolysis:	−
Lipase:	−
DNase:	−
Glucose fermentation:	−
Lactose fermentation:	−
Gelatin hydrolysis:	+
Proteolysis in milk:	+
H_2S in TSI:	−
Rifampin (15 μg):	S
Acids in PYG:	
Butyric acid	+
Succinic acid	+
Phenylacetic acid	−

P. endodontalis

Colony morphology (on BRU/BA):

Form:	circular
Elevation:	convex
Margin:	entire
Color:	colonies develop dark brown or black pigmentation between 7 and 14 days
Surface:	smooth and shiny
Other:	most strains produce colonies that adhere strongly to blood agar plates

Cell morphology (in PYG broth):

0.4-0.6 x 1.0-2.0 μm; bacilli or coccobacilli

Primary tests:

Colonies < 1 mm diam:	−
Brown-black pigmentation:	+*
Red fluorescence:	+⁻
Pitting colonies:	−

Secondary tests:

Catalase:	−
Oxidase:	−
Indole:	+
Motility:	−
Urease:	−

Tertiary tests:

Bile inhibition:	+
Esculin hydrolysis:	−
Lipase:	−
DNase:	−
Glucose fermentation:	−
Lactose fermentation:	−
Gelatin hydrolysis:	+
Proteolysis in milk:	+
H₂S in TSI:	−
Rifampin (15 μg):	S
Acids in PYG:	
Butyric acid	+
Succinic acid	+
Phenylacetic acid	−

P. gingivalis

Colony morphology (on BRU/BA):

Diameter:	1-2 mm
Form:	convex
Color:	form black pigment in 7-10 days

Cell morphology (in PYG broth):

0.5 x 1.0-2.0 μm; most cells are coccbacillary; bacilli up to 5 μm in length may be observed

Primary tests:

Colonies < 1 mm diam:	−
Brown-black pigmentation:	+*
Red fluorescence:	−
Pitting colonies:	−

Secondary tests:

Catalase:	−
Oxidase:	−
Indole:	+
Motility:	−
Urease:	−

Tertiary tests:

Bile inhibition:	+
Esculin hydrolysis:	−
Lipase:	−
DNase:	−
Glucose fermentation:	−
Lactose fermentation:	−
Gelatin hydrolysis:	+
Proteolysis in milk:	+
H₂S in TSI:	−
Rifampin (15 μg):	S
Acids in PYG:	
Butyric acid	+
Succinic acid	+
Phenylacetic acid	+*

Fusobacterium spp.

F. mortiferum

Colony morphology (on BRU/BA):

Diameter:	1-2 mm
Form:	circular
Elevation:	convex or slightly umbonate
Margin:	entire, diffuse, or slightly scalloped
Surface:	smooth
Density:	translucent
Hemolysis:	nonhemolytic

Cell morphology (in PYG broth):

0.8-1.0 x 1.5-10.0 μm; pale, irregularly staining, extremely pleomorphic bacilli with swollen areas, filaments, threads, and large, round bodies

Primary tests:

Colonies < 1 mm diam:	−
Brown-black pigmentation:	−
Red fluorescence:	−
Pitting colonies:	−

Secondary tests:

Catalase:	−
Oxidase:	−
Indole:	−
Motility:	−
Urease:	−

Tertiary tests:

Bile inhibition:	−
Esculin hydrolysis:	+
Lipase:	−
DNase:	−
Glucose fermentation:	+
Lactose fermentation:	+⁻
Gelatin hydrolysis:	−
Proteolysis in milk:	−
Rifampin (15 μg):	R*
Acids in PYG:	
Butyric acid	+
Succinic acid	−⁺
Phenylacetic acid	−

Additional important information:

Because *F. mortiferum* is bile tolerant and esculin hydrolysis-positive, it can be mistaken for a member of the *B. fragilis* group on BBE medium

F. necrophorum

Colony morphology (on BRU/BA):

Diameter:	1-2 mm
Form:	circular
Elevation:	convex to umbonate
Margin:	scalloped to erose
Surface:	often bumpy, ridged, or uneven
Density:	translucent to opaque
Internal structure:	often mosaic when viewed by transmitted light
Hemolysis:	most strains produce either alpha or beta hemolysis on rabbit blood; in general, the beta-hemolytic strains are lipase-positive, and the alpha-hemolytic or nonhemolytic strains are lipase-negative

Cell morphology (in PYG broth):

0.5-0.7 (swellings up to 1.8) x up to 10 μm; pleomorphic bacilli; ends may be round to tapered; range from coccoid bodies to long filaments; filamentous forms with granular inclusions are more common in broth; rods are more common in older cultures and growth on agar

Primary tests:

Colonies < 1 mm diam:	—
Brown-black pigmentation:	—
Red fluoresence:	—
Pitting colonies:	—

Secondary tests:

Catalase:	—
Oxidase:	—
Indole:	+
Motility:	—
Urease:	—

Tertiary tests:

Bile inhibition:	+
Esculin hydrolysis:	—
Lipase:	+*
DNase:	—
Glucose fermentation:	+
Lactose fermentation:	—
Gelatin hydrolysis:	—+
Proteolysis in milk:	V
Rifampin (15 μg):	S
Acids in PYG:	
Butyric acid	+
Succinic acid	—
Phenylacetic acid	—

Additional important information:

F. necrophorum is the only lipase-positive species of *Fusobacterium*

F. nucleatum

Colony morphology (on BRU/BA):

Diameter:	1-2 mm
Form:	circular to slightly irregular
Elevation:	convex to pulvinate
Surface:	some strains produce rough "breadcrumb" colonies
Density:	translucent
Internal structure:	often (but not always) with "flecked" or "ground glass" appearance when viewed by transmitted light
Hemolysis:	nonhemolytic, but may be slightly hemolytic under the area of confluent growth or may produce greenish discoloration of the blood agar upon exposure to oxygen

Cell morphology (in PYG broth):

0.4-0.7 x 3-10.0 μm; pale staining; long, slender, spindle-shaped with sharply pointed or tapered ends ("fusiform"); often have central swellings and intracellular granules; cell length is usually fairly uniform within actively growing cultures; occur in pairs or end-to-end

Primary tests:

Colonies < 1 mm diam:	−
Brown-black pigmentation:	−
Red fluoresence:	−
Pitting colonies:	−

Secondary tests:

Catalase:	−
Oxidase:	−
Indole:	+
Motility:	−
Urease:	−

Tertiary tests:

Bile inhibition:	+
Esculin hydrolysis:	−
Lipase:	−
DNase:	−
Glucose fermentation:	−
Lactose fermentation:	−
Gelatin hydrolysis:	−
Proteolysis of milk:	−
Rifampin (15 μg):	S
Acids in PYG:	
Butyric acid	+
Succinic acid	−
Phenylacetic acid	−

F. varium

Colony morphology (on BRU/BA):

Diameter:	punctiform to 1 mm
Form:	circular
Elevation:	flat to low convex
Margin:	entire
Color:	usually gray-white center with colorless edge ("fried egg colony")
Density:	translucent
Hemolysis:	nonhemolytic

Cell morphology (in PYG broth):

0.3-0.7 x 0.7-2.0 μm; uneven staining; pleomorphic, coccoid, and rod shapes; occurs singly or in pairs

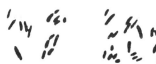

Primary tests:

Colonies < 1 mm diam:	−
Brown-black pigmentation:	−
Red fluoresence:	−
Pitting colonies:	−

Secondary tests:

Catalase:	−
Oxidase:	−
Indole:	+⁻
Motility:	−
Urease:	−

Tertiary tests:

Bile inhibition:	−
Esculin hydrolysis:	−
Lipase:	−
DNase:	−
Glucose fermentation:	+
Lactose fermentation:	−
Gelatin hydrolysis:	−
Proteolysis of milk:	−
Rifampin (15 μg):	R*
Acids in PYG:	
Butyric acid	+
Succinic acid	−
Phenylacetic acid	−

Miscellaneous Asaccharolytic, Nonmotile, Anaerobic, Gram-Negative Bacilli

Bilophila wadsworthia

Colony morphology (on BRU/BA):

Diameter:	0.6-0.8 mm (4 days)
Form:	circular or slightly irregular
Elevation:	raised
Margin:	erose; may be slightly spreading
Color:	gray
Density:	translucent

Cell morphology (on BRU/BA):

0.7-1.1 x 1.0-10.0 μm; pleomorphic bacilli with swollen ends; vacuolelike pale areas, and variable uptake of stain

Primary tests:

Brown-black pigmentation:	−
Pitted colonies:	−
Growth on BBE:	+
	(dark colonies)

Secondary tests:

Catalase:	+
Oxidase:	−
Indole:	−
Desulfoviridin:	+
Urease:	+⁻

Tertiary tests:

Growth in 20% bile:	+
Nitrate reduction:	+
H_2S in TSI:	−
Gelatin hydrolysis:	−

Desulfomonas pigra

Colony morphology (on BRU/BA):

Diameter:	1.0-2.0 mm
Form:	circular to slightly irregular
Elevation:	low convex or umbonate
Density:	translucent
Hemolysis:	nonhemolytic

Cell morphology (in PYG broth):

0.8-1.0 μm; blunt, rounded ends; occur singly or in pairs

Primary test:

Pitted colonies:	—

Tertiary test:

H$_2$S in TSI:	+*

Secondary tests:

Relationship to oxygen:	AN
Catalase:	+$^-$
Oxidase:	—
Indole:	—
Desulfoviridin:	+*
Urease:	—
Flagella:	—*
Penicillin (2 U):	V
Kanamycin (1 mg):	V

Miscellaneous Asaccharolytic, Motile, Anaerobic, Gram-Negative Bacilli

Campylobacter concisus

Colony morphology (on BRU/BA):

Diameter:	1 mm
Form:	entire
Elevation:	convex
Density:	translucent
Other:	no pitting

Cell morphology (in PYG broth):

0.5 x 4.0 μm; curved with rounded ends

Primary test:

Curved bacilli:	+

Secondary tests:

Relationship to oxygen:	MA, AN
Catalase:	—
Oxidase:	+
Indole:	—
Desulfoviridin:	—
Penicillin (2 U):	R
Kanamycin (1 mg):	R
Flagella:	monotrichous

Campylobacter mucosalis

Colony morphology (on BRU/BA):

Diameter:	1.5 mm
Form:	circular
Elevation:	raised with a flat surface
Color:	dirty yellowish
Other:	on moist agar, colonies tend to swarm along the line of inoculation

Cell morphology (in PYG broth):

0.25-0.3 x 1.0-2.8 μm; irregularly curved; in old cultures, coccoid cells and filamentous forms (7-8 μm in length) are seen

Primary test:		**Secondary tests:**	
Curved bacilli:	+	Relationship to oxygen:	AN
		Catalase:	−
		Oxidase:	+*
		Indole:	−
		Desulfoviridin:	−
		Penicillin (2 U):	R
		Kanamycin (1 mg):	V
		Flagella:	monotrichous

Campylobacter sputorum subsp. *sputorum*

Colony morphology (on BRU/BA):

Diameter:	1-2 mm
Form:	circular
Elevation:	low convex
Margin:	irregular, spreading
Color:	gray
Surface:	smooth and shiny
Hemolysis:	some strains weakly alpha-hemolytic

Cell morphology (in PYG broth):

0.3-0.5 x 2.0-4.0 μm; slender, comma-shaped and gull-winged; rounded ends; occasional filaments up to 8 μm in length; some cells from young (10-14-hour-old) cultures have a characteristic darting and corkscrewlike motion

Primary test:		**Secondary tests:**	
Curved bacilli:	+	Relationship to oxygen:	MA*
		Catalase:	−
		Oxidase:	+*
		Indole:	−
		Desulfoviridin:	−
		Penicillin (2 U):	S
		Kanamycin (1 mg):	S
		Flagella:	monotrichous

Appendix E

Desulfovibrio desulfuricans

Cell morphology (in PYG broth):

0.5-1.0 x 3.0-5.0 μm; typically vibroid in shape;
sigmoid forms may occur

Primary test:		Tertiary test:	
Curved bacilli:	+	H$_2$S in TSI:	+*

Secondary tests:

Relationship to oxygen:	AN
Catalase:	+* −
Oxidase:	−
Indole:	−
Desulfoviridin:	+*
Urease:	−
Penicillin (2 U):	V
Kanamycin (1 mg):	V
Flagella:	monotrichous

Additional important information:

Requires media containing sulfates, which blacken if iron salts are present

Cultures have a pronounced smell of H$_2$S

48 hour colonies are < 1 mm diameter

The curved morphology, progressive motility, and positive desulfoviridin test enable presumptive identification of this anaerobic, gram-negative bacillus

Wolinella curva (*Campylobacter curvus*)

Colony morphology (on BRU/BA):

three colony types -
- (1) 1 mm; convex
- (2) up to 5 mm; agar-pitting or corroding
- (3) up to 5 mm; spreading

Cell morphology (in PYG broth):

0.5-1.0 x 2.0-6.0 μm; curved rods; round or tapered ends; helical or straight cells may also exist; rapid, darting motility

Primary test:		Tertiary test:	
Curved bacilli:	+	H$_2$S in TSI:	+−

Secondary tests:

Relationship to oxygen:	MA*, AN
Catalase:	−
Oxidase:	+−
Indole:	−
Desulfoviridin:	−
Urease:	−
Penicillin (2 U):	S
Kanamycin (1 mg):	S
Flagella:	monotrichous

Additional important information:

48 hour colonies are < 1 mm diameter

Appendix E

Wolinella recta (*Campylobacter rectus*)

Colony morphology (on BRU/BA):

gray; translucent; three colony types -
   (1) 1 mm; convex
   (2) up to 5 mm; agar-pitting or corroding
   (3) up to 5 mm; spreading

Cell morphology (in PYG broth):

mainly straight rods

Primary test:		Secondary tests:	
Curved bacilli:	—	Relationship to oxygen:	AN
		Catalase:	—
		Oxidase:	$+^-$
		Indole:	—
		Desuloviridin:	—
		Urease:	—
		Penicillin (2 U):	S
		Kanamycin (1 mg):	S
		Flagella:	monotrichous

Additional important information:
   48 hour colonies are < 1 mm diameter

Miscellaneous Saccharolytic, Motile, Anaerobic, Gram-Negative Bacilli

Anaerobiospirillum succiniciproducens

Colony morphology (on BRU/BA):

Diameter:	0.5-1.0 mm
Form:	circular
Elevation:	convex; some colonies have a raised center
Density:	translucent

Cell morphology (in PYG broth):

0.6-0.8 x 3.0-8.0 μm; helical bacilli; rounded
ends; some cells may reach 20 μm in length;
usually occur singly; straight rods and spherical
forms may occur

Primary tests:

Cellular morphology:	large, helical bacilli
Spreads on BRU/BA:	—

Secondary tests:

Relationship to oxygen:	AN, MA*
Flagella:	bipolar tufts
Oxidase:	V* (key reaction when positive; approx 50% of the time)

Tertiary tests:

Glucose fermentation:	+
Lactose fermentation:	$+^-$
Mannitol fermentation:	—
Major acids in PYG:	A, S

Additional important information:
   48-hour colonies are < 1 mm diameter
   Darting, twisting motility in wet prep

Centipeda periodontii

Colony morphology (on BRU/BA):

On initial isolation -
    Diameter: 2 mm
    Elevation: flat
    Margin: irregular
    Density: transparent
    Hemolysis: nonhemolytic

After subculture -
    Elevation: flat or thinly raised
    Color: gray
    Surface: finely granular
    Density: transparent
    Other: growth spreads rapidly and after 3 days of incubation covers the
                        entire surface of the plate

Cell morphology (in PYG broth):

0.65 x 4.0-17.0 μm (or longer); slightly curved
rods with 3 or more curves; occasional branches;
occur singly or occasionally in chains; 50-250
flagella per cell; flagella are located in a linear
zone that spirals around the cell

Primary tests:

Cellular morphology: Large, straight or curved bacilli
Spreads on BRU/BA: +*

Secondary tests:

Relationship to oxygen: AN
Flagella: peritrichous
Oxidase: —

Tertiary tests:

Glucose fermentation: +
Lactose fermentation: +
Mannitol fermentation: +
Major acids in PYG: A, P, (L), (S)

Selenomonas sputigena

Colony morphology (on BRU/BA):

Diameter:	usually < 0.5 mm, but can be up to 2 mm
Elevation:	usually convex; sometimes flat
Margin:	sometimes irregular
Color:	usually grayish yellow; sometimes grayish
Surface:	smooth
Internal structure:	sometimes mottled, with a more opaque center; sometimes granular

Cell morphology (in PYG broth):

Usually 0.9-1.1 x 3.0-6.0 μm; ends are usually tapered and rounded to give short kidney- to crescent-shaped or vibroid cells; long cells and chains of cells are often helical

Primary tests:

Cellular morphology:	medium-sized, curved bacilli
Spreads on BRU/BA:	—

Secondary tests:

Relationship to oxygen:	AN
Flagella:	subpolar tuft*
Oxidase:	—

Tertiary tests:

Glucose fermentation:	+
Lactose fermentation:	+
Mannitol fermentation:	+
Major acids in PYG:	A, P

Succinivibrio dextrinosolvens

Cell morphology (in PYG broth):

0.4-0.6 x 1.0-7.0 μm; curved rods with pointed ends; helically twisted with less than 1 coil to 3 or more coils per cell; cells may become straight or only slightly curved after maintenance on artificial media

Primary tests:

Cellular morphology:	medium-sized, helical bacilli
Spreads on BRU/BA:	—

Secondary tests:

Relationship to oxygen:	AN
Flagella:	monotrichous
Oxidase:	—

Tertiary tests:

Glucose fermentation:	+
Lactose fermentation:	—
Mannitol fermentation:	V
Major acids in PYG:	A, S

Anaerobic Cocci

Gram-Positive Cocci

Peptostreptococcus spp.

P. anaerobius

Colony morphology (on BRU/BA):

Diameter:	minute to 1.0 mm
Form:	circular
Elevation:	convex
Margin:	entire
Color:	gray to white
Surface:	shiny and smooth
Density:	opaque
Hemolysis:	nonhemolytic
Other:	cultures have a sweet, fetid (putrid) odor

Cell morphology (in PYG broth):

0.5-0.6 μm; cells of young cultures are often elongated; occur in pairs and chains

Secondary tests:

Relationship to oxygen:	AN, MA
Catalase:	—
Indole:	—
Urease:	—

Tertiary tests:

Esculin hydrolysis:	—
Nitrate reduction:	—¹
Glucose fermentation:	+
Lactose fermentation:	—
Major acids in PYG:	A, IB*, IV*, IC*
	(Note the production
	of isoacids)

Additional important information:

An anaerobic, gram-positive coccus that is inhibited by sodium polyanethol sulfonate (SPS)* can be presumptively identified as *P. anaerobius*

P. asaccharolyticus

Colony morphology (on BRU/BA):

Diameter:	minute to 2.0 mm
Form:	circular
Elevation:	low convex
Margin:	entire
Color:	white to buff
Surface:	smooth and slightly shiny
Density:	translucent to opaque
Hemolysis:	nonhemolytic

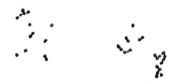

Cell morphology (in PYG broth):

0.5-1.6 μm; occur in pairs, tetrads, or irregular clumps; older cells may stain gram-negative

Secondary tests:

Relationship to oxygen:	AN
Catalase:	—⁺*
Indole:	+*
Urease:	—

Tertiary tests:

Esculin hydrolysis:	—
Nitrate reduction:	—
Glucose fermentation:	—
Lactose fermentation:	—
Major acids in PYG:	A, B

Additional important information:

An SPS-resistant, spot indole-positive, anaerobic, gram-positive coccus from a human specimen can be presumptively identified as *P. asaccharolyticus*

P. indolicus

Colony morphology (on BRU/BA):

Diameter:	0.5-1.0 mm
Form:	circular
Elevation:	convex to slightly pulvinate
Margin:	entire
Color:	grayish to yellow
Surface:	glistening

Cell morphology (in PYG broth):

0.5-0.6 μm; occur singly or in pairs, short chains, tetrads, or small clusters

Secondary tests:

Relationship to oxygen:	AN
Catalase:	—
Indole:	+*
Urease:	—

Tertiary tests:

Esculin hydrolysis:	—
Nitrate reduction:	+
Glucose fermentation:	—
Lactose fermentation:	—
Major acids in PYG:	A, P, B

Additional important information:

Infrequently encountered in human clinical specimens; common in veterinary clinical materials

Coagulase positive*

P. magnus

Colony morphology (on BRU/BA):

Diameter:	minute to 0.5 mm
Form:	circular
Elevation:	raised
Margin:	entire
Surface:	dull and smooth
Hemolysis:	nonhemolytic

Cell morphology (in PYG broth):

0.7-1.2 μm; occur singly and in pairs, tetrads, and clusters

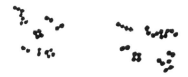

Secondary tests:

Relationship to oxygen:	AN
Catalase:	— + *
Indole:	—
Urease:	—

Tertiary tests:

Esculin hydrolysis:	—
Nitrate reduction:	—
Glucose fermentation:	—
Lactose fermentation:	—
Major acids in PYG:	A

P. micros

Colony morphology (on BRU/BA):

Diameter:	minute to 1.0 mm
Form:	circular
Elevation:	convex
Margin:	entire
Color:	white to translucent gray
Surface:	smooth and shiny
Density:	opaque

Cell morphology (in PYG broth):

0.3-0.7 μm; occur in pairs and chains of 6-20 cells

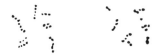

Secondary tests:

Relationship to oxygen:	AN
Catalase:	—
Indole:	—
Urease:	—

Tertiary tests:

Esculin hydrolysis:	—
Nitrate reduction:	—
Glucose fermentation:	—
Lactose fermentation:	—
Major acids in PYG:	A

P. prevotii

Colony morphology (on BRU/BA):

Diameter:	minute
Form:	circular
Elevation:	raised to low convex
Margin:	entire
Color:	white
Surface:	smooth
Density:	semiopaque

Cell morphology (in PYG broth):

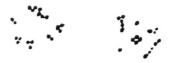

0.6-0.9 μm; occur in tetrads, irregular groups, and occasionally in short chains of 6-8 cells

Secondary tests:

Relationship to oxygen:	AN
Catalase:	−⁺
Indole:	−
Urease:	−

Tertiary tests:

Esculin hydrolysis:	−
Nitrate reduction:	−
Glucose fermentation:	−
Lactose fermentation:	−
Major acids in PYG:	A, P, B

P. tetradius (formerly "*Gaffkya anaerobia*")

Colony morphology (on BRU/BA):

Minute to 1 mm; nonhemolytic

Cell morphology (in PYG broth):

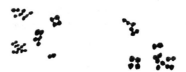

0.8-1.8 μm; occur as pairs, tetrads, short chains, and masses

Secondary tests:

Relationship to oxygen:	AN
Catalase:	−
Indole:	−
Urease:	+*

Tertiary tests:

Esculin hydrolysis:	−
Nitrate reduction:	−
Glucose fermentation:	+
Lactose fermentation:	+
Major acids in PYG:	A, B, L

Sarcina ventriculi

Colony morphology (on BRU/BA):

round colonies with rugged edges

Cell morphology (in PYG broth):

1.8-2.4 μm; occur in packets of 8 to several hundred or more

Secondary tests:		**Tertiary tests:**	
Relationship to oxygen:	AN, MA	Esculin hydrolysis:	+*
Catalase:	V	Nitrate reduction:	+
Indole:	−	Glucose fermentation:	+
Urease:	+*	Lactose fermentation:	+*
		Major acids in PYG:	A

Additional important information:

Forms endospores*

Staphylococcus saccharolyticus

Colony morphology (on BRU/BA):

Diameter:	0.5-2.0 mm
Form:	circular
Elevation:	slightly convex
Margin:	entire
Color:	grayish-white to yellowish
Surface:	smooth and glistening
Density:	opaque

Cell morphology (in PYG broth):

0.6-1.0 μm; occur singly, in pairs, tetrads, or irregular masses

Secondary tests:		**Tertiary tests:**	
Relationship to oxygen:	AN, MA, AT	Esculin hydrolysis:	−
Catalase:	+*	Nitrate reduction:	+
Indole:	−	Glucose fermentation:	+
Urease:	−	Lactose fermentation:	−
		Acids in PYG:	A

Gram-Negative Cocci

Veillonella parvula

Colony morphology (on BRU/BA):

Diameter:	1-3 mm
Margin:	entire
Color:	grayish white
Surface:	smooth
Density:	opaque
Consistancy:	butyrous (butterlike)
Hemolysis:	nonhemolytic
Other:	may fluoresce red under UV light

Cell morphology (in PYG broth):

Tiny (0.3-0.5 μm); occur as diplococci, masses, and short chains

Secondary tests:

Relationship to oxygen:	AN, MA
Catalase:	+*
Indole:	−
Urease:	−

Tertiary tests:

Esculin hydrolysis:	−
Nitrate reduction:	+⁻
Glucose fermentation:	−
Lactose fermentation:	−
Major aids in PYG:	A, P

Additional important information:

A nitrate-positive, anaerobic, gram-negative coccus can be identified as *Veillonella* spp.

References

1. Baron, E. J., P. Summanen, J. Downes, M. C. Roberts, H. Wexler, and S. M. Finegold. 1989. *Bilophila wadsworthia*, gen. nov. and sp. nov., a unique gram-negative anaerobic rod recovered from appendicitis specimens and human faeces. J. Gen. Microbiol. 135:3405–3411.

2. Bartley, S. L. (Centers for Disease Control). 1990. Personal communication.

3. Bryant, M. P. 1984. Genus VI. *Succinivibrio* Bryant and Small, 1956, 22 [AL], p. 644–645. *In* N. R. Krieg and J. G. Holt (eds.), Bergey's manual of systematic bacteriology, vol. 1. Williams & Wilkins, Baltimore.

4. Bryant, M. P. 1984. Genus VII. *Anaerobiospirillum* Davis, Cleven, Brown and Balish 1976, 503 [AL], p. 645–649. *In* N. R. Krieg and J. G. Holt (eds.), Bergey's manual of systematic bacteriology, vol. 1. Williams & Wilkins, Baltimore.

5. Bryant, M. P. 1984. Genus IX. *Selenomonas* Von Prowazek 1913, 36 [AL], p. 650-653. *In* N. R. Krieg and J. G. Holt (eds.), Bergey's manual of systematic bacteriology, vol. 1. Williams & Wilkins, Baltimore.

6. Canale-Parola, E. 1986. Genus *Sarcina* Goodsir 1842, 434 [AL], p. 1100–1103. *In* P. H. A. Sneath, N. S. Mair, and M. E. Sharpe (eds.), Bergey's manual of systematic bacteriology, vol. 2. Williams & Wilkins, Baltimore.

7. Cato, E. P., W. L. George, and S. M. Finegold. 1986. Genus *Clostridium* Prazmowski 1880, 23 [AL], p. 1141–1200. *In* P. H. A. Sneath, N. S. Mair, and M. E. Sharpe (eds.), Bergey's manual of systematic bacteriology, vol. 2. Williams & Wilkins, Baltimore.

8. Cummins, C. S., and J. L. Johnson. 1986. Genus I. *Propionibacterium* Orla-Jensen 1909, 337 [AL], p. 1346-1353. *In* P. H. A. Sneath, N. S. Mair, and M. E. Sharpe (eds.), Bergey's manual of systematic bacteriology, vol. 2. Williams & Wilkins, Baltimore.

9. Dowell, V. R., Jr. 1988. Characteristics of some clinically encountered anaerobic bacteria. Unpublished document that accompanied a lecture presented at the IV National Symposium on Clinical Microbiology of the Sociedade Brasileira de Microbiologia, Rio de Janeiro, Brazil. Centers for Disease Control, Atlanta.

10. Georg, L. K., G. W. Robertstad, S. A. Brinkmann, and M. D. Hicklin. 1965. A new pathogenic anaerobic *Actinomyces* species. J. Infect. Dis. 115:88–99.

11. Hatheway, C. L. 1990. Toxigenic clostridia. Clin. Microbiol. Rev. 3:66–98.

12. Holdeman, L. V., J. L. Johnson, and W. E. C. Moore. 1986. Genus *Peptostreptococcus* Kluyver and van Niel 1936, 401 [AL], p. 1083–1092. *In* P. H. A. Sneath, N. S. Mair, and M. E. Sharpe (eds.), Bergey's manual of systematic bacteriology, vol. 2. Williams & Wilkins, Baltimore.

13. Holdeman, L. V., R. W. Kelly, and W. E. C. Moore. 1984. Genus I. *Bacteroides* Castellani and Chalmers, 1919, 959 [AL], p. 604–631. *In* N. R. Krieg and J. G. Holt (eds.), Bergey's manual of systematic bacteriology, vol. 1. Williams & Wilkins, Baltimore.

14. Howard, B. J., and M. J. Ducate. 1987. Anaerobic bacteria. *In* B. J. Howard, John Klaas II, S. J. Rubin, A. S. Weissfeld, and R. C. Tilton (eds.), Clinical and pathogenic microbiology. The C. V. Mosby Company, St. Louis.

15. Johnson, J. L., W. E. C. Moore, and L. V. H. Moore. 1986. *Bacteroides caccae* sp nov., *Bacteroides merdae* sp. nov., and *Bacteroides stercoris* sp. nov. isolated from human feces. Int. J. Syst. Bacteriol. 36:499–501.

16. Kloos, W. E., and K. H. Schleifer. 1986. *Staphylococcus* Rosenbach 1884, 18 [AL], p. 1013–1035. *In* P. H. A. Sneath, N. S. Mair, and M. E. Sharpe (eds.), Bergey's manual of systematic bacteriology, vol. 2. Williams & Wilkins, Baltimore.

17. Lai, C.-H., B. M. Males, P. A. Dougherty, P. Berthold, and M. A. Listgarten. 1983. *Centipeda periodontii* gen. nov., sp. nov. from human periodontal lesions. Int. J. Syst. Bacteriol. 33:628–635.

18. Moore, L. V. H., J. L. Johnson, and W. E. C. Moore. 1986. Genus *Peptostreptococcus* Kluyver and van Niel 1936, 401 [AL], p. 1083–1092. *In* P. H. A. Sneath, N. S. Mair, and M. E. Sharpe (eds.), Bergey's manual of systematic bacteriology, Vol. 2. Williams & Wilkins, Baltimore.

19. Moore, W. E. C., and L. V. H. Moore. 1986. Genus *Eubacterium* Prevot 1938, 294 [AL], p. 1353–1373. *In* P. H. A. Sneath, N. S. Mair, and M. E. Sharpe (eds.), Bergey's manual of systematic bacteriology, vol. 2. Williams & Wilkins, Baltimore.

20. Moore, W. E. C., L. V. Holdeman, and R. W. Kelly. 1984. Genus II. *Fusobacterium* Knorr 1922, 4 [AL], p. 631–637. *In* N. R. Krieg and J. G. Holt (eds.), Bergey's manual of systematic bacteriology, vol. 1. Williams & Wilkins, Baltimore.

21. Postgate, J. R. 1984. Genus *Desulfovibrio* Kluyver and van Niel 1936, 397 [AL], p. 666–672. *In* N. R. Krieg and J. G. Holt (eds.), Bergey's manual of systematic bacteriology, vol. 1. Williams & Wilkins, Baltimore.

22. Rogosa, M. 1984. Genus I. *Veillonella* Prevot 1933, 118, emend. mut. char. Rogosa 1965, 706 [AL], p. 681–683. *In* N. R. Krieg and J. G. Holt (eds.), Bergey's manual of systematic bacteriology, vol. 1. Williams & Wilkins, Baltimore.

23. Scardovi, V. 1986. Genus *Bifidobacterium* Orla-Jense 1924, 472 [AL], p. 1418–1434. *In* P. H. A. Sneath, N. S. Mair, and M. E. Sharpe (eds.), Bergey's manual of systematic bacteriology, vol. 2. Williams & Wilkins, Baltimore.

24. Schaal, K. P. 1986. Genus *Actinomyces* Harz 1877, 133 [AL], p. 1383–1418. *In* P. H. A. Sneath, N. S. Mair, and M. E. Sharpe (eds.), Bergey's manual of systematic bacteriology, vol. 2. Williams & Wilkins, Baltimore.

25. Schaal, K. P. 1986. Genus *Arachnia* Pine and Georg 1969, 269 [AL], p. 1332–1342. *In* P. H. A. Sneath, N. S. Mair, and M. E. Sharpe (eds.), Bergey's manual of systematic bacteriology, vol. 2. Williams & Wilkins, Baltimore.

Appendix E

26. Shah, H. N., and M. D. Collins. 1988. Proposal for reclassification of *Bacteroides asaccharolyticus*, *Bacteroides gingivalis*, and *Bacteroides endodontalis* in a new genus, *Porphyromonas*. Int. J. Syst. Bacteriol. 38:128–131.

27. Slack, J. M., and M. A. Gerencser. 1975. *Actinomyces*, filamentous bacteria: biology and pathogenicity. Burgess Publishing Company, Minneapolis.

28. Smibert, R. M. 1984. Genus *Campylobacter* Sebald and Veron 1963, 907 [AL]; p. 111–118. *In* N. R. Krieg and J. G. Holt (eds.), Bergey's manual of systematic bacteriology, vol. 1. Williams & Wilkins, Baltimore.

29. Tanner, A. C. R., M. A. Listgarten, and J. L. Ebersole. 1984. *Wolinella curva* sp. nov. : "*Vibrio succinogenes*" of human origin. Int. J. Syst. Bacteriol. 34:275-282.

Most of the diagrams are from the *VPI Anaerobe Laboratory Manual, 4th ed.*, 1977. They are used with the permission of Drs. L. V. H. and W. E. C. Moore, Department of Anaerobic Microbiology, Virginia Polytechnic Institute, Blacksburg. Additional diagrams are from *Bergey's Manual of Systematic Bacteriology, Vol. 1*, 1984. They are used with the permission of Dr. John G. Holt, Bergey's Manual Trust, and Williams and Wilkins, Baltimore.

APPENDIX F

ADDITIONAL READING

The following is an abbreviated listing of significant books, book chapters, and journal articles published since 1980.

Anaerobes in General

Anaerobic Bacteria. K. T. Holland, J. S. Knapp, and J. G. Shoesmith. Blackie, Glasgow, Scotland, 1987 (excellent source of information on anaerobes in general, with limited information regarding anaerobes of medical importance).

Biology of Anaerobic Microorganisms. A. J. B. Zehnder. Wiley–Liss, Inc. (Division of John Wiley & Sons), New York, 1989 (molecular, ecological, and applied aspects of anaerobes).

Clostridia. N. P. Minton and D. J. Clarke (eds.). Plenum, New York, 1989 (taxonomy, phylogeny, physiology, genetics, and biochemistry of the genus *Clostridium*, volume 3 in the Biotechnology Handbook series).

Anaerobes of Medical Importance

Anaerobes in Human Disease. B.I. Duerden and B.S. Draser (eds.). Wiley-Liss, New York, 1991.

Anaerobes Today. J. M. Hardie and S. P. Borriello (eds.). John Wiley & Sons, New York, 1988 (the proceedings of the 5th Anaerobe Discussion Group Symposium, July, 1987).

Anaerobic Bacteria. S. D. Allen (section ed.) In *Manual of Clinical Microbiology*, 4th ed. E. H. Lennette, A. Balows, W. J. Hausler, Jr., and H. J. Shadomy (eds.). American Society for Microbiology, Washington, D. C., 1985, pp. 413–472.

Anaerobic Bacteria. B. J. Howard and M. J. Ducate. In *Clinical and Pathogenic Microbiology*. B. J. Howard, J. Klaas II, S. J. Rubin, A. S. Weissfeld, and R. C. Tilton (eds.). C. V. Mosby, St. Louis, 1987, pp. 371–415.

Anaerobic cultures. J. E. Rosenblatt. In *The Role of Clinical Microbiology in Cost–Effective Health Care.* J. W. Smith (ed.). College of American Pathologists, Skokie, IL, 1985, pp. 349–357.

Anaerobic identification: How far to go. W. J. Martin. Clin. Microbiol. Newsl. 5:135–136, 1983.

Anaerobic Infections. S. M. Finegold, W. L. George, and M. E. Mulligan. Year Book Medical Publishers, Chicago, 1986 (physician–oriented).

Anaerobic Infections. S. M. Finegold and V. L. Sutter. Upjohn, Kalamazoo, MI, 1982.

Anaerobic Infections: Clinical and Laboratory Practice. A. T. Willis and K. D. Phillips. Public Health Laboratory Service, London, UK, 1988.

Anaerobic Infections in Humans. S. M. Finegold and W. L. George. Academic Press, San Diego, CA, 1989 (physician oriented).

Biology of Anaerobes. S. L. Gorbach, J. G. Bartlett, and F. P. Tally. Upjohn, Kalamazoo, MI, 1981.

Clinical Guide to Anaerobic Infections. S. M. Finegold, E. J. Baron, H. M. Wexler, and H. Tester. Star Publishing, Belmont, CA, 1992.

Current relevance of anaerobic bacteriology. S. D. Allen. Clin. Microbiol. Newsl. 6:147–149, 1984.

International Symposium on Anaerobic Bacteria and Bacterial Infections. D. L. Kasper and A. B. Onderdonk (eds.). Rev. Inf. Dis. 12 (Suppl. 2), January–February, 1990.

International Symposium on Anaerobic Bacteria and Their Role in Disease. S. M. Finegold (ed.). Rev. Inf. Dis. 6 (Suppl. 1), March–April, 1984.

Pediatric Anaerobic Infection: Diagnosis and Management, 2nd ed. I. Brook. C. V. Mosby, St. Louis, 1989.

Processing clinical specimens for anaerobic bacteriology: Isolation and identification procedures. Anaerobic gram–positive bacilli. Anaerobic gram–negative bacilli. Anaerobic cocci. M. A. C. Edelstein. In *Bailey & Scott's Diagnostic Microbiology,* 8th Ed. E. J. Baron and S. M. Finegold (eds.). C. V. Mosby, St. Louis, 1990, pp. 477–557.

The anaerobic bacteria. In *Color Atlas and Textbook of Diagnostic Microbiology,* 3rd ed. E. W. Koneman, S. D. Allen, V. R. Dowell, Jr., W. M. Janda, H. M. Sommers, and W. C. Winn, Jr. J. B. Lippincott, Philadelphia, 1988, pp. 393–471.

The Pathogenic Anaerobic Bacteria, 3rd ed. L. D. S. Smith and B. L. Williams. Charles C. Thomas, Springfield, IL, 1984.

Toxigenic clostridia. C. L. Hatheway. Clin. Microbiol. Rev. 3: 66–98, 1990.

Clostridium difficile

Clostridium difficile: Its disease and toxins. D. M. Lyerly, H. C. Krivan, and T. D. Wilkins. Clin. Microbiol. Rev. 1:1–18, 1988.

Clostridium difficile: Its Role in Intestinal Disease. R. D. Rolfe and S. M. Finegold (eds.). Academic Press, San Diego, CA, 1988.

Laboratory Procedures

Rapid methods in microbiology: IV. Presumptive and rapid methods in anaerobic bacteriology. H. B. Moore. Am. J. Med. Tech. 47: 705–712, 1981.

Wadsworth Anaerobic Bacteriology Manual, 4th ed. V. L. Sutter, D. M. Citron, M. A. C. Edelstein, and S. M. Finegold. Star Publishing, Belmont, CA, 1985.

Resistance of Anaerobes to Antimicrobial Agents

Annual incidence, epidemiology, and comparative *in vitro* susceptibilities to cefoxitin, cefotetan, cefmetazole, and ceftizoxime of recent community–acquired isolates of the *Bacteroides fragilis* group. E. J. C. Goldstein and D. M. Citron. J. Clin. Microbiol. 26:2361–2366, 1988.

Beta–lactamase production and susceptibilities to amoxicillin, amoxicillin–clavulanate, ticarcillin, ticarcillin–clavulanate, cefoxitin, imipenem, and metronidazole of 320 non–*Bacteroides fragilis Bacteroides* isolates and 129 fusobacteria from 28 U. S. centers. P. C. Appelbaum, S. K. Spangler, and M. R. Jacobs. Antimicrob. Agents Chemother. 34:1546–1550, 1990.

Comparative activities of newer beta–lactam agents against members of the *Bacteroides fragilis* group. G. J. Cuchural, Jr., F. P. Tally, N. V. Jacobus, T. Cleary, S. M. Finegold, G. Hill, P. Iannini, J. P. O'Keefe, and C. Pierson. Antimicrob. Agents Chemother. 34:479–480, 1990.

Susceptibility of the *Bacteroides fragilis* group in the United States: Analysis by site of isolation. G. J. Cuchural et al. Antimicrob. Agents Chemother. 32:717–722, 1988.

Susceptibility Testing of Anaerobes

Anaerobes: Problems and controversies in bacteriology, infections, and susceptibility testing. S. M. Finegold. Rev. Inf. Dis. 12 (Suppl 2):S223–S230, 1990.

Methods for Antimicrobial Susceptibility Testing of Anaerobic Bacteria, 2nd ed. : Approved Standard. NCCLS Document M11–A2. National Committee for Clinical Laboratory Standards, Villanova, PA, 1990 (see the most recent NCCLS document on this subject).

Revisiting anaerobe susceptibility testing. R. J. Zabransky. Clin. Microbiol. Newsl. 11:185–188, 1989.

Son of anaerobic susceptibility testing: Revisited. E. J. Baron, D. M. Citron, and H. M. Wexler. Clin. Microbiol. Newsl. 12:69–72, 1990.

Susceptibility testing of anaerobes: Fact, fancy, and wishful thinking. E. J. C. Goldstein and D. M. Citron. Clin. Ther. 11: 710–723, 1989.

Susceptibility testing of anaerobic bacteria (a minireview). S. M. Finegold et al. J. Clin. Microbiol. 26:1253–1256, 1988.

Susceptibility testing of anaerobic bacteria: Myth, magic, or method? H.M. Wexler. Clin. Microbiol. Rev. 4:470-484, 1991.

Therapeutic implications of bacteriologic findings in mixed aerobic–anaerobic infections (a minireview). S. M. Finegold and H. M. Wexler. Antimicrob. Agents Chemother. 32:611–616, 1988.

ANSWERS TO SELF-ASSESSMENT EXERCISES

Chapter 1

1. Define the following terms:

   a. **Anaerobiosis: life in the absence of molecular oxygen**

   b. **Capnophile: an organism that grows best in the presence of increased concentrations of carbon dioxide**

   c. **Obligate anaerobe: an organism that grows only in an anaerobic environment, one that will not grow in a microaerophilic environment (5% oxygen), a CO_2 incubator (15% oxygen), or in air (about 21% oxygen)**

   d. **Microaerotolerant anaerobe: an organism that grows in an anaerobic system and a microaerophilic environment but does not grow in a CO_2 incubator or in air**

   e. **Aerotolerant anaerobe: an organism capable of multiplication in atmospheres containing molecular oxygen (such as air), but one that grows best in an anaerobic environment**

2. Respiration involves molecular oxygen as an oxidizing agent and electron acceptor, whereas fermentation does not. **TRUE**

3. The toxicity of oxygen to anaerobes is thought to occur in a two-stage process: a bacteriostatic phase (phase 1) and a bactericidal phase (phase 2). Briefly describe what happens in both phases.

   a. **Phase 1: Electrons are diverted to the reduction of molecular oxygen. Thus, they are not available to the cell for metabolic processes.**

   b. **Phase 2: Toxic reduction products like superoxide anions and hydrogen peroxide kill the cell.**

4. List three products formed during the reduction of molecular oxygen that are capable of causing extensive damage to cellular enzyme systems and cell structure.

   a. superoxide anions

   b. hydrogen peroxide

   c. hydroxyl radicals (singlet oxygen is also an acceptable answer)

5. List two enzymes possessed by bacteria (including some obligate anaerobes) that enable the organisms to counteract or neutralize the toxic effects of products formed during the reduction of molecular oxygen.

   a. superoxide dismutase

   b. catalase

6. List three ecological niches where anaerobes are found.

   a. soil

   b. fresh and salt water sediment (mud)

   c. the bodies of animals (including humans)

7. Define the following terms:

   a. Endogenous anaerobes: anaerobes that comprise part of the indigenous microflora of animals (including humans); also called indigenous anaerobes

   b. Exogenous anaerobes: anaerobes that exist in or arise from locations outside of or other than the bodies of animals; examples include anaerobes that live in soil or mud

   c. Indigenous microflora: the mixture of microorganisms that typically reside on or in (colonize) various anatomic sites of a healthy animal body; such organisms may either be permanent or transient residents

8. List three industrial uses of anaerobes.

   a. anaerobic digestion of sewage

   b. retting process in the textile industry

   c. cheese, vaccine, or vitamin B_{12} production

Chapter 2

1. Give the gram reaction (positive or negative) and morphologic appearance of each of the following anaerobes (e.g., *Propionibacterium acnes* is a gram-positive diphtheroid).

		Gram Reaction	**Morphology**
a.	*Bacteroides fragilis*	gram-negative	bacilli
b.	*Clostridium perfringens*	gram-positive	bacilli (rarely contain spores)
c.	*Fusobacterium nucleatum*	gram-negative	bacilli ("fusiform")
d.	*Peptostreptococcus anaerobius*	gram-positive	cocci
e.	*Veillonella* spp.	gram-negative	cocci
f.	*Wolinella* spp.	gram-negative	bacilli (may be curved)
g.	*Actinomyces* spp.	gram-positive	bacilli (often branched)
h.	*Mobiluncus* spp.	gram-negative (although technically gram-positive)	bacilli (curved)
i.	*Bifidobacterium* spp.	gram-positive	bacilli

2. All sporeforming, anaerobic bacilli are classified as *Clostridium* spp. **TRUE**

3. Two *Clostridium* spp. that routinely stain pink in Gram-stained preparations are *C. ramosum* and *C. clostridiiforme.*

4. It is easy to recognize *Clostridium* spp. in Gram-stained smears of clinical specimens because they always contain spores. **FALSE (Spores are not always seen in clinical specimens when clostridia are present.)**

5. The only *Fusobacterium* species that consistently has cells that are "fusiform" in shape (long, thin rods of variable length, with pointed ends) is *F. nucleatum.*

6. Define the following terms:

Diphtheroid: A gram-positive bacillus resembling *Corynebacterium diphtheriae* in appearance; small, club-shaped bacilli that tend to form various arrangements of cells (e.g., "Chinese letters," "picket fences," etc.)

Pleomorphic: When the cells of a given species of bacterium exist in a variety of forms

7. The two genera of anaerobic, gram-negative bacilli most commonly isolated from clinical specimens are *Bacteroides* and *Fusobacterium.*

8. Name the anaerobe most commonly involved in soft tissue infectious processes and bacteremia; give both genus and species. *Bacteroides fragilis*

9. Most anaerobic, gram-positive cocci isolated from clinical materials are species of *Peptostreptococcus,* and most gram-negative cocci are species of *Veillonella.*

10. List five genera of nonsporeforming, anaerobic, gram-positive bacilli.

   a. *Actinomyces*

   b. *Bifidobacterium*

   c. *Eubacterium*

   d. *Mobiluncus* (although they stain pink when Gram stained)

   e. *Propionibacterium*

   (*Lactobacillus* is also an acceptable answer.)

11. List five species in the *Bacteroides fragilis* group.

   a. *B. distasonis*

   b. *B. fragilis*

   c. *B. thetaiotaomicron*

   d. *B. uniformis*

   e. *B. vulgatus*

   (*B. caccae, B. merdae, B. ovatus,* and *B. stercoris* are also acceptable answers.)

12. Name three species in the genus *Porphyromonas.*

   a. *P. asaccharolytica*

   b. *P. endodontalis*

   c. *P. gingivalis*

Chapter 3

1. Define the following terms:

   a. **Sterile site:** An anatomic site usually devoid of microorganisms (e.g., blood, cerebrospinal fluid, synovial fluid, and healthy tissue)

   b. **Nonsterile site:** An anatomic site usually colonized with members of the indigenous microflora (e.g., skin, oral cavity, and gastrointestinal tract)

   c. **Opportunistic pathogen:** A microorganism (often a member of the indigenous microflora) that causes no harm under ordinary circumstances but can cause serious disease under certain conditions (e.g., if the host becomes immunosuppressed or when the organism gains access to a usually sterile site)

2. List three ways in which endogenous anaerobes are beneficial to the human body.

   a. **They provide a barrier to colonization of mucous membranes by foreign and/or pathogenic organisms.**

   b. **Within the gastrointestinal tract, they synthesize vitamins and cofactors that are utilized by the host and degrade potentially oncogenic compounds.**

   c. **They play a role in maturation of the immune system of neonates.**

3. Which of the following clinical materials would you expect to contain endogenous anaerobes?

   a. **Scrapings from gingival crevices**
   b. **Voided urine**
   c. **Expectorated sputum**
   d. **Cervical swab**
   e. **Throat swab**

   All these materials would contain endogenous anaerobes.

4. What genus of anaerobic, gram-positive bacillus is most prevalent in the indigenous microflora of the skin?

   Propionibacterium

5. Name two species of *Prevotella* that are especially prevalent in the indigenous microflora of the human vagina.

   a. *P. bivia*
   b. *P. disiens*

6. What genus of pigmented, anaerobic, gram-negative bacilli is especially prevalent in the indigenous microflora of the oral cavity?

    Porphyromonas

7. Which genus of anaerobe is most prevalent in the colon?

    Bacteroides

8. The most common species of *Bacteroides* in the colon is *B. fragilis*. **FALSE (*B. vulgatus* and *B. thetaiotaomicron* outnumber *B. fragilis* in the colon.)**

9. Which of the following clinical materials are considered sterile site specimens?

    a. Blood

    b. Catheterized urine

    c. Cerebrospinal fluid

    d. Synovial fluid

    e. Fluid aspirated from an oral lesion

    a, c, and d are sterile site specimens

10. Unless they are carefully collected, specimens from many anatomic locations can become contaminated with endogenous anaerobes. **TRUE**

Chapter 4

1. The term "exogenous anaerobes" refers to anaerobes found on or in the body. **FALSE (It refers to anaerobes found in the external environment.)**

2. Most infectious diseases involving anaerobes involve endogenous anaerobes. **TRUE**

3. Endogenous anaerobes can cause infectious diseases in virtually any tissue or organ of the body. **TRUE**

4. Infectious diseases involving anaerobes are rarely polymicrobial or purulent. **FALSE (Infectious diseases involving anaerobes are usually polymicrobial and purulent.)**

5. Foodborne botulism is acquired by ingesting food containing **spores** of *Clostridium botulinum,* whereas infant botulism results from the ingestion of food containing **toxins** of *C. botulinum.* **FALSE (The reverse is true.)**

For questions 6-9, indicate whether the organism is considered an endogenous or exogenous anaerobe.

6. *Bacteroides fragilis*   **endogenous**

7. *Clostridium tetani*   **exogenous**

8. *Fusobacterium nucleatum*   **endogenous**

9. *Clostridium botulinum*   **exogenous**

10. Define the term "intoxication" as it refers to diseases caused by anaerobic bacteria. **A poisoning that results from ingestion or *in vivo* production of clostridial toxins**

11. Give five examples of conditions that predispose an individual to infections by endogenous anaerobes:

    a. **Aspiration of oral contents into the lungs**

    b. **Tooth extraction**

    c. **Oral surgery**

    d. **Gastrointestinal tract surgery**

    e. **Genitourinary tract surgery**

12. Name the anaerobe most often involved in each of the following infectious processes/diseases:

    a. Actinomycosis (give genus and species)
       Actinomyces israelii

    b. Bacterial vaginosis (give genus)
       Mobiluncus (especially *Mobiluncus curtisii* subsp. *curtisii*)

    c. Gas gangrene (give genus and species)
       Clostridium perfringens

    d. Infectious processes in soft tissues (give genus and species)
       Bacteroides fragilis

    e. Antibiotic-associated diarrhea and pseudomembranous colitis (give genus and species)
       Clostridium difficile

13. Give three indications of possible involvement of anaerobes in an infectious process, other than predisposing factors:

    a. **The specimen has a foul odor.**

    b. **The specimen contains a large quantity of gas.**

    c. **The specimen has a black discoloration and/or fluoresces brick-red under long-wave ultraviolet (UV) light.**

14. List four examples of virulence factors possessed by anaerobes:

    a. **Exotoxins (Many specific examples may be listed.)**

    b. **Endotoxins**

    c. **Capsules**

    d. **Pili that enable organisms to adhere to mucosal surfaces**

15. Differentiate between the terms "infection" and "infectious disease" as they are used in this book. **An infection is the invasion of tissue by microorganisms and their subsequent multiplication within that tissue. An infection may or may not result in clinical signs and symptoms. An infection that results in clinical signs and symptoms is called an infectious process or infectious disease.**

16. Why is the presence of a *Clostridium difficile* carrier of special importance in a hospital setting? **An individual whose intestinal tract is colonized with *C. difficile* could serve as a reservoir of nosocomial (hospital-acquired) infectious diseases such as antibiotic-associated diarrhea, colitis, or pseudomembranous colitis.**

Chapter 5

1. The following specimens have been received in the laboratory for anaerobic culture. Indicate which are acceptable (A) and which are unacceptable (U) for routine anaerobic bacteriology:

U	a. Voided urine
A	b. Blood
A	c. Cerebrospinal fluid
U	d. Feces
U	e. Catheterized urine
A	f. Synovial fluid
U	g. Swab containing material from oral abscess
U	h. Swab containing pus from skin lesion
A	i. Bone marrow
A	j. Tissue biopsy specimen

    Specimens considered unacceptable are those apt to be contaminated with endogenous anaerobes.

2. In general, the best types of clinical specimens to submit to the anaerobic bacteriology laboratory are those that have been collected with a **needle and syringe. Such specimens are less likely than swab specimens to be contaminated with indigenous microflora.**

3. In general, the worst types of clinical specimens to submit to the anaerobic bacteriology laboratory are those that have been collected with a **swab. Specimens collected using swabs are apt to be contaminated with indigenous microflora.**

4. An elderly, alcoholic patient is suspected of having aspiration pneumonia. Which one of the following specimens is recommended for isolation of any anaerobic bacteria that might be contributing to the infectious process?

   a. Bronchial washings

   b. Expectorated (coughed) sputum

   c. **Percutaneous transtracheal aspirate**

   d. Specimen collected by nasopharyngeal tube

   e. Sputum induced by nebulization

   The answer is c. The other specimens will be contaminated with anaerobes that are part of the upper respiratory tract flora.

5. A female patient is suspected of having endometritis. Which of the following specimens is recommended for isolation of any anaerobic bacteria that might be contributing to the infectious process?

   a. Vaginal swab

   b. Cervical swab

   c. **Specimen collected by culdocentesis**

   The answer is c. Vaginal and cervical swabs will be contaminated with anaerobes usually present in vaginal and cervical secretions.

6. You have been asked to present a short in-service to nursing personnel regarding important considerations in the **transport** of clinical specimens for anaerobic bacteriology. List three of the important considerations you will include in your presentation.

   a. **Specimens must be transported rapidly. Any organisms present in the specimens were removed from a 37°C environment (their optimum growth temperature) and are at about 25°C during transit.**

   b. **Specimens must be protected from oxygen during transport. This can be accomplished by using special anaerobic transport containers.**

   c. **Specimens must not dry out during transport. Containers used for transport must incorporate some means of keeping specimens moist.**

7. The clinical microbiology section of a laboratory continues to receive specimens for anaerobic bacteriology that are unacceptable. Describe three ways clinical laboratory professionals could improve the situation.

   a. **Participate in in-services and rounds, providing guidelines for proper selection, collection, and transport of specimens for anaerobic bacteriology.**

b. Prepare and disseminate written laboratory directives that contain guidelines for proper selection, collection, and transport of specimens for anaerobic bacteriology.

c. Develop and enforce criteria for rejection of specimens.

8. What is the meaning of the term "PRAS media?"

Media that were prereduced and anaerobically sterilized during manufacture.

Chapter 6

1. List four types of solid culture media recommended in this chapter for primary isolation of anaerobes from clinical specimens.

   a. Brucella blood agar (BRU)

   b. Bacteroides bile esculin (BBE) agar

   c. Kanamycin-vancomycin-laked blood (KVLB) agar

   d. Phenylethyl alcohol (PEA) agar

2. While examining a Gram-stained smear of material aspirated from a festering leg wound, a microbiologist observes numerous neutrophils, large, gram-positive bacilli, gram-negative coccobacilli, and gram-positive cocci. In addition to the four plates routinely used for primary isolation of anaerobes, what additional plated medium or test should be included in the primary isolation setup? Why?

   An egg yolk agar (EYA) plate and/or a direct Nagler test. The large, gram-positive bacilli could be *Clostridium* species. The EYA plate and direct Nagler test are sometimes useful in making presumptive identifications of clostridia.

3. List the three most commonly used methods employed by clinical microbiology laboratories to incubate inoculated plates anaerobically.

   a. Anaerobic chambers

   b. Anaerobic jars

   c. Anaerobic bags or pouches

4. Match the media used in the primary isolation setup for anaerobes with their intended purpose:

a. Bacteroides bile esculin (BBE) agar **2**

1. To enhance the brown-black color of colonies of pigmented *Prevotella* spp.

b. Phenylethyl alcohol (PEA) agar **3**

2. For rapid presumptive identification of organisms in the *Bacteroides fragilis* group

c. Kanamycin-vancomycin-blood agar (KVA) or kanamycin-vancomycin-laked blood (KVLB) agar **1**

3. To suppress the growth of any *Enterobacteriaceae*, including swarming *Proteus* species

5. A properly selected, collected, and transported specimen was received in the laboratory for anaerobic culture. Appropriate plated and tubed media were inoculated and incubated within an anaerobic jar. No anaerobes were recovered from any of the plated media, but a *Bacteroides* species was recovered from the thioglycollate broth. Give two possible explanations for these findings.

    a. **The patient may have been receiving antimicrobial agents at the time the specimen was collected. The concentration of these agents in the specimen could have inhibited growth on the plated media. However, the antimicrobial agents were diluted in the broth, thus permitting growth of the *Bacteriodes* species present in the specimen.**

    b. **The plated media that were used might have been stored under aerobic conditions prior to use. Toxic substances (e.g., superoxide anions and hydrogen peroxide), produced during reduction of molecular oxygen, could have accumulated on the surface of the media. Such substances could have killed the anaerobes present in the specimen.**

    c. **Because an anaerobic jar was used, anaerobic conditions may not have been obtained due either to "poisoned" catalyst pellets or a crack somewhere in the jar, lid, or "O-ring."**

6. What selective and differential media could be used when attempting to recover *Clostridium difficile* from fecal specimens?

    CCFA, CMA, or CMBA

7. What solid medium is used to determine an organism's ability to produce the enzymes lecithinase and lipase?

    Egg yolk agar

8. What do the initials "CCFA" stand for?

    Cycloserine-cefoxitin-egg-fructose agar

9. Because most anaerobic infectious processes involving anaerobes are pyogenic, it would be impractical to culture a wound specimen for anaerobes when no leukocytes are observed on the Gram stain of the specimen. **FALSE (Although anaerobic infections are usually pyogenic, intact leukocytes may not be observed in Gram-stained material from cases of clostridial myonecrosis. The appearance of any leukocytes present in such specimens may have been distorted by clostridial toxins; thus, the specimen should be processed whether or not leukocytes are observed.)**

10. To obtain anaerobic incubation conditions, a laboratory is using anaerobic jars containing reusable palladium-coated alumina catalyst pellets. Indicate whether the following statements are TRUE or FALSE.

    a. Anaerobic bacteria produce gases (including H_2S) that can "poison" the catalyst pellets, rendering them ineffective. **TRUE**

b. Following incubation, the presence of a water vapor condensate on the inner walls of the anaerobe jar is an indication that the pellets are "poisoned" and not operating effectively. **FALSE (Water vapor is an indication that the catalyst is functioning properly.)**

c. To "rejuvenate" catalyst pellets, keep them in a 160°C oven for a minimum of two hours. **TRUE**

d. It is recommended that catalyst pellets should be "rejuvenated" after every use. **TRUE**

e. If catalyst pellets are properly "rejuvenated," it is not necessary to include an anaerobic indicator strip (e.g., a methylene blue strip) each time a jar is set up. **FALSE (An indicator strip is required each time a jar is set up.)**

11. A wound aspirate containing five organisms (*Escherichia coli, Bacteroides fragilis,* a pigmented species of *Prevotella, Staphylococcus aureus,* and *Peptostreptococcus anaerobius*) is inoculated onto plates to be incubated in a CO_2 incubator (blood agar, CA, MAC, and PEA) and plates to be incubated anaerobically (PRAS BRU/BA, PRAS PEA, PRAS BBE, and PRAS KVLB). State which of the five organisms would be expected to grow on each of the plates.

    Plates incubated in the CO_2 incubator:

Blood agar	*Escherichia coli, Staphylococcus aureus*
CA	*Escherichia coli, Staphylococcus aureus*
MAC	*Escherichia coli*
PEA	*Staphylococcus aureus*

    Plates incubated anaerobically:

PRAS BRU/BA	**All five organisms**
PRAS PEA	**All except** *Escherichia coli*
PRAS BBE	*Bacteroides fragilis*
PRAS KVLB	*Bacteroides fragilis,* **the pigmented species of Prevotella**

Chapter 7

1. Name the anaerobe (genus and species) that produces a characteristic double zone of hemolysis on BRU/BA.

   Clostridium perfringens

2. Which two genera of anaerobes produce colonies on BRU/BA and/or KVLB agar that fluoresce brick-red under long-wave (366 nm) ultraviolet light?

   Porphyromonas **and** *Prevotella*

3. The main reason for including PRAS PEA agar in the primary isolation setup is to inhibit the growth of what specific facultative anaerobe?

   Swarming *Proteus* species

4. The Nagler test is useful for presumptive identification of *Clostridium perfringens* because *C. perfringens* is the only organism that produces a positive result with this test. **FALSE (Four species of *Clostridium* produce a positive Nagler test result.)**

5. What characteristic odor is associated with colonies of *Clostridium difficile*?

   A horse stable odor

6. No anaerobes are capable of growing on a chocolate agar plate incubated at 35-37°C in a CO_2 incubator. **FALSE (Certain aerotolerant anaerobes could grow.)**

7. Which disk is especially important to add to the pure culture/subculture plate of an anaerobic, gram-positive coccus?

   An SPS (sodium polyanethol sulfonate) disk

8. What rapid test is useful in differentiating a *Bacillus* species from an aerotolerant species of *Clostridium*? Why?

   Catalase test because *Bacillus* spp. are catalase-positive and *Clostridium* spp. are catalase-negative

9. Cite three reasons why identifying anaerobic isolates is important.

   a. To guide antimicrobial therapy

   b. Possibly to indicate the source of the infection

   c. To generate a data base of information regarding pathogenic organisms

10. Which of the bacteria in the following list would be able to grow on or in the following anaerobically incubated media? (Place the letters of the correct answers on the lines following the types of media.)

    a. Virtually all gram-positive anaerobes

    b. Virtually all gram-negative anaerobes

    c. Facultative, gram-positive organisms

    d. Facultative, gram-negative organisms

    e. Virtually all *Bacteroides* and *Prevotella* spp.

    f. Members of the *Bacteroides fragilis* group

BRU/BA	a, b, c, d, e, f
PEA	a, b, c, e, f (not d)
EYA	a, b, c, d, e, f
BBE	f (This medium is selective and differential for members of the B. fragilis group.)
KVLB	e, f
THIO broth	a, b, c, d, e, f (Virtually all bacteria will grow in THIO; it is not selective for anaerobes.)

Chapter 8

1. Presumptively identify the following anaerobes using Table 8-3 and the appropriate flow charts. The first one was worked through for you as an example.

   a. **Presumptive identification:** *Fusobacterium necrophorum.* Poor growth on BBE agar and no growth on KVLB agar indicate that the organism is probably not a *Bacteroides* species. The disk results indicate a *Fusobacterium* species. *F. necrophorum* is the only *Fusobacterium* species that is lipase-positive.

   b. **Presumptive identification:** *Clostridium perfringens.* The major clues are the double zone of hemolysis and the positive lecithinase result.

   c. **Presumptive identification:** *Propionibacterium acnes.* An anaerobic, gram-positive diphtheroid that is catalase-positive and spot indole-positive can be presumptively identified as *P. acnes.*

   d. **Presumptive identification:** *Veillonella* sp. An anaerobic, gram-negative coccus that is nitrate-positive can be presumptively identified as *Veillonella* sp. However, some species of *Veillonella* are nitrate-negative.

   e. **Presumptive identification:** *Prevotella intermedia.* The major clues are good growth on KVLB agar, brown pigmentation starting to occur, and brick-red fluorescence, which together provide presumptive evidence of a pigmented species of *Prevotella.* *P. intermedia* is both lipase-positive and indole-positive.

   f. **Presumptive identification:** *Peptostreptococcus anaerobius.* The major clue is the SPS result. An anaerobic, gram-positive coccus that is SPS-sensitive can be presumptively identified as *P. anaerobius.*

   g. **Presumptive identification:** *Fusobacterium nucleatum.* No growth on BBE agar or KVLB agar rules out most *Bacteroides* species. The disk results indicate a *Fusobacterium* species. *F. nucleatum* is the only *Fusobacterium* species that is not pleomorphic. All cells are fusiform in appearance. Both smooth and rough colonies are possible, with rough colonies resembling the "molar tooth" colonies of *Actinomyces israelii.*

   h. Presumptive identification: *Peptostreptococcus asaccharolyticus*. Although *P. indolicus* is also an anaerobic, gram-positive coccus that is SPS-resistant and indole-positive, it is rarely encountered in human clinical specimens. However, it is commonly encountered in veterinary specimens.

   i. Presumptive identification: *Fusobacterium mortiferum*. The disk results indicate a *Fusobacterium* species, but this organism is growing on the BBE plate. The *F. varium/F. mortiferum* group of fusobacteria are bile-tolerant. *F. varium* is spot indole-positive, whereas *F. mortiferum* is spot indole-negative.

2. Differentiate between the terms "presumptive identification" and "definitive identification."

   A presumptive identification of a microorganism is based upon simple colony and Gram stain observations and the results of relatively simple and inexpensive tests. Although a presumptive identification is believed to be valid, the possibility exists that it is not. A definitive identification is a valid identification of a microorganism, regardless of the method(s) used to make the identification.

3. List five colony observations/characteristics that are of value in making presumptive identifications of anaerobes.

   a. Color/pigment
   b. Hemolysis/double zone of hemolysis
   c. Pitting of the agar
   d. Odor
   e. Swarming
   Other acceptable answers are surface, density, consistency, form, elevation, and margin.

4. List ten simple test results or reactions that are of value in making presumptive identifications of anaerobes.

   a. Aerotolerance testing
   b. Fluorescence
   c. Special potency antimicrobial disks
   d. SPS disk, nitrate disk, bile disk
   e. Catalase test
   f. Spot indole test
   g. Motility test
   h. Lecithinase activity
   i. Lipase activity
   j. Urease test

5. When used in conjunction with other characteristics, a long-wave (366 nm) ultraviolet light source (a Wood's lamp) is useful in the anaerobic bacteriology laboratory for presumptive identification of certain anaerobic isolates. Answer the following questions about fluorescence.

   a. An anaerobic, gram-positive bacillus is producing large, ground-glass colonies on CCFA, is turning the usually pink medium a yellow color, and has a "horse stable" odor. What color fluorescence would you expect if a BRU/BA plate containing the same organism was held under UV light?

      Chartreuse (The organism is *Clostridium difficile*. CCFA is selective for *Clostridium difficile*. The CCFA plate is not used to check for fluorescence because the medium is fluorescent.)

   b. An anaerobic, gram-negative coccobacillus is growing well on KVLB agar, and some of the colonies have started to turn brown-black. What color fluorescence would you expect if the KVLB plate was held under UV light?

      Brick-red (Many strains of pigmented *Prevotella* fluoresce brick-red [red-brown to red-orange]. Other strains can fluoresce different colors, including yellow, orange, pink-orange, pink, and brilliant red.)

   c. An anaerobe that produces grayish-white, bread crumb colonies on BRU/BA appears as fusiform, gram-negative bacilli on Gram stain. What color fluorescence would you expect if the BRU/BA plate was held under UV light?

      Chartreuse (The organism is *Fusobacterium nucleatum*.)

   d. A nitrate-positive anaerobe is growing well on BRU/BA and appears as tiny, gram-negative cocci and diplococci on Gram stain. What color fluorescence would you expect if the BRU/BA plate was held under UV light?

      Red (The organism is a *Veillonella* sp. Some strains of *Veillonella* produce a red fluorescence that is weaker than that of the pigmented species of *Porphyromonas* and *Prevotella* and fades rapidly upon exposure to air.)

6. In some laboratories, it is not necessary to definitively identify every anaerobic isolate. **TRUE (To definitively identify every anaerobic isolate would be expensive and unnecessary. Many frequently isolated anaerobes can be satisfactorily identified by combining colony and Gram stain observations with the results of simple, rapid, and inexpensive test procedures.)**

7. Coccobacillary, gram-negative bacilli were seen on the Gram-stained smear of a positive blood culture. The contents of the blood culture bottle were inoculated to four different plates. The organism did not grow on a CO_2-incubated chocolate agar plate. Although the organism grew well on an anaerobically incubated BRU/BA plate, it did not grow at all on the anaerobically incubated KVLB or BBE plates. A Gram stain of a BRU/BA plate colony confirmed that the organism was a gram-negative coccobacillus, and aerotolerance testing demonstrated that it was truly an anaerobe. Young, nonpigmented colonies on the BRU/BA plate fluoresced brick-red under UV light; and, with age, the colonies became brown-black. This anaerobic, gram-negative coccobacillus was presumptively identified as a pigmented *Porphyromonas* species. Indicate whether the following statements are TRUE or FALSE.

a. There is insufficient information to presumptively identify this organism as a *Porphyromonas* species. **FALSE**

b. The Gram stain results are inconsistent with a PID of a *Porphyromonas* species. **FALSE**

c. The fluorescence and pigmentation results are consistent with a PID of a *Porphyromonas* species. **TRUE (However, these results are also consistent with a PID of a pigmented species of *Prevotella*.)**

d. The primary purpose for using a KVLB plate is to enhance the brown-black pigmentation of the pigmented species of *Prevotella*. **TRUE**

e. A possible explanation for failure of the organism to grow on the KVLB plate is that it is a vancomycin-sensitive strain of *Porphyromonas*. **TRUE (Because the organism did not grow on KVLB agar, it was presumptively identified as a *Porphyromonas* species rather than a *Prevotella* species. But this is only a presumptive identification. To speciate this organism, you must inoculate some type of biochemical- or enzyme-based identification system.)**

8. Presumptively identify (to genus and species) the following anaerobes **without** referring to the flow charts. Use the flow charts only if you are unable to identify the organisms without them.

a. Anaerobic, gram-positive diphtheroids have been isolated from several blood culture bottles of a patient suspected of having subacute bacterial endocarditis. Simple test procedures have demonstrated that the organism is catalase-positive, spot indole-positive, and nitrate-positive.

Presumptive identification: *Propionibacterium acnes*

b. Very pleomorphic, anaerobic, gram-negative bacilli have been recovered from an aspirate that was carefully collected from a liver abscess. Special potency anti-microbial disks have demonstrated that the organism is vancomycin-resistant, colistin-sensitive, and kanamycin-sensitive. Simple test procedures have shown that the organism is spot indole-positive, lecithinase-negative, and lipase-positive.

Presumptive identification: *Fusobacterium necrophorum*

c. Anaerobic, gram-positive cocci have been recovered from a polymicrobic infectious process that developed following GU surgery. Simple test procedures have demonstrated that the organism is catalase-negative, SPS-resistant, spot indole-positive, and nitrate-negative.

Presumptive identification: *Peptostreptococcus asaccharolyticus*

d. Large, rectangular, anaerobic, gram-positive bacilli have been recovered from a wound resulting from puncture of the patient's foot by a rusty nail. No spores were observed in the Gram-stained preparations of either the specimen or the pure culture. A double zone of hemolysis was seen on the BRU/BA subculture plate. The organism was lecithinase-positive and lipase-negative. Nagler and reverse CAMP tests were both positive.

Presumptive identification: *Clostridium perfringens*

e. Somewhat pleomorphic, anaerobic, gram-negative bacilli were recovered from a brain abscess that developed subsequent to GI surgery. The organism grew well on BBE, turned the medium brown, and produced a stippling in the medium around areas of heavy growth. The organism grew well on KVLB, but the colonies were nonpigmented and did not fluoresce under UV light. The organism was catalase-positive and indole-negative.

Presumptive identification: ***Bacteroides fragilis***

f. An anaerobic, gram-positive coccus is recovered in pure culture from material that was carefully aspirated from a cervical abscess. The organism is catalase-negative, SPS-sensitive, spot indole-negative, and nitrate-negative.

Presumptive identification: ***Peptostreptococcus anaerobius***

Chapter 9

1. Explain what is meant by a conventional or traditional system for definitive identification of anaerobes.

   A conventional or traditional approach to identifying anaerobes employs a variety of tubed biochemical test media (either PRAS or non-PRAS media) and gas-liquid chromatographic analysis of acid end products produced by anaerobes in broth culture.

2. Identify two advantages and one disadvantage of biochemical-based minisystems for anaerobe identification, when compared to a conventional identification system.

   Advantages: (a) Biochemical-based minisystems are easier and faster to inoculate than a conventional system. (b) They generate code numbers, which can be looked up in a manufacturer-supplied code book.

   Disadvantage: Data bases from which code books are developed do not contain all the anaerobes that can potentially be isolated from clinical specimens.

3. Describe four advantages and one disadvantage of preexisting enzyme-based minisystems for anaerobe identification, when compared to a conventional identification system.

   Advantages: (a) Preexisting enzyme-based minisystems are easier and faster to inoculate than a conventional system. (b) They do not require anaerobic incubation. (c) They require only four hours of incubation (compared to 24-48 hours or more with the conventional system). (d) They generate code numbers, which can be looked up in a manufacturer-supplied code book.

   Disadvantage: Data bases from which code books are developed do not contain all the anaerobes that can potentially be isolated from clinical specimens.

4. Identify two advantages of preexisting enzyme-based minisystems for anaerobe identification, when compared to biochemical-based minisystems.

   Biochemical-based minisystems require anaerobic incubation and 24-48 hours or more before results are apparent. Preexisting enzyme-based systems (a) can be incubated in air and (b) require only four hours of incubation before test results can be read.

5. State three components of the Hungate technique or VPI system for isolation and identification of anaerobes.

   Major components of the Hungate technique are (a) an oxygen-free gassing system employing cannula, (b) media produced under oxygen-free conditions (PRAS media), and (c) solid media contained in tightly stoppered tubes ("roll tubes") in place of plated media.

6. Explain the principles of gas-liquid chromatography; include the meaning of "motile phase" and "stationary phase."

   Like other types of chromatographic techniques, GLC uses the chemical and physical properties of a particular component to separate it from other components in a mixture. All chromatographic techniques have a mobile phase and a stationary phase. In GLC, an inert gas serves as the mobile phase, and a liquid serves as the stationary or separating phase. An unknown mixture is usually prepared for separation, volatilized, and carried as a gas through a packed chromatographic column. Components of the mixture are then separated, and they elute from the column, in order, based on their chemical and physical properties. As the individual components elute from the column, they are detected by one of several types of detecting devices (detectors). A graphic display of the separated components is then printed on a recorder in the form of a chromatogram. Identification of an unknown component is based on its position on the chromatogram (i.e., its relative retention time) as compared to a standard. Quantitation of an unknown component is based on the area under the peak with respect to a known standard.

7. Describe two ways in which gas-liquid chromatography can be used to identify anaerobes.

   Gas-liquid chromatography can be used (a) to analyze metabolic end products (volatile and nonvolatile acids) produced by anaerobes in broth culture and (b) to analyze cellular fatty acids.

8. List five genera of anaerobes that contain flagellated species.

   Any of the following genera: *Anaerobiospirillum, Butyrivibrio, Campylobacter, Clostridium, Desulfovibrio, Mobiluncus, Selenomonas, Succinimonas, Wolinella.*

9. Name two desulfoviridin-producing genera of anaerobes.

   ***Desulfomonas, Desulfovibrio,* or *Bilophila* are acceptable answers.**

10. Briefly discuss laboratory confirmation of the following clostridial diseases: botulism, tetanus, *C. perfringens* food poisoning, *C. difficile*-induced diarrhea.

Botulism:	Laboratory confirmation of foodborne botulism requires isolating *C. botulinum* from the stool specimen or detecting botulinal toxin in the serum, stool, or the epidemiologically implicated food. Laboratory confirmation of wound botulism requires isolating *C. botulinum* from a wound specimen or detection of botulinal toxin in the patient's serum or stool.

Diagnosis of infant botulism can be confirmed by detecting *C. botulinum* and/or botulinal toxin in the stool of infants who are demonstrating the characteristic signs and symptoms of paralyzing botulinal toxin.

Tetanus: Laboratory results are rarely required to confirm the diagnosis. When *C. tetani* is isolated from the wound(s) of patients with tetanus, toxicity and neutralization tests can be performed by intramuscular injection of culture supernatant or whole culture into untreated mice and mice protected with tetanus antitoxin.

C. perfringens
food poisoning: Laboratory confirmation requires two or more of the following findings:

- $>10^5$ colony forming units (CFU) of *C. perfringens* per gram of implicated food

- A median *C. perfringens* spore count of $>10^6$ per gram of stool from affected patients (such counts have been found in healthy individuals)

- Isolation of the same serotype of *C. perfringens* from stools of affected patients as isolated from the suspected food (isolates are not always serotypable)

- Demonstration of the presence of *C. perfringens* enterotoxin in stool specimens from ill persons and its absence in stool specimens from well persons

C. difficile-
induced diarrhea: A tissue culture cytotoxin assay (CTA) is the best method for laboratory diagnosis/confirmation of *C. difficile*-induced diarrhea. Other methods include CIE, ELISA, latex agglutination, and a dot immunobinding assay. Positive latex results should be confirmed by CTA, and stools yielding negative dot immunobinding assay results should be further tested using CTA.

Chapter 10

1. Give two valid reasons why antimicrobial susceptibility testing should **not** be performed routinely on all anaerobic isolates.

Any of the following are acceptable: (a) Not all anaerobic isolates are clinically significant. (b) The cost is high. (c) Problems exist with all currently available methods for susceptibility testing of anaerobes.

2. Cite three clinical situations in which antimicrobial susceptibility testing of anaerobic isolates **should** be performed.

   (a) Usual therapeutic regimens fail, and the infection persists.

   (b) Antimicrobial agents play a pivotal role in determining the outcome.

   (c) Making empiric decisions based on precedent is difficult (i.e., no precedent may exist).

3. Name three species of anaerobes for which susceptibility testing **should** routinely be performed, whenever the isolates are deemed clinically significant.

   Any members of the *B. fragilis* group, *B. gracilis*, pigmented species of *Porphyromonas* and *Prevotella*, *C. perfringens*, *C. ramosum*, and certain *Fusobacterium* spp.

4. List three antimicrobial agents that should be included in susceptibility testing of clinically significant anaerobic isolates.

   Any of the following are acceptable: Penicillin G, one or more of the broad-spectrum antipseudomonal penicillins (e.g., ticarcillin or piperacillin), clindamycin, cefoxitin, and certain other cephalosporins with known activity against anaerobes.

5. Name three antimicrobial agents that are essentially always effective against anaerobes.

   Any of the following are acceptable: Chloramphenicol; combinations of a beta-lactam drug and a beta-lactamase inhibitor (e.g., amoxicillin plus clavulanic acid, ampicillin plus sulbactam, and ticarcillin plus clavulanic acid), except for many strains of *Bilophila wadsworthia*; imipenem; and metronidazole (except against most nonsporeforming, gram-positive bacilli and some strains of *Peptostreptococcus* spp.

6. List three methods of antimicrobial susceptibility testing of anaerobes that employ agar plates as part of the procedure and three methods that do not.

   Agar plate methods include agar dilution, spiral gradient endpoint, Epsilometer, and disk diffusion. Of these, only agar dilution is currently recommended by the NCCLS. Broth methods include broth dilution, broth microdilution, and broth-disk elution. The latter is currently not recommended by the NCCLS.

7. To what method of antimicrobial susceptibility testing of anaerobes should all other methods be compared whenever other methods are being evaluated?

   The NCCLS agar dilution method.

8. Briefly describe three mechanisms by which anaerobes may become resistant to penicillins.

   The three mechanisms are (a) production of enzymes called beta-lactamases, (b) alteration of the number or type of penicillin-binding proteins, and (c) blocked penetration of the drug into the active site via alteration of the bacterial outer membrane pores or "porins."

9. What are the three major approaches to treating anaerobe-associated infectious processes?

   (a) Create an environment in which anaerobes cannot proliferate by removing dead tissue (debridement), draining pus, eliminating obstructions, decompressing tissues, releasing trapped gas, and improving circulation and oxygenation of tissues. (b) Arrest the spread of anaerobes into healthy tissues with antimicrobial agents. (c) Use specific antitoxins to neutralize toxins produced by the anaerobes when such toxins are present.

10. There is no NCCLS standardized "Kirby-Bauer type" of disk diffusion susceptibility testing for anaerobes. **TRUE**

11. The broth-disk elution method of susceptibility testing provides a minimal inhibitory concentration (MIC) for each antimicrobial agent being tested. **FALSE (Only a single concentration [the breakpoint concentration] of each antimicrobial agent is used in the broth-disk elution method.)**

12. The most practical method for susceptibility testing of anaerobes in small clinical microbiology laboratories is the NCCLS standardized agar dilution technique. **FALSE (The agar dilution method is labor-intensive and not practical for testing small numbers of clinically significant anaerobic isolates. This technique is probably the least practical method for small labs.)**

13. All clinically significant anaerobic isolates should be tested for beta-lactamase production. **FALSE (One should automatically assume that certain anaerobic isolates [e.g., members of the *B. fragilis* group] are beta-lactamase producers and perform some type of susceptibility test on them.)**

14. Broth microdilution methods for susceptibility testing of anaerobes are popular in clinical laboratories because they are commercially available, can be stored frozen or lyophilized for extended periods of time, and, in some cases, results can be read by instruments. **TRUE (All these statements about broth microdilution methods are true.)**

Chapter 11

1. Explain the term "quality assurance" (QA) as it relates to anaerobic bacteriology.

   Quality assurance refers to any and all actions taken by health care professionals to assure the overall quality of clinical anaerobic bacteriology services within a given health care institution. Inclusive in the term are actions that improve the quality of (a) specimens sent to the laboratory, (b) procedures performed within the laboratory, and (c) events that occur after results leave the laboratory.

2. Explain the term "quality control" (QC) as it relates to anaerobic bacteriology.

   Quality control refers to any and all actions taken by laboratory personnel to assure the quality of procedures performed within the anaerobic bacteriology section of the microbiology laboratory. Quality control is a very important, but not the sole, component of QA.

3. Differentiate between the terms "accuracy" and "precision" as they relate to laboratory tests.

   Accuracy is a measure of the "truth" or validity of a result. Precision is the expression of the variability of analysis or an indication of the amount of random error that exists in an analytical process.

4. List three important considerations pertaining to the **quality** of anaerobic bacteriology specimens submitted to the laboratory.

   a. **Is it an appropriate specimen for anaerobic bacteriology?**

   b. **Was the specimen collected in a manner that would minimize contamination with organisms of the indigenous microflora (including anaerobes)?**

   c. **Was the specimen properly transported to the laboratory (i.e., rapidly and in a container designed to minimize exposure of the specimen to oxygen)?**

5. Describe three specific actions that laboratory personnel may take to improve the overall quality of anaerobic bacteriology specimens submitted to the laboratory.

   a. **Write and distribute guidelines for the proper selection, collection, and transport of specimens for anaerobic bacteriology**

   b. **Ensure the availability of appropriate specimen transport containers**

   c. **Participate in educational programs (e.g., rounds and in-services), describing proper techniques for selection, collection, and transport of specimens for anaerobic bacteriology**

6. List ten QA/QC considerations pertaining to events that occur **within** the anaerobic bacteriology section of the laboratory.

   a. **Education of microbiology laboratory personnel**

   b. **Assessment of specimen quality**

   c. **Speed with which the specimen is processed**

   d. **Objective criteria for presumptive identification of organisms observed in Gram-stained smears**

   e. **Speed with which the preliminary report is sent**

   f. **Choice and QC of media**

   g. **Monitoring of the system being used for anaerobic incubation**

   h. **Written QC procedures for stains, reagents, etc.**

   i. **Choice and QC of the susceptibility testing method**

   j. **Participation in proficiency surveys**

7. Describe two actions that laboratory personnel may take to ensure the quality of events that occur **after** anaerobic bacteriology results leave the laboratory.

    a. Calculate the overall time that elapses between specimen collection and posting of laboratory results to patients' charts

    b. Work with physicians to determine if results from the anaerobic bacteriology laboratory are being interpreted properly and whether or not they influence the course of treatment

8. List five common sources of error in the anaerobic bacteriology laboratory.

    a. Contamination of clinical specimens with endogenous anaerobes

    b. Failure to process specimens rapidly

    c. Sole reliance on a liquid medium, such as thioglycollate broth, for recovery of anaerobes from specimens

    d. Use of inappropriate/inadequate plated media

    e. Failure to use QC organisms to check isolation and differential media, media supplements, reagents, ID systems, and susceptibility testing methods

    (Note: There are other acceptable answers.)

9. Laboratories isolating large numbers of *Bacteroides* and *Clostridium* spp. and relatively small numbers of fusobacteria, *Porphyromonas* spp., and anaerobic cocci can pride themselves on their anaerobic bacteriology capabilities. **FALSE (Many species of *Bacteroides* and *Clostridium* are relatively aerotolerant. Laboratories isolating large numbers of the more fastidious fusobacteria, *Porphyromonas* spp., and anaerobic cocci may take pride in themselves.)**

Chapter 12

1. State two of the most exasperating problems facing microbiologists in the anaerobic bacteriology laboratory.

    a. How to improve the quality of specimens that are submitted to and processed by the laboratory

    b. How far to go in the workup of anaerobic cultures

2. Identify three ways in which microbiologists could interact with physicians and/or nurses to reduce costs associated with anaerobic bacteriology.

    a. Microbiologists can participate in rounds, in-services, and other educational activities. They can publish and distribute laboratory directives that clearly define appropriate and inappropriate specimens, provide information concerning rejection of inappropriate/poor quality specimens, and furnish guidelines regarding repeat cultures.

b. Microbiologists can communicate with clinicians to clarify questions regarding the extent to which the clinicians wish particular specimens to be worked up or isolates to be identified.

c. Microbiologists can ensure that laboratory reports contain sufficient information to enable clinicians to interpret culture results.

3. Name four parameters that should be considered when determining the extent of anaerobic workup for a given specimen.

   Anaerobic isolates should be identified only to a level appropriate for

   a. the specimen source (i.e., is the specimen apt to contain indigenous microflora contaminants or not?)

   b. the types and numbers of different anaerobes present

   c. the patient's status (i.e., is the patient recovering or is his or her condition deteriorating?)

   d. the needs of the clinician (i.e., how extensive a workup does the clinician feel is necessary?)

4. State at least two important cost containment guidelines regarding each of the following: specimens submitted for anaerobic bacteriology, request slips accompanying specimens for anaerobic bacteriology, media used in the anaerobic bacteriology laboratory.

   a. Specimens submitted for anaerobic bacteriology should be (1) collected in a manner that will minimize contamination with endogenous anaerobes, (2) submitted rapidly, and (3) transported in special containers designed to minimize exposure of the specimen to oxygen.

   b. Request slips should (1) state the patient's condition and the clinician's diagnostic considerations, (2) clearly indicate the type of specimen and anatomic site from which it was collected, and (3) be signed by the attending physician.

   c. In general, media used in the anaerobic bacteriology laboratory should (1) be of high quality, (2) support the growth of fastidious anaerobes, and (3) be in date and properly stored.

5. Give an important cost containment guideline regarding each of the following: *Clostridium difficile* cultures, beta-lactamase testing, reports sent from the anaerobic bacteriology laboratory.

   a. *Clostridium difficile* cultures (1) should be performed only on watery or bloody diarrheal specimens and (2) should not be performed on specimens from patients known to have pathognomonic pseudomembranes or microabscesses.

   b. Beta-lactamase testing should not be performed (1) routinely on all anaerobic isolates, (2) on *B. fragilis* group isolates (which one may assume are beta-lactamase producers), or (3) on *C. perfringens* isolates (which one may assume for now are not beta-lactamase producers).

c. Reports sent from the anaerobic bacteriology laboratory should contain sufficient information to enable a clear understanding/interpretation of the results.

6. Cite at least three clinical situations in which susceptibility testing of anaerobic isolates would be warranted.

   a. The organism is known to exhibit variability in its susceptibility patterns (e.g., *B. fragilis* group, other anaerobic, gram-negative bacilli, and *Clostridium*).

   b. The organism has been isolated in pure culture.

   c. The organism has been isolated from a seriously ill patient.

   d. The organism has been isolated from a patient who has undergone long-term therapy.

   e. The organism has been isolated from a patient who has failed to respond to empiric therapy or is relapsing.

Chapter 13

1. Names three anaerobes of significance in veterinary medicine that are of little or no significance in human medicine.

   Any of the following are acceptable answers: *Actinomyces bovis, A. hordeovulnaris, A. suis, Bacteroides nodosus, Clostridium chauvoei, Fusobacterium russii, Peptostreptococcus indolicus, Prevotella ruminicola,* **and** *Treponema (Serpula) hyodysentariae*

2. Name an anaerobe associated with each of the following diseases and an/the animal species in which each disease occurs:

Big head	*C. novyi* or *C. sordellii*	rams
Black disease	*C. novyi* type B	sheep and cattle
Black leg	*C. chauvoei*	cattle and sheep
Lockjaw	*C. tetani*	horses, other herbivores, pigs, dogs, cats
Lumpy jaw	*A. bovis*	cattle
Malignant edema	*C. septicum*	cattle
Overeating disease	*C. perfringens* type C	cattle
Red water disease	*C. haemolyticum*	cattle
Swine dysentery	*Treponema (Serpula) hyodysenteriae*	swine

3. What are the three types of *Clostridium botulinum* of greatest importance in veterinary medicine.

   Types B, C$_1$, C$_2$, and D cause most animal cases of botulism.

4. Name the three types/subtypes of *Clostridium perfringens* of greatest importance in veterinary medicine.

   Types B, C, and D are of major clinical importance in animals.

5. Identify the essential elements in selection, collection, and transport of veterinary clinical specimens for anaerobic bacteriology.

   Specimens must contain few (preferably no) endogenous anaerobes. They must be collected in a manner that minimizes their contamination by indigenous microflora that might be present at or near the collection site. Specimens must be transported to the laboratory as rapidly as possible in transport containers designed to minimize their exposure to oxygen. They should not be refrigerated or allowed to dry out en route.

6. List the three major components in the treatment of anaerobe-associated infectious processes in animals.

   Debridement, drainage, and appropriate antimicrobial therapy

INDEX

NOTES: Page numbers in *italics* refer to the definitions lists at the beginning of each chapter. Page numbers in **boldface** refer to the "Compendium of Clinically Encountered Anaerobes" in Appendix E.

Page numbers followed by *t* refer to tables, and page numbers followed by *f* refer to figures.